CHARLES M. SALTER ASSOCIATES INC.

ACOUSTICS

ARCHITECTURE · ENGINEERING · THE ENVIRONMENT

WILLIAM STOUT PUBLISHERS · SAN FRANCISCO

This book was set in Monotype Bembo, Univers, and Gill Sans by Pick Design; the editor was Rachel Dangermond of Writing by Design; the designers were Dylan Jhirad and Henry Rollmann; the art was drawn by Michael Flynn, Dylan Jhirad, and Henry Rollmann; the publisher was William Stout Publishers, San Francisco; printed and bound in Hong Kong by Global Interprint, Inc.; Robert J. Richardson was the indexer.

Library of Congress Card Catalog Number 97-62505
ISBN 0-9651144-6-5

CONTENTS

Introduction

Charles M. Salter, P.E.

THE idea for this book developed from my teaching experience at UC Berkeley's College of Environmental Design and from our firm's consulting projects over the last twenty years. The purpose in putting together this information is to lend an understanding of acoustics and audio/video design to anyone without a background or formal training in acoustical engineering.

Acoustics play a significant role in our daily lives—whether at rest, work, or play, we are all aware of sound and how it affects us. In this book, the nature and control of acoustics are discussed in the context of history, physics, hearing, buildings, environment, psychology, law, and planning. These are issues our firm encounters during the course of consulting activities.

We have taken our most frequently asked questions and attempted to answer them with physical explanations and practical examples throughout the book. The text is supplemented with case studies that provide a wide range of first-hand information, already tested in practice.

The book is intended as a reference for practicing design professionals, building owners, attorneys, and people with an interest in this field. You can read the preliminary chapters to understand the whole picture or skip around to specific topics with which you want to become familiar. The information will help you learn the right questions to ask about noise abatement and offer an understanding of how proper acoustical conditions can make your project more successful.

Units and Equations

In this book, all of the equations employ both the International System of Units (metric) and the English Engineering System (English). All metric units are converted from their English counterparts.

The following is a list of all the units used in the book, their abbreviations, and their conversion factors between metric and English systems. The list is broken up into different categories (length, area, etc.). Within each section, the units are listed in alphabetical order, with the metric units always listed first.

LENGTH

1 meter (m) = 1×10^3 mm = 3.281 ft. = 39.37 in. = 1×10^{-3} km = 6.214×10^{-4} mi.

1 millimeter (mm) = 1×10^{-3} m = 3.281×10^{-3} ft. = 3.937×10^{-2} in. = 1×10^{-6} km

1 foot (ft.) = 0.3048 m = 304.8 mm = 12 in. = 1.894×10^{-4} mi.

1 inch (in.) = 2.540×10^{-2} m = 25.4 mm = $^{1}/_{12}$ ft.

1 kilometer (km) = 1×10^3 m = 1×10^6 mm = 3.281×10^3 ft. = 0.6214 mi.

1 mile (mi.) = 1.609×10^3 m = 1.609×10^6 mm = 5.280×10^3 ft. = 1.609 km

TIME

1 second (sec.) = $^{1}/_{3600}$ hr. = $^{1}/_{60}$ min.

1 hour (hr.) = 3.6×10^3 sec. = 60 min.

1 minute (min.) = 60 sec. = $^{1}/_{60}$ hr.

MASS

1 kilogram (kg) = 1×10^3 g = 2.205 lb

1 gram (g) = 1×10^{-3} kg = 2.205×10^{-3} lb

1 pound (lb) = 0.4536 kg = 453.6 g

AREA

1 square meter (m^2) = 1×10^6 mm^2 = 10.76 sq. ft. = 1.550×10^3 in.^{2} = 1×10^{-6} km^2 = 3.861×10^{-7} sq. mi.

1 square millimeter (mm^2) = 1×10^{-6} m^2 = 1.076×10^{-5} sq. ft. = 1.550×10^{-3} in.^{2} = 1×10^{-12} km^2 = 3.861×10^{-13} sq. mi.

1 square foot (sq. ft.) = 9.290×10^{-2} m^2 = 9.29×10^4 mm^2 = 144 in.^{2} = 2.296×10^{-5} acre = 3.587×10^{-8} sq. mi.

1 square inch (sq. in.) = 6.452×10^{-4} m^2 = 645.2 mm^2 = $^{1}/_{144}$ sq. ft.

1 acre = 4.047×10^3 m^2 = 4.356×10^4 sq. ft. = $^{1}/_{640}$ sq. mi.

1 square kilometer (km^2) = 1×10^6 m^2 = 1×10^{12} mm^2 = 1.076×10^7 sq. ft. = 0.3861 sq. mi.

1 square mile (sq. mi.) = 2.590×10^6 m^2 = 2.590×10^{12} mm^2 = 2.788×10^7 sq. ft. = 640 acre = 2.590 km^2

VOLUME

1 cubic meter (m^3) = 35.31 cu. ft.

1 cubic foot (cu. ft.) = 2.83×10^{-2} m^3

DENSITY

1 kilogram per cubic meter (kg/m^3) = 6.243×10^{-2} lb/cu. ft.

1 pound per cubic foot (lb/cu. ft.) = 16.02 kg/m^3

SPEED

1 meter per second (m/s) = 196.9 fpm = 2.23 mph

1 foot per minute (fpm) = 5.102×10^{-3} m/s = 40.91 mph

1 mile per hour (mph) = 0.4470 m/s = 88.02 fpm

POWER

1 watt (W) = 1×10^{-3} kW = 284.3×10^{-6} tons of refrigeration

1 kilowatt (kW) = 1×10^3 W = 0.2843 tons of refrigeration

1 ton of refrigeration = 3.517×10^3 W = 3.517 kW

PRESSURE

1 pascal (Pa), or **newton per square meter** (newton/m^2) = 4.0186×10^{-3} in. of H$_2$O = 2.089×10^{-2} psf = 1.450×10^{-4} psi

1 inch of H$_2$O (in. H$_2$O)* = 248.8 Pa = 5.197 psf = 3.609×10^{-2} psi

1 pound per square foot (psf) = 9.929×10^5 Pa = 3.990×10^3 in. H$_2$O = $^{1}/_{144}$ psi

1 pound per square inch (psi) = 6.895×10^3 Pa = 27.71 in. H$_2$O = 144 psf

VOLUME FLOW RATE

1 cubic meter per hour (m^3/hr.) = 0.588 cfm

1 cubic foot per minute (cfm) = 1.7 m^3/hr.

GALVANIZED SHEET METAL THICKNESS TOLERANCES

Gage	Nominal Thickness	
	(mm)	(in.)
33	0.2286	0.0090
32	0.3404	0.0134
31	0.3607	0.0142
30	0.3988	0.0157
29	0.4369	0.0172
28	0.4750	0.0187
27	0.5131	0.0202
26	0.5512	0.0217
25	0.6274	0.0247
24	0.7010	0.0276
23	0.7772	0.0306
22	0.8534	0.0336
21	0.9296	0.0366
20	1.006	0.0396
19	1.158	0.0456
18	1.311	0.0516
17	1.461	0.0575
16	1.613	0.0635
15	1.803	0.0710
14	1.994	0.0785
13	2.372	0.0934
12	2.753	0.1084
11	3.132	0.1233
10	3.510	0.1382
9	3.891	0.1532
8	4.270	0.1681

ACOUSTICS

1 metric sabin = 10.76 sabin

1 sabin = 9.290×10^{-2} metric sabin

1 hertz (Hz) = 1 cycle per second (cps)

Notes: Conversions taken from three sources: (1) *ASHRAE Handbook: Fundamentals*. American Society of Heating, Refrigerating and Air-Conditioning Engineers, Inc. (I-P Edition) (Atlanta, 1993), 35.1–35.2; (2) Hans C. Ohanian, *Physics*. 2d ed. (W.W. Norton & Company, 1989), Appendix 7; and (3) *HVAC Duct Construction Standards: Metal and Flexible*. Sheet Metal and Air Conditioning Contractors National Association, Inc., 1st ed. Appendix-1 (Virginia, 1985).

* in. of H$_2$O ref. 60°F

List of Illustrations

Figure 7.5 [STC contour]
Reprinted from ASTM E413.

Figure 7.6 [comparative STC values for various wall constructions]
Adapted from Michael Rettinger, *Acoustic Design and Noise Control* (Chemical Publishing Company, 1973).

Figure 7.7 [IIC contour]
Reprinted from ASTM E989.

Table 7.2
[STC ratings of wood frame floor/ceiling constructions]
Adapted from J.B. Grantham and T. B. Heebink, "Sound attenuation provided by several wood-framed floor-ceiling assemblies with troweled floor toppings," *Journal of the Acoustical Society of America*, vol. 54, no. 2 (1973), pp. 353–360.

Table 7.3.1
[IIC ratings of wood frame floor/ceiling constructions]
Adapted from J.B. Grantham and T. B. Heebink, "Sound attenuation provided by several wood-framed floor-ceiling assemblies with troweled floor toppings," *Journal of the Acoustical Society of America*, vol. 54, no. 2 (1973), pp. 353–360.

8

Figure 8.1
[threshold of human perception for vibration]
Adapted from the International Organization for Standardization (ISO) 2631-2 (1989), p. 11.

Figure 8.2
[generic criteria for vibration-sensitive equipment]
Adapted from "Generic criteria for vibration-sensitive equipment," Vibration Control in Microelectronics, Optics, and Metrology vol. 1619 (proceeding of The International Society of Optical Engineering Conference, November, 1991), p. 71.

9

Table 9.1
[recommended background noise levels in buildings]
Adapted in part from Chapter 43, *ASHRAE Application Handbook* (1995).

Table 9.2 [maximum air speeds in ductwork]
Adapted from Chapter 43, *ASHRAE Application Handbook* (1995).

Table 9.3
[recommended air speed in ductwork near diffuser]
Adapted from Chapter 43, *ASHRAE Application Handbook* (1995).

Table 9.5 [transformer sound ratings]
NEMA Standard ST 20 (1986).

Figure 9.19 [transmissibility]
Adapted from J. P. Den Hartog, *Mechanical Vibrations* (Dover Publications, Inc., 1985).

10

Figure 10.3 [pressure measured using the hydrophone with and without control]
Adapted from M.J. Brennan, S.J. Elliott, and R.J. Pinnington, "Active Control of Fluid Waves in a Pipe." *ACTIVE 95*, p. 383–394.

13

Figure 13.2 [plaintiffs' bedroom]
Charles M. Salter, photographer.

17

Figure 17.1 [Hollywood Bowl]
David R. Schwind, photographer.

Figure 17.2 [Camp Snoopy]
Charles M. Salter, photographer.

Figure 17.3 [Monterey Bay Aquarium]
Peter Dodge, EHDD, photographer.

Figure 17.4 [Airtouch Boardroom]
Chas McGrath, photographer.

Figure 17.5 [KQED]
Chas McGrath, photographer.

Figure 17.6 [Dwinelle Distance Learning Center]
Alan Geller, photographer.

Figure 17.7 [Harris Concert Hall]
Timothy Hursley, photographer.

Figure 17.8 [A/V Pacific Bell Corporate Television]
Chas McGrath, photographer.

Figure 17.9 [Cashman Complex]
Photographer unknown.

Figure 17.10 [Pleasant Hill City Hall]
Charles W. Callister, Jr., photographer.

Figure 17.11 [Oakland Federal Building]
Cesar Rubio, photographer.

Figure 17.12 [Skywalker Ranch]
Doug Salin, photographer.

Figure 17.13 [Treasure Island]
Courtesy of the Roma Design Group.

Figure 17.14 [Lucille Salter Packard Children's Hospital at Stanford]
Chas McGrath, photographer.

Figure 17.15 [Sutton Place]
Photographer unknown.

Figure 17.16 [Portside]
Lance Keimig, photographer.

Figure 17.17 [private residence]
Lance Keimig, photographer.

Figure 17.18 [Oceanside Water Pollution Control Plant]
Stefan Curl, photographer.

Figure 17.19 [University of Washington Biomedical Sciences Research Building]
Steve Keating, photographer.

Figure 17.20 [Newport Beach Public Library]
Nick Merrick, Hedrich Blessing, photographer.

Figure 17.21 [SFMOMA]
Russell Abraham, photographer.

Figure 17.23 [presentation studio at CMSA]
John Sutton, photographer.

Figure 17.24 [Off Planet]
Chas McGrath, photographer.

Figure 17.25 [Mudd's]
Russell Abraham, photographer.

Figure 17.26 [Stanford Shopping Center]
Charles M. Salter, photographer.

Figure 17.27 [Todd-AO]
Karen Heyman, photographer.

Figure 17.28 [Dolby]
Doug Salin, photographer.

Figure 17.29 [One Market Plaza]
Timothy Hursley, photographer.

Figure 17.30 [Fairfield Center for Creative Arts]
Timothy Hursley, photographer.

Figure 17.31 [Stanford Theater]
Russell Abraham, photographer.

Figure 17.32 [Moraga Valley Presbyterian Church]
Perry Johnson, photographer.

18

Table 18.1
[California building code requirements during design]
Reprinted from *California Building Code*, UBC Section 1208A.

Table 18.2
[minimum California building after construction]
Reprinted from *California Building Code*, UBC
Section 1208A.

Table 18.3 [minimum noise isolation requirements]
HUD Minimum Property Standards: Multi-family
Housing, 1973.

Table 18.4 [The relationship between STC of a partition
and audibility of speech and music for various back-
ground noise levels]
Taken from Charles M. Salter and Richard B. Rodkin,
"Community Noise and Residential Acoustics,"
California Builder Magazine (Jan–Mar 1980).

Table 18.6 [typical noise criteria levels]
Adapted from Chapter 43, *ASHRAE Application
Handbook*, 1995.

19

Table 19.6 [degree of satisfaction/dissatisfaction]
Adapted from W. J. Cavanaugh, W. R. Farrell, P. W.
Hirtle, and B.G. Watters, Bolt Beranek & Newman,
"Speech Privacy in Buildings," *Journal of the Acousti-
cal Society of America*, vol. 34 (1962), pp. 475–492.

1

History

Rachel E. Dangermond

The Bible	6th century BC	325 BC	AD 27
Exodus 26:7 God details how the tabernacle is to be constructed, with acoustical instructions	Pythagoras investigates musical tuning and temperament	Aristotle writes about the production and reception of sound, and the cause of echoes	Marcus Vitruvius Pollio writes *De Architectura*

IN the last quarter of the 20th century, acoustical design has improved the quality of many buildings, and moved beyond, to influence most of our living environments. However, the desire to understand and control acoustics is not a product of the modern age—the behavior of sound has been contemplated since the dawn of civilization. One of the earliest written acoustical instructions appears in the Old Testament, in Exodus 26:7, where God explains to Moses how to construct the tabernacle.[1] These instructions detail the size, construction, and way in which curtains of goat hair should be hung over the tabernacle, prompting historians to speculate that the advice was acoustical in nature.

As early as the 6th century BC, Pythagoras had started investigating musical tuning and temperament. By using a monochord, a type of single-string instrument, he was able to measure the mathematical equivalents of musical intervals. By 325 BC, Aristotle had begun writing about the production and reception of sound, and the cause of echoes. The ancient Greeks were particularly involved in acoustical research and development, applying their empirical findings to the design of outdoor theaters (see

Mid-16th century	1650	Mid-17th century	1673
Adrian Willaert works with antiphonal composition for the Cathedral San Marco	Athanasius Kircher designs a parabolic horn	Robert Boyle observes that sound waves require a medium for transmission	Athanasius Kircher writes *Phonurgia Nova*

Figure 1.1 The Greek Amphitheater.

Figure 1.1). The design elements that shaped the sound quality of these theaters were (1) good line of sight to the stage; (2) use of a reflecting wall; and (3) low ambient noise. Greek amphitheaters achieved an excellent sound quality and many are still being used today.

The ancient Romans were equally involved in the development of acoustical design principles. In Marcus Vitruvius Pollio's famous ten-volume treatise, *De Architectura*, he suggested the use of resonant vessels to enhance sound in theaters (ca. AD 27). Archaeologists have found examples of these sounding vessels in ancient theaters throughout the Greco-Roman world.

Perhaps the most significant example of acoustical design was the Renaissance cathedral and its influence on liturgical ritual and music. The long reverberation times found in cathedrals contributed to the overall grandeur of these architectural spaces. About the middle of the 16th century, composers had begun tailoring music to take advantage of this rich reverberance. Composer Adrian Willaert's music was inspired by the unusual layout of the Cathedral San Marco in Venice (Figure 1.2), with its facing balconies and unique double organ loft. San Marco's basilica has a rich, natural reverberant sound that makes separate singing by small choirs overlap in time and lends an interesting, spatial element to music.

Figure 1.2 Cathedral San Marco.

Early 18th century	1725	1735	1816
Joseph Sauveur introduces the term acoustics to define the science of sound	Brooke Taylor determines a formula for the relationship between the length, tension, and mass of stretched string with respect to sound frequency	Sir Isaac Newton attempts to determine the speed of sound in the air	Pierre Simon Laplace formulates the equation for calculating the speed of sound

By the mid-17th century, the phenomenon of sound reflection and echoes was explained by most scientists as being similar to the manner in which light beams reflect from a mirror (as shown in Figure 1.3). In 1673, Athanasius Kircher published *Phonurgia Nova*, which described the way sound is reflected, concentrated, and dispersed by the shape of its enclosure.[2] (See Figures 1.4 and 1.8.) Kircher had earlier designed a parabolic horn that served as both a hearing aid and speaking trumpet. Another discovery came from Robert Boyle, who realized that sound does not travel through a vacuum, but instead requires a medium (air or solid structure) for transmission.[3]

In the early part of the 18th century, Joseph Sauveur introduced the term acoustics to define the science of sound.[4] Sauveur used acoustics to

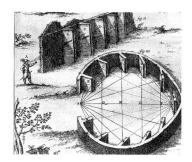

Figure 1.3 The geometric concentration of sound reflection is illustrated in this seventeenth century drawing.

understand and appreciate music; he pioneered studies on the relationship of physical measures of frequency and perceived pitch. The study of musical acoustics continued with Brooke Taylor's expansion on Pythagoras's work. Taylor used the monochord to determine a formula for the rela-

Figure 1.4 Athanasius Kircher's interpretation of what a built theater might look like, using Vitruvius's design.

1822

Architect Charles Bullfinch applies stretched fabric across the dome of the Hall of Representatives to mitigate echoes

1830

Felix Savart fashions a device to measure the limits of human hearing • The siren is used to ascertain the frequency of a musical note by ear

1837

Architect Robert Mills raises the floor of the Hall of Representatives

1839

John Scott Russell presents the principle of the isacoustic curve

Figure 1.5 Drawing of the Hall of Representatives chamber in the U.S. Capitol, following its reconstruction after the war of 1812.

tionship between the length, tension, and mass of stretched string with respect to the frequency of sound.[5] In 1735, Sir Isaac Newton put forth equations to determine the speed of sound in the air, but not until 1816 was the correct equation for calculating the speed of sound discovered by Pierre Simon Laplace.[6]

Research in the 19th century made substantial contributions to the understanding of how sound behaves in architectural spaces. This research was prompted by the growing need for public assembly spaces and the development of orchestral music. After the war of 1812, architect Charles Bullfinch was hired to mitigate the poor acoustics in the Hall of Representatives (see Figure 1.5). Realizing that fabric absorbed sound, Bullfinch

1847	1850	1862	1876
Russell delivers lecture, "On the Arrangement of Buildings with Reference to Sound," to Royal Institute of British Architects	Joseph Henry discovers the precedence effect	Helmholtz publishes *On the Sensations of Tone*	Alexander Graham Bell invents the microphone

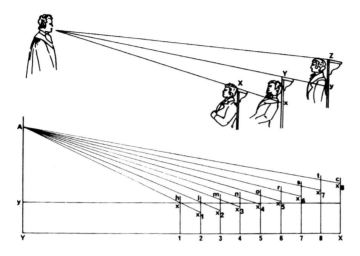

Figure 1.6 The isacoustic curve, as presented by John Scott Russell.

applied stretched fabric across the entire dome of the hall. This technique was also applied in concert halls as a remedy for echoes, as in the work of Gustav Lyon in France.

Bullfinch's treatment did not completely solve the acoustical problems of the Hall of Representatives. However, several years later, architect Robert Mills thought to raise the floor of the Hall by four feet, shifting the focal point of the dome below the listening plane and significantly improving the overall sound quality.[7] In 1839, John Scott Russell, a professor of natural philosophy at the University of Edinburgh, published an article that presented the principle of the isacoustic curve—equal sound coverage across an audience (see Figure 1.6). Russell's theories were published in three issues in the then-popular *Building News*.[8] The methods and theories used by Bullfinch, Mills, and Russell are all early examples of acoustical treatments for improving speech intelligibility.

Around this early part of the century, a number of inventions and applications helped with our understanding of acoustics. Felix Savart fashioned a device to measure the limits of human hearing, using an instrument, similar to a siren, as a measuring device to ascertain frequency.[9] (See Figure 1.7.)

Perhaps the greatest contribution of the time came from Joseph Henry and his discovery of the precedence effect, which he called the *limit of perceptibility*. Using the lawn facing the west wall of the old Smithsonian Building, Henry determined the distance from the wall where he could stand and still detect an echo when he clapped his hands. Standing less than 30 feet from the wall, he observed no echo or prolongation of the sound. He concluded that a total to and fro distance of 60 feet, corresponding to

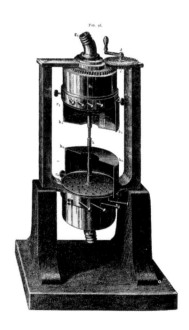

Figure 1.7 A measuring device similar to a siren.

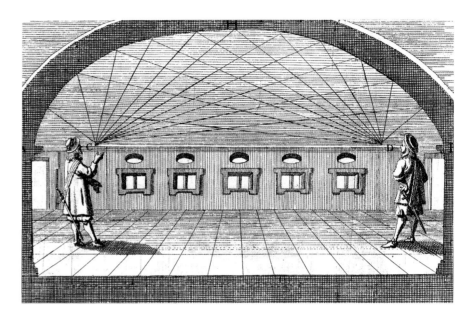

Figure 1.8 An early illustration of the way curved surfaces focus sound.

the time delay interval of approximately .05 seconds was the upper limit for when a direct sound, followed by a delayed sound, was not observed as an echo. When less than .05 seconds elapsed, the direct sound blended with the delayed sound to produce a single acoustical image. When the time delay between direct and reflected sound exceeded this limit, an echo was heard.[10] Henry also correctly identified the role of room volume, shape, and surface absorption on the behavior of sound waves. He is considered the first researcher to conduct quantitative studies in architectural acoustics and to apply his research to the design of buildings.

Henry's contemporary, Hermann Helmholtz, worked to develop acoustical principles into a science. Helmholtz is known for his exploration of the mathematics of musical tones and his definition of timbre. He wrote an extensive treatise on acoustics and hearing called *On the Sensations of Tone* (1862).[11] In 1876, Alexander Graham Bell invented the microphone, an important instrument that fostered the growth of electroacoustics. The English physicist, Lord Rayleigh, published *The Theory of Sound,* which marked a turning point in acoustics and is often cited as the beginning of modern acoustical theory (1877). It is still used as a reference by scientists, engineers, and musicians interested in the mathematical basis of sound and vibration.[12]

While many acoustical concepts presented during this time added to the growing knowledge of acoustics, some were simply ill-conceived. Architect Alexander Saeltzer proposed in his 1872 book, *A Treatise on*

Acoustics in Connection with Ventilation Both Ancient and Modern, that bad, gloomy, and poisonous air was really to blame for poor acoustics. He claimed proper ventilation was the key to fixing the problem. In his treatise, Saeltzer suggests that sound "…not knowing which way to turn; poison on all sides, ever anxious to do its duty, full of natural vitality… becomes disheartened, leaves first its battle field…and at last, exhausted, looks up as high as possible to attain rest in the strata of warmer more flexible air, air more congenial to its nature, which is always found at the highest point."[13]

Toward the end of the 19th century, an important contributor to the acoustical canon was Wallace Clement Sabine, a young physics instructor. Sabine was commissioned to design a solution to the acoustical problems in Harvard University's Fogg Art Museum lecture hall (see Figure 1.9). The reverberation in the hall made it unusable for lectures. Without the benefit of a sound level meter, Sabine imaginatively measured the reverberation time in the room by using an organ pipe. In his experiments, he would sound the pipe and release it, at which point he would start a clock, and stop it when the sound became completely inaudible. He recorded the time it took for the sound to decay and plotted the test results on a graph that showed how various materials absorb sound.

Sabine was assigned a laboratory at Harvard to continue conducting acoustical experiments and to analyze many other spaces on the campus.[14] Five years later, he developed the mathematical formula for the prediction of reverberation time in large rooms. In 1895, he was hired to work on the acoustics of the proposed Boston Symphony Hall with architect Charles McKim. Sabine encouraged McKim to use a rectangular plan for the hall, rather than a semicircular shape, patterning it after the famous Veriesaal Music Hall in Vienna[15] (see Figure 1.10). Boston Symphony Hall is considered one of the finest concert halls in the world today, but because it was significantly larger than any other hall at that time (with over 2,600 seats), the first touring European orchestras were quick to criticize the acoustics. Their criticism led Harvard to deny Sabine tenure.

Sabine then moved to Illinois, where, with the financial backing of a wealthy textile mogul, he built the world's first acoustical laboratory. Riverbank Acoustical Labs opened in 1918 and was dedicated to full-scale measurement of the sound absorption of materials and sound transmission of wall structures. Sabine is now considered to be the first acoustical consultant. The sabin, the unit for acoustic absorption, was named in his honor.

The first acoustical tile was developed and patented by Sabine and R. Gustavino. During the twenties, a number of acoustical materials were

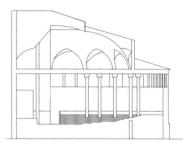

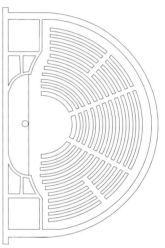

Figure 1.9 The Lecture Hall in the Fogg Art Museum, Harvard University.

1926

Warner Brothers presents the
first full-length motion picture
w/ synchronized sound

1927

Floyd Watson builds and equips
the first anechoic (no echo)
chamber

1929

Acoustical Society of America
established

Pre 1930

Acoustone made by the United
States Gypsum Company

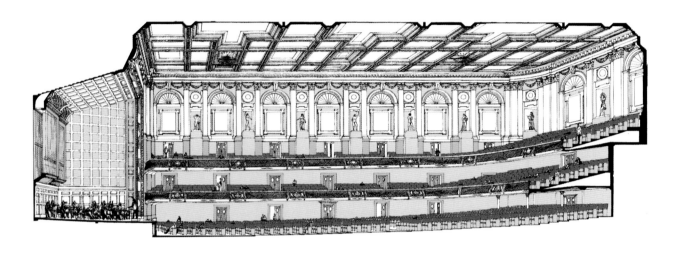

Figure 1.10 Boston Symphony Hall.

developed by various manufacturers: cabots quilts (made from seaweed);
balsam wood (composed of wood fibers felted into a low-density blanket);
and Flaxlinum (an insulation board) are just a few examples. Acousti-
Celotex, the first practical acoustical tile, was developed around this same
time. The United States Gypsum Company introduced Acoustone, a form
of fissured mineral tile, which was flame resistant and more attractive than
its predecessors.

Acoustical explorations in the early part of the century continued
in the same direction that Sabine had begun. In 1923, Floyd Watson published *Acoustics of Buildings,* a practical guide to architectural acoustics. His
work included a table of absorption coefficients for materials and examples of how architects can design rooms that are acoustically desirable.
Edgar Buckingham created a theoretical formulation for sound transmission loss in 1925, developing the formula based on his measurements of
decaying sound fields in a room; as with Sabine's earlier work, the only
instruments available to Buckingham at the time were a stopwatch and his
ears. Two years later, Floyd Watson was the first to build and equip an anechoic chamber.[16]

Films with sound were being introduced around this time, presenting new acoustical design challenges for movie theaters. Theaters had originally been designed to accommodate the Wurlitzer organs that played during silent films. In 1926, Warner Brothers presented the first full-length
motion picture, with synchronized sound accompaniment, using equipment from Bell Laboratories.

Radio broadcasts were widely popularized by this time, helping to advance the acoustical quality and speech perception in recording rooms. Carl Eyring addressed this subject in his paper, "Reverberation in Dead Rooms," wherein he generalized Sabine's reverberation formula to fit broadcast studios and motion picture sound stages (1930).

Around the thirties, the first sound level meter was invented by Paul Sabine, Wallace Sabine's cousin. The sound level meter allowed an objective measure of sound pressure level. In 1929, The Acoustical Society of America was established, and, at the instigation of manufacturer Wallace Waterfall, the leading materials manufacturers formed the Acoustical Materials Association in 1933. They selected Riverbank Acoustical Labs as their official test site to rate sound-absorbing wall and ceiling materials.

In 1931, the Johns-Manville Company built the first reverberation room to develop sound-absorbing products. Owens-Corning Fiberglas Corporation developed a new type of acoustical material in the forties: a glass fiber mat bonded with a phenolic resin that could be formed into a blanket, board, or finished tile. The material was not only flame resistant, but resistant to the effects of high humidity, a problem with many of the other materials. Glass fiber materials are now commonly used in a wide variety of acoustical applications.

Spanning the forties and fifties, surveys of the acoustical quality of multi-family housing were underway in England and other European countries. A series of technical papers summarized this testing and correlated the results of airborne sound isolation and impact noise with a survey of people's complaints. The most common complaints were of sound from doors shutting, radios, televisions, footsteps, children playing, plumbing noise, adult voices, and lifts (elevators). This airborne and impact insulation data were organized into three groups.[17] The findings about people's response to noise intrusion are similar to those criteria recommended by the Department of Housing and Urban Development (HUD) for sound control in low-cost, market-rate, and luxury housing (as discussed in Chapter 18).

Also during this time, cinematic innovations such as CinemaScope and Cinerama had begun using multi-channel sound to create an acoustic analogue to wide screen formats. In the forties, Disney's landmark film, *Fantasia,* used a system called *Fantasound* that featured as many as 96 speakers to play back 7 tracks of sound.

In the fifties, Erwin Meyer and his associates at Göttingen University attempted to correlate desirable sound quality in a room with acoustical measurements. They determined that the earlier, short-term portion of the

1940s

Disney's *Fantasia* uses seven channels located behind the projection screen

1940s & 1950s

Surveys of the acoustical quality of multi-family housing are underway in England and other European countries

1950

Erwin Meyer and his associates at Göttingen University attempt to correlate a desirable sound quality in a room to early reflection patterns

1950s

Karl Kryter publishes *The Effects of Noise on Man*

reverberation process was important for the perception of sound quality in concert halls. Up until then, the applied science of concert hall acoustics had focused on correcting disturbing echoes. At this time, Karl Kryter wrote *The Effects of Noise on Man,* a groundbreaking work that distinguished between psychological and environmental effects of noise levels.

At the introduction of electronic devices for recording and playback, Bell Laboratories in New Jersey began studying the nature of room acoustics for recording studios and theaters. In more than fifty years of pioneering research in this area, Bell Labs became associated with many notable inventions and famous scientists, including Harvey Fletcher and Vern Knudsen. During the sixties, researchers at Bell Labs had already started using computers to analyze room acoustics. Bishnu Atal and Manfred Schroeder, among others, demonstrated the utility of computers for both prediction and simulation of room acoustics. This field has continued to the present day in the form of auralization, "techniques for virtual acoustic simulation of an enclosure." The Japanese acoustician Yoichi Ando has proposed an auralization system whereby concertgoers can selectively compare different seats in a concert hall before purchasing tickets.

In 1965, Dolby Laboratories introduced noise reduction techniques for audio recordings. Dolby continues to make considerable technological improvements in all aspects of sound recording and reproduction quality in theaters. Twenty years after Dolby, Lucasfilm's Tomlinson Holman, founder of THX sound, extended these improvements to the cinema by setting acoustical standards for equipment and the acoustical quality of the theater.

In 1972, mounting evidence on how acoustics affect our lives led the U.S. Congress to enact the Noise Control Act. The U.S. Environmental Protection Agency followed two years later with a document entitled "Information on Levels of Environmental Noise Requisite to Protect Public Health and Welfare with an Adequate Margin of Safety" ("EPA Levels Document"). The EPA Levels Document addresses hearing safety, activity interference, and the issue of annoyance, and forms the basis of many modern noise-impact criteria.[18]

The field of acoustics has progressed from a combination of logic and conjecture into a science, where physical laws are described by mathematics and increasingly enriched by technology. Physics, mathematics, architecture, environmental science, psychology, anatomy, electricity, and computer science have all contributed to our knowledge of acoustics. This historical account offers an overview of the important events that helped shape the acoustical principles and applications discussed throughout this book.

1965	1972	1974	1980s
Dolby Laboratories introduces noise reduction techniques for audio recordings	US Congress enacts the Noise Control Act	The US Environmental Protection Agency publishes *EPA Levels Document*	Lucasfilm's Tomlinson Holman, founder of THX sound, sets acoustical standards for cinema equipment and the acoustical quality of the theater

Notes

1. Ex 26:7a, *New Oxford Annotated Bible.*

2. Cecil D. Elliot, *Technics and Architecture* (Massachusetts Institute of Technology, 1992), p. 408.

3. R. Bruce Lindsay, et al., "Acoustics: Historical and Philosophical Development." *Benchmark Papers in Acoustics*, ed. R. Bruce Lindsay (Dowden, Hutchison & Ross, Inc., 1973), p. 11.

4. Joseph Sauveur, "Système Général des Intervales du Son." *Acoustics: Historical and Philosophical Development,* trans. and ed. R. Bruce Lindsay, p. 88.

5. Lindsay, *Acoustics*, p. 95.

6. Ibid., p. 13.

7. Elliot, *Technics*, pp. 411–412.

8. Ibid., pp. 419–420.

9. Frederick V. Hunt, *Origin of Acoustics* (The Acoustical Society of America through the American Institute of Physics, 1992), p. 136.

10. Robert S. Shankland, "Architectural Acoustics in America to 1930." *Journal of the Acoustical Society of America*, 61.2 (February 1977), pp. 250–251.

11. Hermann Helmholtz, trans. Alexander J. Ellis, *On the Sensations of Tones* (Dover Publications, Inc., 1954), pp. 152–153.

12. Robert Bruce Lindsay, *Introduction to the Theory of Sound, Lord Rayleigh* (Dover Publications, 1945), p. 25.

13. Alexander Saeltzer, *A Treatise on Acoustics in Connection with Ventilation* (D. Van Nostrand, 1872).

14. Elliot, *Technics*, pp. 423–427.

15. Leo L. Beranek, *How They Sound: Concert and Opera Halls* (The Acoustical Society of America through The American Institute of Physics, 1996), p. 79.

16. Hale J. Sabine, "Building Acoustics in America, 1920–1940." *Journal of the Acoustical Society of America*, 61.2 (February 1977), p. 255.

17. A. Chapman, "A Survey of Noise in British Homes." *National Building Studies*, Technical Paper No. 2 (1948); P. H. Parkin and E. F. Stacy, "Recent Research on Sound Insulation in Houses and Flats." *Journal of the Royal Institute of British Architects*, 61, 9 (June 1954), p. 372.

18. U. S. Environmental Protection Agency, *Information on Levels of Environmental Noise Requisite to Protect Public Health and Welfare with an Adequate Margin of Safety*, Rept. 550/9-74-004, Washington, D.C.:GPO (1974).

2

Fundamentals

Durand R. Begault, Ph.D.

Wave Frequency • Sound Pressure and Sound Pressure Level • Predicting Sound Pressure Levels from Multiple Sources • Directivity • Diffuse Sound Fields • The Inverse Square Law

A *sound wave* is a physical disturbance of molecules within a medium—air, water, or solid—that can be detected by a listener. Most sound waves result from a vibrating object. Look around and you'll see countless objects in a state of vibration: the windows in your house when a truck drives by, a guitar when its strings are plucked, or tree branches in the wind. Each of these are examples of a sound source. These different waves combine and reach a listener via numerous direct and indirect pathways. The listener's inner ear contains organs that vibrate in response to these molecular disturbances, converting the vibrations into changing electrical potentials that are sensed by the brain—allowing the phenomenon of hearing to occur.

Acoustical analysis involves not only the sound source but also who is hearing it (receiver) and everything in between (the path). The path is made up of the environment encompassing both sound source and receiver. The medium of transmission can either be air, or a combination of mediums, involving a conversion to vibration and then back to sound, through solid objects such as walls and floors. Figure 2.1 shows an example of the chain of events from sound source to listener and a generalized source-path-receiver model.

Figure 2.1 *(top)* A generalized source-path-receiver model. The source is an entity that causes acoustical vibration. The path is the environmental context and the transformational aspects of the medium. The receiver can refer to a human listener or a microphone.

(bottom) An example of the source-path-receiver model. The circles indicate sound radiation; their darkness indicates relative intensity. A listener (receiver) is listening to music from a record player; this is a desired sound source. In an adjoining utility room, a machine emits sound and infrasonic vibration; this is a noise sound source. The two rooms and their surfaces (path) transform the acoustical vibration from the sources before they reach the listener. The sound from the loudspeaker will reach the ears of the listener via a direct path and indirect paths from reflections off of walls; the latter is termed *reverberation*. The machine is heard and felt through the wall as a result of *sound transmission* (the wall itself becomes a vibrating source) and possibly due to sound *leakage* through cracks in the construction.

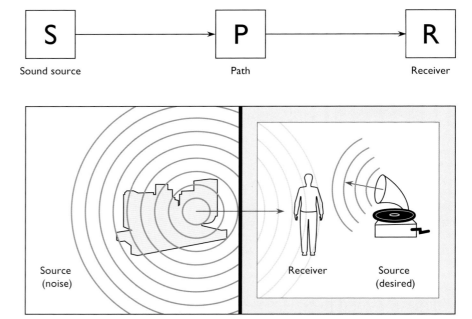

The perception of a listener can be influenced by the treatment of either the path or the source. For instance, we can enhance the intelligibility of speech in a conference room by electronically amplifying the spoken voice, or the sound output from a power plant can be reduced to limit the disturbance in a community. Note the distinction in these examples between the spoken voice and the sound sources. The spoken voice is desirable, while sound from a power plant is not; we refer to undesirable sound as noise.

Wave Frequency

The molecular disturbance caused by an acoustic source involves a series of high and low pressure areas (termed *compression* and *rarefaction*). Figure 2.2 shows five discrete moments of time that comprise a single wave cycle. An equivalent illustration, indicating pressure variation continuously over time, is shown in Figure 2.3.

A sound's *frequency* is defined in terms of the number of wave cycles that occur during one second. The unit used for describing frequency is *hertz* (Hz). For higher frequencies, *kilohertz* (kHz) is used to indicate the number of oscillations times 1,000 that occur within a second. For example, 1.68 kHz (1.68 x 1000 (kilo) Hz) is the same frequency as 1,680 Hz.

Figure 2.2 Compression and rarefaction of air molecules at five discrete moments of time. The "+" indicates compression (an increase in pressure) and the "–" indicates rarefaction (a decrease in pressure). This represents a single cycle of pressure variation.

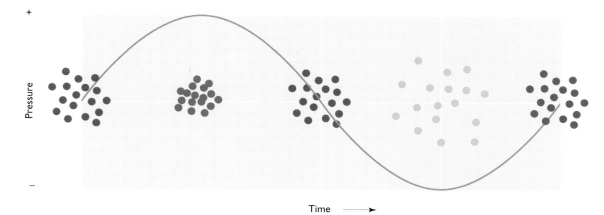

Time ⟶

If you drop a rock into the middle of a lake, ripples propagate outward from the point of contact. These circular ripples are comparable to sound waves traveling through air. If you count the number of wave ripples that pass a single point on the lake during one second, you can calculate the wave's frequency. For example, the frequency of the red wave in Figure 2.4 is five times greater than the frequency of the black wave.

Waves that have a repeated pattern of oscillation are called *periodic waves*. Figures 2.3 and 2.4 show the simplest type of periodic wave, the *sine wave*. Sine waves (also called "pure tones") have a single constant frequency, obtainable only from electronic devices.

How do these frequencies relate to hearing? When the frequency is in the range of roughly 20 Hz to 20 kHz, the waves are heard as sound waves; these are termed audio frequencies. Human speech contains frequencies that lie between 200 Hz and 5 kHz; the sound of an orchestra can contain frequencies between 25 Hz to 13 kHz or even higher. Frequencies below 20 Hz are sensed as vibration, are not audible to most people, and are termed *infrasonic*. Frequencies above 20,000 Hz are termed *ultrasonic*.

Figure 2.5 shows the typical frequency range for various sound sources. Many situations encountered in buildings involve a combination of both audio and infrasonic frequencies; that is, sound and vibration. For instance, at frequencies up to around 100 Hz, such as those produced by a pipe organ, it is possible to simultaneously hear sound and feel vibrations.

Real-world waves are not as periodic as those just described; in fact, most waves usually contain a mixture of many frequencies. While a sine wave is considered technically to be a "simple" wave, in actuality, almost all waves in nature are "complex," in that they contain multiple frequencies. The reason a violin and a viola sound different from each other is because each has a different combination of frequencies, which is referred to as the sound source's *spectrum*. The interaction and behavior of the different frequencies within a spectrum can be quite complex, and are in fact responsible for the rich palette of sound colors that we experience daily. Using the treble and bass tone controls of a home audio system is an everyday example of how a sound's spectrum can be changed by selectively emphasizing some frequency components and de-emphasizing others.

Figure 2.3 A single cycle (wavelength period) of a continuously repeating wave (here, a sine wave) is shown as a continuous function of time on the x axis, with pressure on the y axis shown in both positive and negative directions from the center line.

The speed of sound through air depends on a number of environmental factors such as temperature and humidity: a good approximation is 344 m/sec (1,128 ft/sec). The speed will vary depending on the propagation medium; for instance, the speed of sound is faster through water than through air.

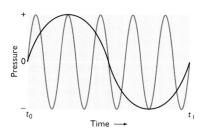

Figure 2.4 Two sine waves with different frequencies. The red sine wave has a frequency that is five times the frequency of the black sine wave, because there are five repetitions of the wave over the time span t_0–t_1.

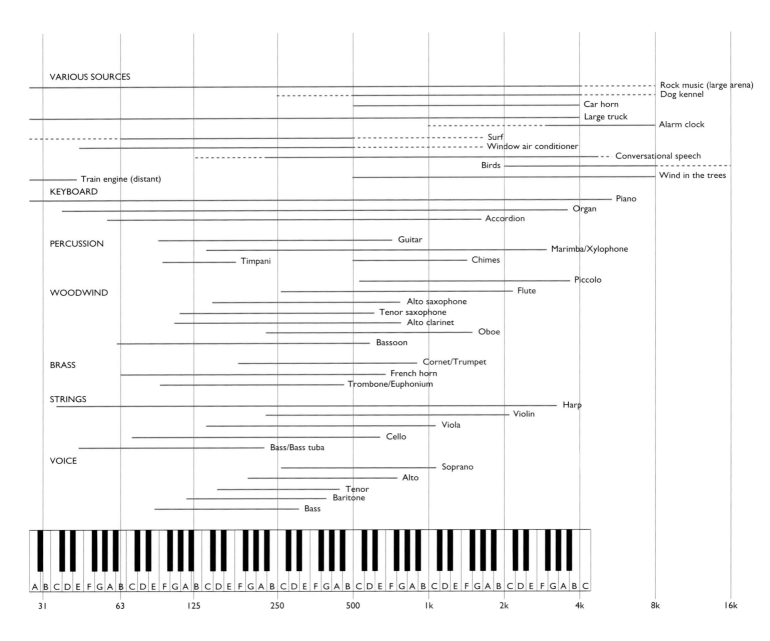

Figure 2.5 Frequency range of typical sound sources.

The wave length of a sound (λ) is related to the speed of sound (c) and the frequency (f) by the following formula: $\lambda = c/f$

Although noise was previously defined as "undesirable sound," it technically refers to sound waves with no periodic frequency. Figures 2.6 and 2.7 show example plots of noise—noise consists of random frequencies that vary from moment to moment and sound like a waterfall. If the noise is concentrated in a narrow range of frequencies, it is termed *tonal noise*. Most sound sources in our environment are composed of a combination of these types of periodic and random signals. For instance, Figure 2.8 shows a speech signal of the word "left." Speech is made up of a sequence of sound events with unique frequency characteristics called phonemes. The phoneme "e" is periodic—we could sing a song with just this "e" sound by changing its frequency. On the other hand, the phoneme "f" is noisy, caused by the air that passes between the teeth and lower lip. Try this out for yourself by saying the word "left" very slowly and extending the "f."

Sound Pressure and Sound Pressure Level

The concept of *sound pressure* is basic to an understanding of sound waves. Figure 2.3 shows how sound pressure represents an increase and decrease above and below the atmospheric air pressure we normally experience. A variation in sound pressure is perceived as a change in loudness; loudness is discussed in Chapter 3.

The range of sound pressures that humans can detect is enormous. The quietest sound a typical young person can hear is equivalent to 20 *micropascals* (.00002 pascals), while the most intense sound that humans can tolerate is equivalent to a sound pressure of around 200 *pascals* (Pa). This is a change in magnitude of 10,000,000 to 1! By using a particular logarithmic unit known as the *decibel* (dB), a wide range of pressure measurements are compressed onto a logarithmic scale. The dB scale is easy and convenient to use when describing sound. The range of decibels most commonly encountered in acoustics extends from 0 to 140 dB—0 dB corresponding to the threshold of hearing, and 140 dB corresponding to the threshold of pain. Within these limits is the dynamic range of the auditory system. A sound pressure expressed using the dB scale is termed the *sound pressure level* (SPL) and is the most frequently used metric in acoustics. In order to go from sound pressure to SPL, there are three steps:

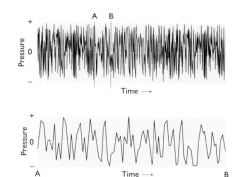

Figure 2.6 *(top)* A noise wave: contrasting the sine wave, a noise wave is completely aperiodic.

Figure 2.7 *(bottom)* A close-up of the noise shown in Figure 2.6.

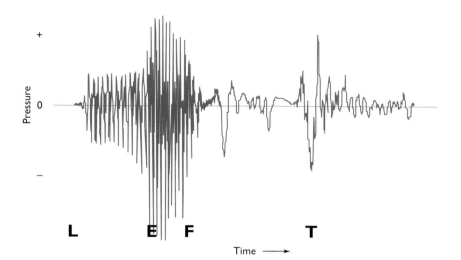

Figure 2.8 A speech wave of the word "left." Note how this wave varies over time. The "e" portion of the wave is more periodic than the "f" portion, which is noisy.

(1) convert the sound pressure at successive instantaneous values into an average sound pressure over a particular time period; (2) express this average value as a ratio to a reference level that is based on the threshold of hearing; and (3) convert to a decibel scale by multiplying 20 times the logarithm of that ratio. Table 2.1 shows comparative sound pressure and sound pressure level values for common sound sources.[1]

Predicting Sound Pressure Levels from Multiple Sources

In many cases, it is desirable to predict how the sound pressure level would change by adding additional sound sources. For example, consider the sound of a single pump in a mechanical equipment room. An engineer

Technically, sound pressure is a measure of acoustic force over a unit area, measured in newtons per meter squared (newton/m^2). One newton/m^2 is equivalent to one pascal (Pa); it's easier to say "2 pascals" than "2 newtons per meter squared." A *newton* is the amount of force needed to accelerate a mass of one kilogram one meter per second per second.

Sound pressure (Pa)	Sound pressure level (dB)	Example sound source
200.0	140	Threshold of pain
20.0	120	Near a jet aircraft engine
2.0	100	Near a jackhammer
0.2	80	Typical factory
0.02	60	Normal speech level
0.002	40	Quiet living room
0.0002	20	Quiet recording studio
0.00002	0	Threshold of hearing

Table 2.1 Comparison of sound pressure and dB SPL for typical sound sources.

The mathematical definition of sound pressure level:

$$\text{dB SPL} = 20 \log (P_1/P_0)$$

Where the value of P_1 is the average pressure of the wave, and P_0 is ambient atmospheric pressure (equivalent to the threshold of hearing). An international standard sets P_0 to a sound pressure level of .00002 newtons/m^2. When P_1 is equal to P_0, the equation works out to be equal to the standardized reference level of 0 dB.

Difference between two sound levels	Add to higher sound level
0 or 1 dB	3 dB
2 or 3 dB	2 dB
4–9 dB	1 dB
10 dB or more	0 dB

Table 2.2 Decibel Addition

Newtons? Pascals? Hertz? Decibels? These scientific units are based on the names of prominent figures in science. Hertz honors the 19th century scientist Heinrich Hertz. Newtons are named after Sir Isaac Newton, who sat under apple trees and invented differential calculus. Pascal was a famous French mathematician of the 17th century. And the decibel is named after Alexander Graham Bell, the father of telephony.

measures the sound at a reference distance (typically 1 m) and obtains a level of 65 dB. The engineer wants to know what the sound level would be if several more pumps were added in the room. How does an engineer predict the sound level of multiple sound sources? In our example, adding an additional pump in the room would not double the sound pressure level; dB values are not additive. A simple calculation of the total resulting sound pressure level can be made by using the following shortcut for decibel addition: (1) if the difference between two sound levels is 0 or 1 dB, add 3 dB to the higher level; (2) if the difference between two sound levels is 2 or 3 dB, add 2 dB to the higher level; (3) if the difference between two sound levels is 4 to 9 dB, add 1 dB to the higher level; (4) if the difference between two sound levels is 10 dB or more, the result is the higher of the two sources; and (5) to combine more than two levels, first add the two lowest together according to the above rules; then add the next two lowest levels together until only two values are obtained. Then the above rules are applicable (see Table 2.2).

Referring back to the problem, we have one pump measured at 65 dB. Two pumps with the same level will result in an additional 3 dB (or 68 dB total). For three pumps, the two lower sound levels are added together: since they are the same level, we get 68 dB + 65 dB, resulting in 70 dB. For four pumps, 65 dB + 65 dB = 68 dB for pumps one and two; 65 + 65 dB = 68 dB for pumps three and four; and therefore add 68 + 68 dB, which results in an overall level 71 dB. With five pumps, the total is 72 dB (71 dB + 65 dB). Thus, five pumps would be 7 dB louder than one pump.

To combine multiple sound sources together of the same intensity, the formula is SPL + 10 log(N) = Total sound level, where SPL is the sound level of one sound and N is the number of sources. To check our answer above: 65 + 10 log(5) = 72 dB.

Directivity

The spatial properties of either a sound source or a receiver at various frequencies and directions constitute its directional pattern or *directivity*. When a sound source radiates energy evenly in all directions it is called omnidirectional. Therefore, an omnidirectional microphone would be equally sensitive to sounds from all directions, but most sound sources emit more power in some directions than others. A jet engine, for instance, is much louder on its exhaust side than on its intake side. Each frequency that makes up the sound source will have its own directivity. Figure 2.9 shows the directional pattern for various frequencies emitted by a loudspeaker.

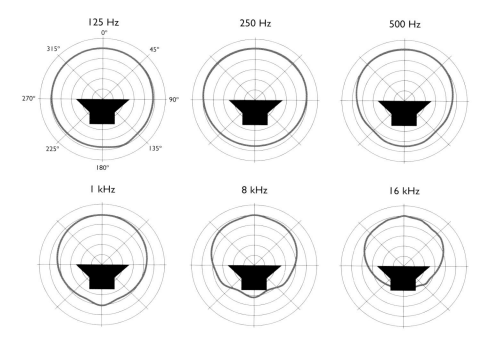

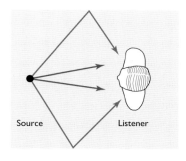

Environmental context

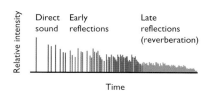

Figure 2.9 Directional properties of a loudspeaker. Each plot shows the directivity for a different frequency. Note how the sound becomes more directional with increasing frequency.

The lower frequencies are less directional than higher frequencies. In general, when a wavelength of a sound is larger than the source generating it, the sound pattern has an omnidirectional characteristic.

Diffuse Sound Fields

Direct sound is the sound wave that reaches the listener via a direct path, without having bounced off a reflecting surface. A *diffuse sound field*, on the other hand, refers to the energy from a sound source that reaches the listener indirectly, after reflecting off surrounding surfaces. The buildup of diffuse sound over time is known as *reverberation*. Reverberation is a collection of time-delayed versions of a sound that have decayed in intensity over time as they arrive at the listener. A representation of the reverberation process is shown in Figure 2.10.

While reverberation is most often heard in enclosed spaces, sound reflections also occur in outdoor settings. Only in anechoic chambers or in atypical environmental locations such as on a mountain summit is sound ever free of reflections. This is the definition of a *free sound field*, "a medium where only the direct sound reaches the receiver." In most rooms, the direct and time delayed sounds arrive so quickly in succession that they are perceived as one sound source, arriving from a single location defined by the direct sound. However, if the reflection arrives late enough in time and has a significantly high amplitude, it is heard separately as an echo.

Figure 2.10 (*top*) A simplified plot of a direct sound (blue) and two early reflections (red) from a sound source to a listener. (*bottom*) Reflectogram showing direct sound (blue), early reflections (red), and reverberation (green). The early and late reflections taken together constitute the diffuse sound field.

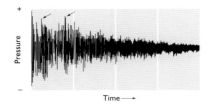

Figure 2.11 This impulse response was obtained by popping a balloon in a room and recording the results. The arrows indicate significant early reflections.

Figure 2.11 shows an impulse response of a room, obtained by recording a balloon pop. A room impulse response is a graphic representation of the moment-to-moment variation of sound pressure in a diffuse field. The room impulse response is equivalent to the reflectogram shown at the bottom of Figure 2.10. Two possibly significant early reflections that might be heard as echoes are indicated by arrows in Figure 2.11. Chapter 6 treats the topics of reverberation, echoes, and diffuse sound fields in greater depth.

The Inverse Square Law

The inverse square law expresses the decrease in sound pressure as a function of distance. Each doubling of distance from a reference point translates into a 6 dB loss in sound pressure level as shown in Figure 2.12.

The inverse square law primarily pertains to point sound sources out-of-doors. Examples approximating point sound sources include window air conditioners and loudspeakers. A line source on the other hand radiates sound cylindrically. Unlike the point source, the sound pressure level for a line source will be reduced by 3 dB for every doubling of distance. Water passing through a pipe approximates a line source.

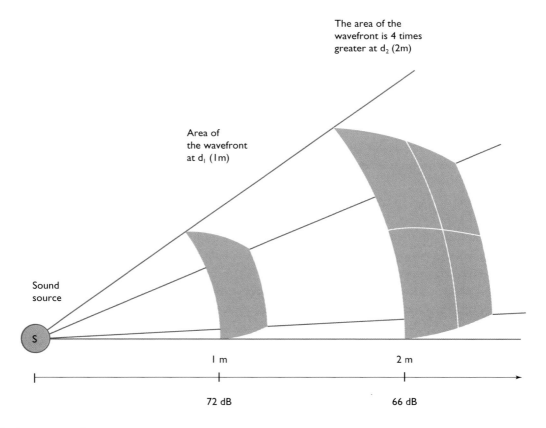

Figure 2.12 The inverse square law. Because the area of the wave is four times as large for each doubling of distance, there is a 6 dB loss.

The inverse square law can be very useful for estimating the fall-off in direct sound level from an outdoor source. However, for a sound source in a reverberant room, the inverse square law does not apply. This is because the reverberation contributes to the overall level; the fall-off in the direct sound level is compensated for by reverberant energy that builds up within a room.

Conclusion

In this chapter, the fundamental concepts of acoustics were introduced: frequency, sound pressure level, spectrum, directivity, and reverberation. All of these concepts are measurable in a physical sense, but Chapter 3, "Psychoacoustics and Hearing," covers how we interpret these fundamental concepts. Acoustical measurements make up a large part of the engineering efforts in acoustics. Chapter 4 discusses techniques used for measuring the frequency and amplitude of noise.

Notes

1. It is possible to refer to the *sound power* (W) of a source, independent of the distance and directivity of the source, unlike the sound pressure level. A sound source radiates sound waves whose total power can be measured in *watts,* a standard scientific unit for measuring energy, work, or the quantity of heat. Like sound pressure level, the sound power level is expressed mathematically as a dB ratio to a reference level. In this case, the reference is 10^{-12} watts (1 picowatt):

Sound Power Level
(dB) $= 10 \log_{10} (W_1/W_0)$

Where W_1 is the power in watts of the sound source, and W_0 is the reference power level of 1 picowatt.

Sound intensity (I) refers to the rate of flow of sound energy per unit area in a specified direction; it is therefore a measurement of not only sound pressure but molecular air particle velocity. As with SPL, the *sound intensity level* is measured as a ratio to a reference quantity. In a *free sound field,* an open field, or other environmental context where reflected sound is effectively not present, the values obtained for SPL and sound intensity level are the same:

Sound intensity level
dB $= 10 \log_{10} (I_1/I_0)$

Where I_1 is the power measured in watts/m^2 and I_0 is a reference value of 10^{-12} watts/m^2.

3

Psychoacoustics & Hearing

Durand R. Begault, Ph.D.

The Auditory Mechanism • Perceptual Interpretation of Physical Cues • Pitch • Loudness • Timbre and Spatial Location • The Precedence Effect • Psychoacoustic Measures

HUMAN hearing can be separated into physiological and perceptual aspects. The physiology of hearing refers to aspects of the auditory mechanism that respond directly to acoustical events, while perception refers to processing of acoustic events by the brain. The connection between physical measurements of sound, the perception of the listener, and legal or scientific standards is illustrated in Figure 3.1.

The Auditory Mechanism

Traditionally, the auditory mechanism is subdivided into the outer, middle, and inner ear. Figure 3.2 and this text provide a thumbnail sketch of the auditory mechanism, along with a description of their function. Sound first enters the auditory mechanism via the *pinna* (the visible portion of the outer ear). The pinna acts as a filter whose frequency response depends on the incidence angle of sound. Because of this, the pinna is considered to function as a cue to auditory localization. Following the pinna, incoming sound is transformed by the effects of the *meatus* (or "ear canal"). The meatus can be approximated by a tube 6 mm (0.2 in.) in diameter and 27 mm (1.0 in.) long, with a resonant frequency of around 3.5 kHz.

The end of the ear canal marks the beginning of the middle ear, which consists of the *eardrum* and the *ossicles* (the small bones popularly

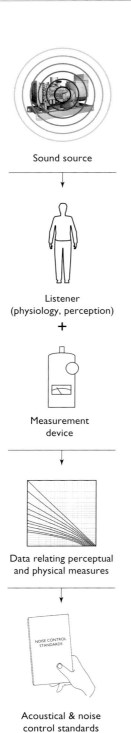

Sound source

Listener
(physiology, perception)

+

Measurement
device

Data relating perceptual
and physical measures

Acoustical & noise
control standards

Figure 3.1 The relationship between objective physical measures and subjective perceptual measures form the basis for many acoustical and noise control standards.

termed the "hammer-anvil-stirrup"). Sound is transformed at the middle ear from acoustical energy at the eardrum to mechanical energy at the ossicles. The ossicles convert the mechanical energy into fluid pressure within the inner ear (the *cochlea*) via motion at the oval window. The fluid pressure causes frequency-dependent vibration patterns along the approximately 30 mm (1.0 in.) long basilar membrane within the inner ear. These vibration patterns cause numerous fibers protruding from auditory hair cells (*cilia*) to bend at certain locations along the basilar membrane.

High frequency sound activates the basilar membrane near its connection beneath the oval window. With lower frequencies, the vibration occurs farther along the membrane. These cilia in turn activate electrical potentials within the neurons of the auditory system, resulting in aural perception and cognition.

Hearing loss or damage to the hearing mechanism can be caused by either brief or long-term exposure to appropriately high sound levels. Damage to the hearing mechanism or to health in general is termed a physiological effect of noise, and can result from both unsafe work conditions and loud recreational activities such as listening to music through headphones or firing guns. The hearing loss that occurs naturally in aging is known as *presbycusis*.

Perceptual Interpretation of Physical Cues

Under controlled conditions, a measurement of a physical aspect of sound is repeatable, allowing an accurate prediction of its variables. By contrast, the measurement of human perceptual response to sound is less predictable, and has a non-linear relationship to physical measurements. For these reasons, a distinction is made between acoustics and *psychoacoustics*. Psychoacoustics refers to the scientific study of human auditory perception. The non-linear relationship between physical and psychoacoustical aspects of sound can be made by the following analogy to cooking. A chef can add 1, 1¼, or 4 teaspoons of oregano to a sauce, but the sauce with 4 teaspoons will not taste "four times as spicy," and it may be impossible to notice the difference between a sauce with 1 and 1¼ teaspoons. Taste is the perceptual dimension, while the amount of spice added to the sauce is the physical dimension. The change in the physical dimension does not correspond to the same proportional change in the perceptual dimension.

Psychoacoustic studies (colloquially referred to as "listening tests") are conducted in order to establish a standardized relationship between physical and perceptual phenomena, for instance, between sound pressure level and loudness. Carrying the cooking analogy further, a gastronomic experiment could establish the relationship between "teaspoons of oregano" and "perceived spiciness." A similar procedure is used for establishing relationships between physical and perceived magnitudes of sound. Table 3.1 identifies equivalent physical and psychoacoustical parameters. The relationship between physical and perceptual parameters have been incorporated into noise control standards. Studies have investigated the role of noise in disturbing sleep; as a consequence, noise control standards allow

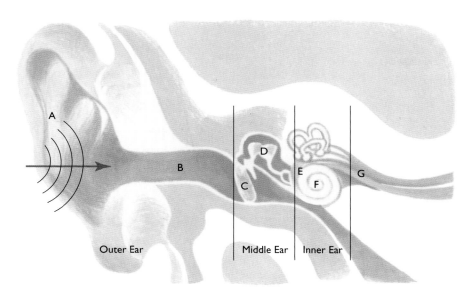

Outer Ear

Middle Ear | Inner Ear

less noise at night than during the day. Most people have experienced how background sound can influence daily activity. Similarly, there are many documented studies on how work performance or learning can be adversely affected by noise. These are called behavioral effects of noise, in contrast to the physiological noise effects discussed earlier.

Pitch

Frequency is a measurable quantity, whereas pitch refers to the perception of frequency. *Pitch* is also a term used by musicians to refer to musical notes; someone with "perfect pitch" is skilled at matching a pitch to an exact frequency. One term common to both musicians and acousticians is the *octave*. An octave relationship is a frequency interval between two sounds whose ratio is 2. Thus, 100 Hz to 200 Hz is an octave; as is 31.5 Hz to 63 Hz.

Pitch can be important for describing certain types of problems in noise control applications. Generally, a noise is most disturbing when it is concentrated in a narrow frequency range; this is termed tonal noise. The sound from a machine can have a specific pitch due to the frequency of the motor's oscillation; forced air can whistle across a vent. We are all familiar with the hum of the ballasts in a fluorescent light system. Broadband noise, conversely, would be exemplified by the sound of distant freely flowing traffic.

Loudness

Scientific tests have determined the relationship between sound pressure level and the perception of *loudness*. The equal loudness contours (also termed "Fletcher-Munson curves") in Figure 3.3 show this relationship. The graph's contours indicate levels in terms of phons, which represent equal loudness for a given pure tone SPL referenced to 1 kHz. For instance, the red dots on the contour line for 40 phons show that 62 dB at

Physical Terminology	Perceptual Terminology
Frequency	Pitch
Sound pressure level	Loudness
Spectrum	Timbre (tone color)

Table 3.1 Physical versus perceptual terminology.

Figure 3.3 Equal-loudness contours for pure tones (Fletcher–Munson Graph). The contour lines indicate equal loudness levels (phons) relative to a 1 kHz frequency. Refer to text for discussion of red dots on the 40 phon contour line.

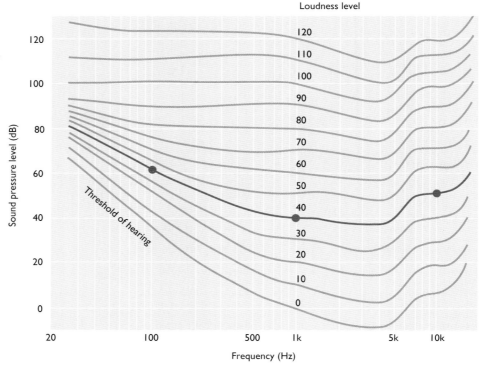

The equivalence between pitch and frequency becomes more complicated with real sound sources. For example, musicians commonly modulate frequency over time using a technique known as vibrato. Although the frequency is varied as much as a semitone at a rate of 5 to 8 Hz, a single pitch is perceived.

100 Hz sounds equally loud as 40 dB at 1 kHz, and 50 dB at 10 kHz. The frequency weighting filters built into sound level meters described in Chapter 4 physically approximate these contours.

Not surprisingly, the contours that exhibit maximal sensitivity are at those frequencies associated with speech (approximately 200 Hz to 5 kHz). At medium and high sound levels, the contours are relatively linear, while at lower levels, the contours indicate that sensitivity to low frequencies is less than at high frequencies. Thus, the relationship between physical and perceptual scales is dependent on both frequency and sound pressure level. Another measurement of loudness is the *sone* scale. A sound with a loudness of 40 phons is equal to 1 sone. This is an arithmetic scale such that a doubling in sones is equivalent to a doubling of loudness. The formula for relating sones and phons is: sones $= 2^{(\text{phons} - 40)/10}$.

For community noise assessment, certain sound sources that are considered to be noisy by one group of people may not be a problem for another group. A general procedure in such assessments is that the average person's level of annoyance needs to be considered. About 10 percent of any population can be expected to object to any noise not of their own making. This group is referred to as hypersensitive. About 25 percent are practically imperturbable. This group is insensitive to noise. The remaining two-thirds group are considered people who have what is called normal sensitivity. Some people will object to certain noises through association; for example, the fear of having an aircraft crash into one's house can motivate objection to aircraft sound, while another type of sound at the same sound level may not be perceived as disturbing.

One type of psychoacoustic measure is known as a *just noticeable difference* (JND). An example of a JND as applied to environmental acoustics is in Table 3.2, which shows the expected response to an increase in noise level.

Timbre and Spatial Location

The spectrum of a sound source is largely responsible for the perceptual quality of *timbre,* or "tone color." Timbre is sometimes defined in terms of what it is not, for example, "the quality of sound that distinguishes it from other sounds of the same pitch and loudness." Our ability to discriminate between different timbres is very complicated and not fully understood, but the main cues seem to involve the change in a sound's spectrum over time.

Spatial location is also an important perceptual quality of sound. The audible difference in level between the ears as the location of a sound moves relative to a listener is termed an *interaural level difference*, and is the same cue manipulated by a stereo sound system. For instance, if you snap your fingers to the right of your head, the level will be louder at the right ear than at the left ear. High frequencies above 1.5 kHz are shielded from the opposite ear by the head. Another cue for spatial hearing is the *interaural time difference*. The wave reaches the right ear before the left ear, since the path length to that ear is shorter. This time difference cue is most effective for frequencies below 1.5 kHz.

Besides level and time differences, another cue for localization is the spectral modification caused by the outer ears (the pinnae). For every sound source position relative to a listener, the pinnae cause a unique spectral modification that acts as an acoustic signature, as shown in Figure 3.4. These spectral modifications are especially important in perceiving the up/down and front/back locations of a sound source.

The Precedence Effect

The *precedence effect* (also called the "Haas effect") explains an important inhibitory mechanism of the auditory system that allows one to hear

Increase in noise level (dB)	Expected response
1	Possibly detectable under laboratory conditions
6	Some individual comment and reaction is expected but no group action is likely
10	Perceived as twice as loud
20	Perceived as four times as loud

Table 3.2 Expected response to increase in noise level.

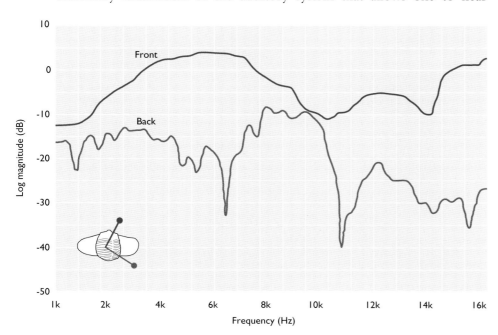

Figure 3.4 Spectral modification caused by the pinnae for two positions. Notice how the sound is brighter in front and more attenuated in back. *Inset:* overhead view showing sound source positions.

sounds in the presence of reverberation. It is also important for understanding the disturbance of speech intelligibility in rooms. Although we are bombarded with multiple sound reflections in a reverberant environment, our hearing system interprets the sound as located at one point. Up to about 40 msec, we perceive both sound reflections and the direct sound as a single, integrated sound source. In other words, the direct sound takes precedence over the later sounds. This is also true for sound reflections after about 40 msec, if the amplitude of the reflections is low enough. But if the amplitude of a reflection is sufficiently high, and occurs after about 40 msec, we hear the reflection as a separate sound source, or echo, because the precedence effect no longer operates. This is the same type of echo experienced when shouting into a canyon and the sound reflection off the walls in the distance is heard. Echoes can be very disturbing if heard during a music performance or during a lecture; as a result, echo mitigation is an important part of room acoustics design.

Psychoacoustic Measures

Many of the noise criteria discussed in this book are based on psychoacoustic measures. In other words, an attempt is made to relate a physical measurement of a quantity to a perceptual quantity, in order to predict human response to a given acoustical phenomenon.

Some specific psychoacoustic measures that are commonly used are listed in Table 3.3. As an example, Figure 3.5 illustrates Noise Criteria (NC) curves, which are used for relating background noise to the octave

Table 3.3 Relationship between some psychoacoustic measures and applications.

Psychoacoustic-based measure	Description & Application
Articulation index (AI)	Estimate of speech intelligibility in noisy contexts
Speech interference level (SIL)	Simplified AI method
A-weighted sound levels (dBA)	dB levels adjusted for a particular equal loudness contour. Widely used as "general measure," hearing conservation (OSHA), and community noise ordinances
Noise criteria (NC) curves	Frequency-based value used to describe maximum allowable background noise, used for continuous (as opposed to time-varying) noise
Noise criteria-A (NCA) curves	Like NC, but curves allow more low-frequency noise
Noise criteria-B (NCB) curves "balanced noise criterion"	An improved version of NC curves, accounting for speech interference level by HVAC systems
Preferred noise criteria (PNC)	Like NC, but curves adjusted for characterizing a "blander" background noise
Room criteria (RC)	Like NC, but curves extend to lower frequencies; designed to be more sensitive to "rumble" and "hissiness" from HVAC systems
Perceived noise level (PNL)	A rating of aircraft "noisiness" used in assessment of aircraft flyover disturbance
Night average sound level (NL) Community noise equivalent level (CNEL) Day/night average sound level (DNL)	Used for assessing annoyance by noise to a community, including various types of day v. nighttime sensitivity weightings
Bels (equal to 10 dB)	Used to rate the noise of some ventilation fans

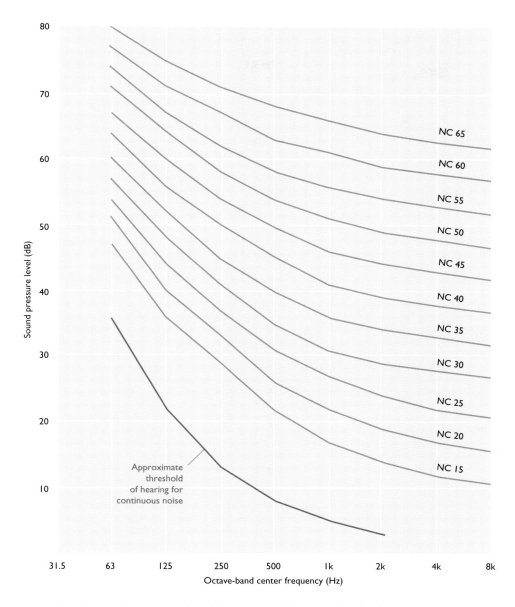

Figure 3.5 Noise Criteria (NC) curves. These are used to describe the target criteria for the level of background noise in a room. Like the equal-loudness curves shown in Figure 3.3, the NC curves compensate for the fact that human hearing is less sensitive to lower frequencies than to higher frequencies. In measurement applications (see Chapter 4), a particular curve is indicated on the basis of the loudest frequency region of the background noise.

band sound pressure level in rooms. These methods for rating noise are periodically revised or modified as researchers develop new insights into the effects of noise and vibration on human physiology and perception.

Conclusion

In this chapter, it was pointed out that physical measurements of sound form the basis of psychoacoustic descriptors. Psychoacoustics is a joint field of physics and psychology that deals with acoustical phenomena as related to audition. The relationship between physical acoustic variables and human response is not linear and cannot be precisely predicted. A wide variety of psychoacoustic measures are used to correlate the physical measurement of a sound with people's subjective response, depending on the specific application.

4

Measurements

Durand R. Begault, Ph.D.

Measurement Types • Sound Level Measurements • Octave-Band and One-Third Octave-Band Measurements • Diffuse versus Free Field Incidence • Fourier Analysis

As a part of acoustical design, it is often necessary to quantify aspects of an existing environment and to predict the result of adding an acoustical treatment. Quantification and prediction are two aspects of acoustical measurement; quantification involves the use of instrumentation, while prediction involves calculations based on previously made measurements. By forming a predictive model, it is possible to develop a design that can be tested against a target criteria. Because methods of acoustical quantification are standardized, a model can be created that is both accurate and replicable from situation to situation. Figure 4.1 shows a simple illustration of the relationships just described. The difficulty of accurately predicting a result obviously increases with the complexity of the acoustical context.

Sound Level Measurements

In Chapter 2, the concept of sound pressure level was introduced as a way of characterizing the amplitude of a sound source. Table 2.1 in Chapter 2 lists sound levels between the threshold of hearing (0 dB) and the threshold of pain (140 dB).

The sound level meter is essential to quantifying a sound level. Figure 4.2 shows a generic sound level meter and indicates its major features. These are (1) the use of an omni-directional microphone as a sensor; and (2) the meter display, which indicates the sound level. The microphone

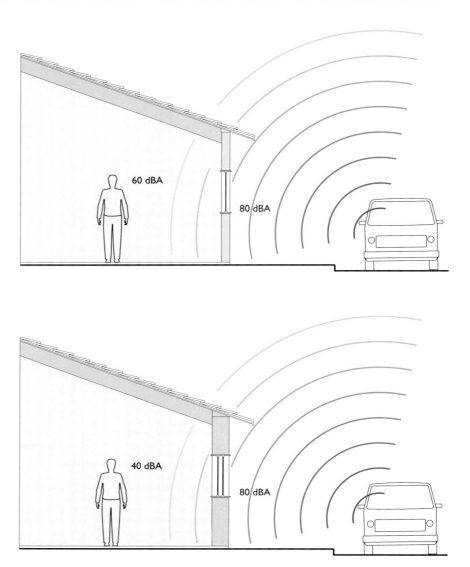

Figure 4.1 A description of the relationship between acoustic quantification and prediction. *Problem*: A house had too much outside noise intrusion from vehicular traffic. *Solution*: (1) measure noise level in the house; (2) specify target criteria for noise reduction; (3) design noise control construction; and (4) verify predicted noise reduction of improved wall and window constructions.

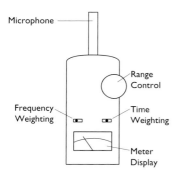

Figure 4.2 Generic sound level meter.

and the meter display are the respective "input" and "output" components of the device. Another characteristic common to sound level meters is the range control for adjusting the dynamic range.

The sound level meter indicates sound pressure levels, however, we are usually more interested in adjusting the measurement to correspond to perception. The averaging of sound pressure level over time is referred to as a *time-weighted average measurement*. An average level is often more important than the moment-to-moment fluctuations in level. We can also apply different types of *frequency weightings* in order to make sound pressure levels better correspond with human perception of loudness.

Typically, a sound level meter has two types of time weightings: fast and slow response. This refers to the time period over which the level is averaged and subsequently indicated at the output. The time periods correspond to 125 milliseconds for "fast" and 1000 msec (one second) for "slow." Consider the sound level of a slow-moving truck, as shown in Figure 4.3; both slow and fast response values are indicated. Let's say that there's a bump in the road that is of interest; the slow measurement will hide the event, but the fast response will indicate it quite clearly as both truck axles pass over the bump. This is because the truck passby noise has a high degree of variation over time. To measure the level of a steady non-varying source, such as an air conditioner, the slow and fast settings will indicate the same sound level.

In many applications, it is desirable to measure an average sound pressure level over a particular duration. This is accomplished with an integrating/averaging sound level meter; the obtained sound level is termed the *equivalent continuous sound level* or *average sound level* (L_{eq}). The L_{eq} is the steady sound level whose sound energy is equivalent to that of varying sound in the measured period. Often, many L_{eq} measurements need to be obtained over an extended duration. For example, over a 24-hour period, an acoustic analysis might include the L_{eq} for each one-hour segment.

The meter in Figure 4.2 shows a frequency weighting selection; typically, the level can be adjusted to measure with either a *C-weighting* or an *A-weighting*. Although we can obtain a "flat response" (sometimes termed a "linear measurement") from a sound level meter, weightings are usually applied in order to make the measurements better correspond to actual loudness. The equal loudness contours in Figure 3.3 (Chapter 3) indicate how perceptual sensitivity varies as a function of frequency. Figure 4.4 illustrates the A and C-weightings. The C-weighting is very close to a flat response over the audible frequencies; it has a gentle roll-off below 50 Hz and above 5 kHz. The A-weighting contrasts this in that frequencies below 1000 Hz are de-emphasized: the roll-off is more severe. For the meter to read 70 dB with an A-weighting, a 100 Hz frequency tone would need to have a linear level 20 dB more than if the tone were at 3 kHz.

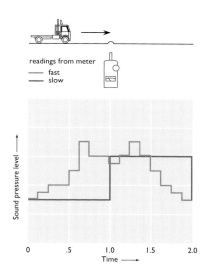

Figure 4.3 Truck passby; difference in slow and fast sound level readings. Note that it is possible to detect when the wheels pass over the bump in the road with the fast setting, but not with the slow setting.

Figure 4.4 Frequency weightings for dBA and dBC.

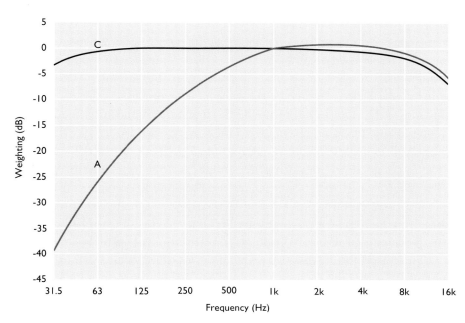

A-weighting forms the basis for most standards related to noise. C-weighting is sometimes used instead of "flat" to avoid the influence of low frequencies in the measurement.

Fc (Hz)	octave	1/3 octave
20		x
25		x
31.5	x	x
40		x
50		x
63	x	x
80		x
100		x
125	x	x
160		x
200		x
250	x	x
315		x
400		x
500	x	x
630		x
800		x
1000	x	x
1250		x
1600		x
2000	x	x
2500		x
3150		x
4000	x	x
5000		x
6300		x
8000	x	x
10000		x
12500		x
16000	x	x

Table 4.1 Center frequencies (F_c) for octave-band and one-third-octave-band analysis.

When reporting levels from a sound level meter, the American National Standard Institute (ANSI Y10.11) recommends that the time and frequency weighting be indicated by adding a subscript to the level. For example, "L_{AF} = 70 dB" refers to an A frequency-weighted, fast time-weighted, sound pressure level of 70 dB. According to the standard, if no subscript for frequency weighting is given, then A-weighting is to be assumed. However, it is common practice to attach the frequency weighting indication after the decibel, although the decibel itself is not weighted: for instance, one commonly sees "70 dBA" or "62 dBC."

The instantaneous *peak sound level* (L_{pk}) is another attribute that can be useful to measure. This is simply the highest sound pressure level that occurs during a measurement. Since slow and fast time weightings are based on averaging, these sound level values will always be lower than L_{pk}. But for some applications, such as for hearing risk assessment, it is important to know the highest level of an individual acoustic event over time. In Figure 4.5, L_{pk} would be 72 dB, up until the sound event of 84 dB occurs; therefore, L_{pk} equals 84 dB during the entire measured period. It is also possible to measure the highest level using both time and frequency weightings. The maximum A-weighted sound level (LA_{max}) uses a fast time averaging. LA_{max} of the signal shown in Figure 4.5 would be less than L_{pk} as a result of the time averaging and frequency weighting. The maximum A-weighted level using no time weighting is called the peak A-weighted sound level (LA_{pk}).

Octave-Band and One-Third-Octave-Band Measurements

Chapter 2 demonstrated that the waves of most sounds are composed of multiple frequencies. These waves are complex, in that they contain numerous frequencies that make up the sound's spectrum. But our hearing system tends to integrate spectral information within different ranges of frequency, as opposed to evaluating individual frequencies. Sound level measurements are sometimes made by analyzing the amount of energy within *octave-bands* or *one-third-octave-bands*. The word *band* refers to the bandwidth of the filters used to divide the incoming sound into frequency ranges. These filters are analogous to the way a prism divides an incoming beam of light into different color bands. The spectrum of the sound source is obtained by measuring the sound pressure level within each bandwidth. Octave-band measurements usually provide enough information about a sound spectrum, but one-third-octave-bands are closer to the manner in which the ear processes sound. Table 4.1 lists the standard center frequencies used in octave-band and one-third-octave-band analyses.

The reason for making octave-band or one-third-octave-band measurements is so that the frequency content of a sound source can be accurately characterized. This allows a more effective application of acoustical treatments in many situations. For instance, consider two different sound sources A and B with the same unweighted sound level. Figure 4.6 compares the acoustical energy distribution of the two sources. The type of treatment in these situations would need to be matched to results of the

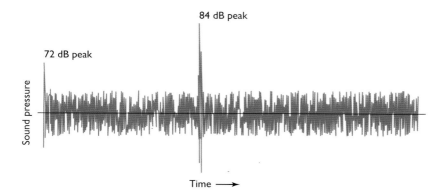

Figure 4.5 The maximum (peak) level of this signal is 72 dB, until an event that causes an 84 dB peak.

84 dB peak

72 dB peak

Sound pressure

Time ⟶

octave-band analysis. This is because the sound absorption of materials and the sound isolation of constructions vary with frequency. The potential disturbance for a particular activity can also be assessed from an octave-band analysis. For example, because speech intelligibility is highly dependent on frequencies between 200 Hz to 5 kHz, sound source B would interfere with speech more than sound source A.

The majority of acoustical measurements made for building acoustics, environmental acoustics, and industrial noise purposes utilize A-weighted, octave-band, or one-third-octave-band analysis. However, some applications demand an even narrower frequency band. Where people are complaining about tonal noise, it is necessary to measure in narrow bands or individual frequencies to determine the cause of the problem. The investigation involves measurements at the location of the complaint and close to suspect sources of the tonal noise. By identifying the spectral components, it is possible to determine the frequencies of sound that people are complaining about and the sources of those frequencies.

Diffuse versus Free Field Incidence

In Chapter 2, the concept of a diffuse sound field was contrasted to a free field. The basic difference is that in a free field, sound arrives by a direct path only; in a diffuse field, sound arrives not only along a direct path but from indirect paths of reflections and reverberation as well. Because the diffuse field arrives from potentially any direction relative to a receiver, the sound field is termed as having a *random incidence;* the free field condition is sometimes termed *frontal incidence.* Some sound level meters have a switch to choose between the two types of measurements. Typically, free field conditions occur out-of-doors and diffuse field conditions occur indoors. The switch activates an appropriate equalization for the system. Special microphones or adapters can also be used for each purpose. In making these measurements, it can be important to determine the influence of the diffuse sound field on a particular measurement location. If care is not taken in the placement of a sound level meter with reference to the sound source, the measurements may be unduly influenced by reverberation or reflections.

Fourier Analysis

We can measure the spectral content of a sound source over a fixed period of time by performing a *Fourier analysis* on the wave. Fourier analysis is

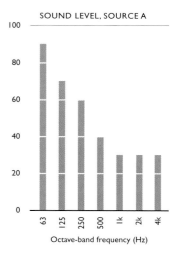

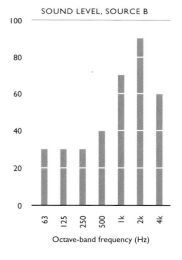

Figure 4.6 Sound levels measured in octave-bands, for two sound sources A and B. While both sources have the same overall SPL, the octave-band measurements immediately show that sound source A has energy primarily in the lower frequency range, especially around 63 Hz, and that sound source B has energy predominately in higher frequencies around 2 kHz. The frequency information is important since the acoustical treatment and predictive modeling for a particular situation is frequency-dependent.

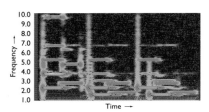

Figure 4.7 A spectrogram of a sound with three acoustic "events." A spectrogram is useful for post-analysis of a recording, in order to view the time-varying nature of a sound source. The color scale indicates the relative intensity of the spectral energy at a particular frequency.

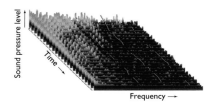

Figure 4.8 A "waterfall" display of a piano note decay. Each time-slice indicates a separate Fourier analysis, which collectively show the spectral evolution of a sound. Note how the higher frequencies die out faster than the lower frequencies.

often abbreviated FFT, for "Fast Fourier Transform." The result of an FFT is very similar to the octave-band and one-third-octave-band analyses described above, except that the time period over which the analysis is performed can be varied during the post-measurement phase. FFT analyses can be very useful in areas such as audio forensics. Fourier analysis works by mathematically separating a complex wave into a series of sine waves whose levels can then be determined. To determine the temporal evolution of the harmonic structure of a sound, multiple FFTs of a wave can be taken over successive time periods. These can then be displayed using a spectrogram as shown in Figure 4.7; the relative intensity in each frequency corresponds to the color scale value. Another way to view the time-varying nature of the spectral energy of a sound is to take multiple FFTs and then arrange them on the z axis of a 3-D graph, as shown in Figure 4.8. This is also termed a perspective (time-level-frequency) graph, or a *waterfall display.*

Conclusion

Acoustical measurement techniques provide a means of quantifying sound levels in an environment. Measuring A-weighted sound levels is common. Also common are octave-band, one-third-octave-band, and narrow-band measurements. Sound level measurements allow calculation, analysis, and prediction of various acoustical factors included in this book, such as measurements of background noise, of sound isolation, and of reverberation and acoustical absorption in a room.

5

Environmental Noise

Alan T. Rosen

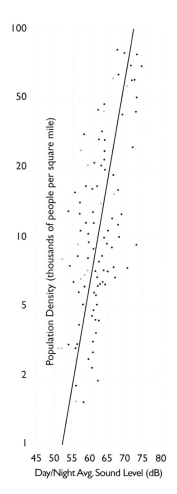

Figure 5.1 Day/Night Average Sound Level as a function of population density.

Terminology • Noise Measurements • Noise Sources and Prediction Methods • Sound Propagation • Noise Impact Assessment • Noise Impact Studies • Mitigation Measures • Outdoor Noise Control • Indoor Noise Control

ANNOYANCE from unwanted sound has been a problem since the dawn of civilization. Julius Caesar was said to have expressed great displeasure at having his sleep disturbed by noisy chariots.[1] He solved the problem by restricting travel around his court during the early morning hours. In medieval times, horse-drawn buggies annoyed patients when they traveled near hospitals. Straw was laid on the roads to lessen the noise of the iron wheels hitting the cobblestones.

Concern over environmental noise continues into the present day. By the early 1970s, the United States introduced federal legislation that mandated research on environmental noise and its effects on people.[2] Much of the early research was sponsored by the U.S. Environmental Protection Agency (EPA). This research led governmental agencies to require noise impact studies. Figure 5.1 is a graph from an EPA study that shows how noise levels increase with population.

This chapter is divided into sections which focus on the fundamentals of environmental noise and impact assessment. Terminology is discussed first. Information about how to measure environmental noise follows. Sources of noise and prediction methodologies are then presented. Finally, noise impact assessment and mitigation are covered.

Terminology

This section reviews the more commonly used environmental noise descriptors. These descriptors regularly employ A-weighting (dBA) to approximate the way the human hearing mechanism responds to sound.

Equivalent or Average Sound Level (L_{eq}). Most environmental sounds vary with time. The most common method of quantifying the loudness of time-varying environmental sound is to average the level over a specified period. The descriptor for this averaging is the equivalent or average sound level (L_{eq}). The average level allows comparison of different sources that change with time. One can compare time-varying traffic noise having an average level of 70 dBA to steady fan noise with an average level of 70 dBA and say that they have similar loudness.

Figure 5.2 shows the relationship between average, maximum, and minimum sound level. The maximum level (L_{max}) is used in studies of sleep disturbance. The L_{max} for an event such as an aircraft flyover is often called the *single event* sound level. The minimum level (L_{min}) can be used to assess audibility of sounds.

Figure 5.2 Environmental noise varies with time. L_{eq} is the average sound level; L_{max} is the maximum; L_{min} is the minimum.

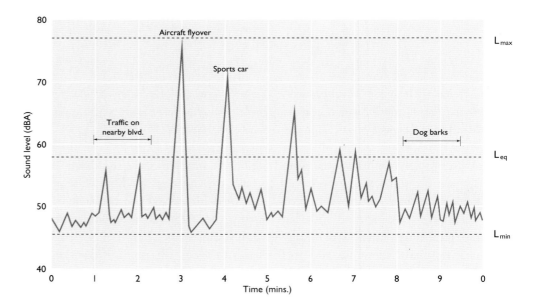

DNL and CNEL. The DNL (Day/Night Average Sound Level) and CNEL (Community Noise Equivalent Level) are single number ratings that describe the noise environment at a site. The DNL is determined by averaging the daytime and nighttime levels (logarithmically) over a 24-hour period. A 10 dB "penalty" is added to the nighttime level (10 PM to 7 AM) and included in the DNL calculation to account for the increased sensitivity of people at night. The CNEL is identical to the DNL except that it includes a 5 dB penalty for evening noise levels (7:00 PM to 10:00 PM). Table 5.1 shows how the DNL and CNEL are calculated using hourly L_{eq}.

Many governmental agencies have adopted standards for community noise exposure based upon the DNL and CNEL. For common sources such as traffic on a roadway, they vary by less than 1 dB. Therefore, the DNL and CNEL are often used interchangeably.

Statistical Descriptors. The average sound level quantifies the loudness of time-varying sources, but it does not describe how the level changes with time. To help provide this information, a category of noise metrics called statistical descriptors are used. The most common statistical descriptors are the L_{10}, L_{50}, and L_{90}.

The L_{10}, L_{50}, and L_{90} correspond to the sound level exceeded 10, 50, and 90 percent of the time. The L_{10}, L_{50}, and L_{90} are used infrequently nowadays since most standards are based on the average sound level, which can be readily measured or predicted.

Time of day	Measured L_{eq} (dB)	Penalty added to L_{eq} for DNL	L_{eq} with DNL penalties (dB)	Penalty added to L_{eq} for CNEL	L_{eq} with CNEL penalties (dB)
5–6 PM	61	0	61	0	61
6–7 PM	62	0	62	0	62
7–8 PM	60	0	60	+5	65
8–9 PM	58	0	58	+5	63
9–10 PM	56	0	56	+5	61
10–11 PM	55	+10	65	+10	65
11–mid	55	+10	65	+10	65
mid–1 AM	54	+10	64	+10	64
1–2 AM	52	+10	62	+10	62
2–3 AM	51	+10	61	+10	61
3–4 AM	50	+10	60	+10	60
4–5 AM	52	+10	62	+10	62
5–6 AM	55	+10	65	+10	65
6–7 AM	58	+10	68	+10	68
7–8 AM	62	0	62	0	62
8–9 AM	61	0	61	0	61
9–10 AM	60	0	60	0	60
10–11 AM	58	0	58	0	58
11–noon	59	0	59	0	59
noon–1 PM	61	0	61	0	61
1–2 PM	61	0	61	0	61
2–3 PM	60	0	60	0	60
3–4 PM	61	0	61	0	61
4–5 PM	61	0	61	0	61

Calculated DNL is 62 dB   Calculated CNEL is 62 dB

Table 5.1 Calculation of DNL and CNEL from a table of hourly L_{eq} measurements.

DNL = L_{dn} =
$10 \log_{10} \frac{1}{24} \{15(10^{L_d/10}) + 9(10^{(L_n+10)/10})\}$

Where L_d = daytime average sound level (7 AM–10 PM) and L_n = nighttime average sound level (10 PM– 7 AM).

CNEL = L_{den} =
$10 \log_{10} \frac{1}{24} \{12(10^{L_d/10}) + 3(10^{(L_e+5)/10}) + 9(10^{(L_n+10)/10})\}$

Where L_d = daytime average sound level (7 AM–7 PM), L_e = evening average sound level (7 PM–10 PM), and L_n = nighttime average sound level (10 PM–7 AM).

Noise Measurements

There are two basic types of environmental noise measurements: short-term and long-term. Short-term measurements usually last 15 minutes while long-term ones can last 24 hours or more. At least one long-term measurement is done for most environmental noise studies. When evaluating a large site, additional short-term measurements can supplement a long-term measurement to help accurately document the noise throughout the site.

When the goal of a measurement is to determine the noise level at a specific receiver, then it is preferable to measure at that specific location. If measurements cannot be made at the receiver, then they can be made at other locations and estimated using standard prediction methods (described later in this chapter).

Short-Term Measurements. During short-term measurements, an observer is usually present. The meter displays the A-weighted level of events such as car or truck passbys and the maximum and minimum levels are noted. At the end of the measurement, the meter can provide other data such as the average level.

The following information should be documented for each measurement: (1) date and time of day; (2) duration; (3) weather (temperature, wind, humidity, barometric pressure, if feasible); (4) major and minor noise sources; (5) distances from meter to source(s) and landmarks; and (6) acoustic shielding such as fences, berms, and topography. Figure 5.3 shows a short-term measurement data sheet.

Long-Term Measurements. A long-term measurement is accomplished using a programmable meter capable of storing many hours of data. Modern sound level meters can store several days of data. The results can be used to determine the DNL. Some meters can be programmed to store the level of the loudest event each hour or the number of events exceeding a preset level. This feature is useful in studies of single event noise, such as aircraft flyovers or train passbys.

The data from a long-term measurement provides a variety of other information such as differences between daytime and nighttime or weekday and weekend levels. The average level during the "noisiest hour" of a day can also be determined. Current federal highway criteria are based on the "noisiest hour."

Long-term measurements are made by placing a meter in a box on a tree or utility pole near the noise source, but high enough above the ground to minimize the chance of the meter being tampered with or stolen. With any unattended measurement, there is a possibility that an extraneous event, such as a bird near the microphone or radio interference, can skew the results. If knowing the precise source of a noise is important, tape recordings can be made. There are programmable monitoring systems that can start a tape recorder when the sound level exceeds a preset threshold. This technique helps to identify sources that occur infrequently.

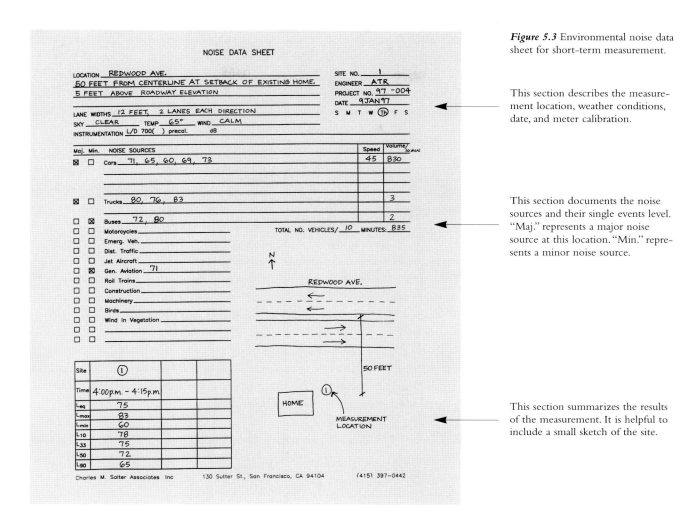

Figure 5.3 Environmental noise data sheet for short-term measurement.

This section describes the measurement location, weather conditions, date, and meter calibration.

This section documents the noise sources and their single events level. "Maj." represents a major noise source at this location. "Min." represents a minor noise source.

This section summarizes the results of the measurement. It is helpful to include a small sketch of the site.

During one project we tried to measure the sound of distant traffic. The data showed surprisingly high nighttime levels. Our tape recording revealed that some of the noise was not from traffic but from teenagers partying near our noise monitor. Figure 5.4 shows the hourly data from our 24-hour measurement and the effect of party noise on the measurement.

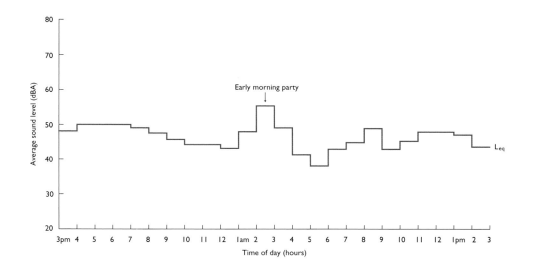

Figure 5.4 Effect of party noise on 24-hour noise levels. DNL = 54 dB with party, 49 dB without party.

Spectral Analysis. An A-weighted level does not provide any "spectral" information. In other words, there is no indication whether the sound is mainly high frequency or low frequency, or whether it contains annoying tones. Spectral data is usually obtained by recording the sound of interest and playing it through a spectrum analyzer. Some ordinances require an evaluation of the tonal content of noise.

  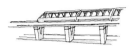 

Noise Sources and Prediction Methods

Environmental sounds can be man-made or natural. Natural sounds are usually considered desirable and rarely the source of complaints. Some, such as dogs, can be annoying. Table 5.2 shows typical levels of some natural sources.

Common sources of annoyance are vehicular traffic, construction, mechanical ventilation equipment, industrial machinery, and amplified music. Table 5.3 summarizes these source levels. A discussion of common sources and their prediction methods follows.

Vehicular Traffic. Noise from trucks, cars, buses, and motorcycles comes from the engine, exhaust, tires, and wind turbulence. The overall level depends on the loudness of each of these components which, in turn, depends on such factors as speed and road composition. On a smooth road at low speed, most of the noise from a car emanates from the engine and exhaust. At high speed, the dominant noise is wheels on the pavement.

Traffic noise prediction modeling began in the late 1960s. The Federal government funded most of this research. A traffic noise prediction model uses various input parameters including hourly traffic volume, speed, number of trucks, distance to road, ground type, topography, and presence of barriers.

If the noise level of traffic is known, changes based on increases in traffic can be estimated assuming that the vehicular speed does not change. For example, if the amount of traffic doubles, a 3 dB increase results. A quadrupling of traffic results in a 6 dB increase.

Aircraft. Aircraft flyovers are one of the most significant sources of community annoyance. In the late 1950s, the introduction of turbojet aircraft dramatically increased community noise exposure around commercial airports.

A principal component of early turbojets was "jet noise," the characteristic roar from the turbulent mixing of air from engine exhaust. These early turbojet engines are designated as Stage 1 aircraft by the Federal Aviation Administration (FAA), which certifies aircraft according to their noise emissions.[3] In the 1970s, commercial aviation manufacturers replaced Stage 1 aircraft with Stage 2 turbofan aircraft, which are about 10 dB quieter.

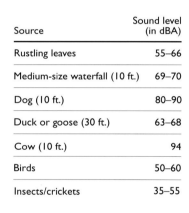

Source	Sound level (in dBA)
Rustling leaves	55–66
Medium-size waterfall (10 ft.)	69–70
Dog (10 ft.)	80–90
Duck or goose (30 ft.)	63–68
Cow (10 ft.)	94
Birds	50–60
Insects/crickets	35–55

Table 5.2 Natural sounds.

Source	Noise Level (in dBA)
Jet aircraft (200 ft.)	112
Live music	110
Subway train (30 ft.)	100
Printing plant	90
Truck/bus (50 ft.)	85
Vacuum cleaner (10 ft.)	70
Automobile (50 ft.)	65
Normal conversation (3 ft.)	65
Whisper (3 ft.)	42

Table 5.3 Man-made sounds.

In the 1980s, jet engine manufacturers began building turbofan Stage 3 aircraft, which are about 10 dB quieter than Stage 2 aircraft. With Stage 3 aircraft, the jet exhaust noise is minimal and the primary sources are turbomachinery associated with the various rotating components of the engines. Stage 3 noise reduction also comes from increased power, which allows for steeper climbs during departure.

Aircraft noise prediction models are periodically modified as aircraft designs change. These models require the following input: aircraft volume, time of day/night, flight tracks, climb profiles, and meteorological data. The computer programs establish a grid system, calculate the composite noise exposure at grid locations, and extrapolate from the data to draw noise contours.

Trains (Heavy Rail and Rapid Transit). Heavy rail consists of large diesel-electric locomotives and freight or passenger cars. These trains have three main noise sources: the engine, cars, and whistle. The locomotive engine is usually the dominant source. Noise emanates from both the engine casing and the exhaust stack. The noise output of the engine is virtually independent of the train speed but is affected by grade.

Freight and passenger car noise comes from the wheel/rail interaction. Factors affecting the wheel/rail noise are jointed versus welded tracks, presence of grade crossings, wheel irregularities, passage over bridges, and short radius curves.[4] Figure 5.5 shows a time history of a train passby.

Rapid rail and light rail refer to urban and interurban transportation systems that use electric engines. Rapid rail includes transit systems such as the Bay Area Rapid Transit (BART) in northern California, the Chicago Transit Authority (CTA), and the New York City Transit Authority (NY-CTA). Light rail refers to street cars or traditional trolley systems such as the Municipal Railway System (MUNI) in San Francisco.

For rapid rail and light rail vehicles, the noise during travel is primarily from the wheel/rail interaction. Other sources are the engines, brakes, and horns. Noise levels increase with speed. At a stop, the dominant source is auxiliary equipment such as air conditioning systems.[5]

In addition to airborne noise, rail activity generates vibration in the ground. This vibration creates low-frequency rumbling noise and is usually felt as shaking in buildings.

There are several models available to predict train noise. The models generally use the following input: number of engines, number of cars, grade, speed, rail construction (jointed or continuously welded), grade crossings (horn blasts), and presence of elevated structures. Table 5.4 shows the noise levels for various rail transit components.

Construction. Construction is often recognized as a disturbance and assessed as a part of environmental impact studies. Common sources of construction noise are bulldozers, trucks, pile drivers, and compressors. Table 5.5 lists the noise levels of construction equipment.

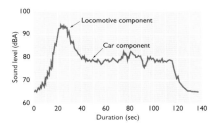

Figure 5.5 A-weighted time history of heavy rail passby, measured 100 ft. from track.

Transit sources	Typical sound level (in dBA) @ 50 ft.
Rail cars	80
Locomotives, diesel	88
Locomotives, electric	86
Monorail	80
Locomotive horns	105
Transit car horns	90
Auxiliary equipment	65
Locomotive idling	80
Rail transit idling	70
Buses idling	75
Curve squeal	100
Crossing signals	73
Substations	63

Table 5.4 Sound levels for various transit sources.

Figure 5.6 Temperature affects sound propagation by creating shadow zones where sound levels are reduced…

Equipment	Typical sound level (in dBA) @ 50 ft.
Dump truck	88
Portable air compressor	81
Concrete mixer (truck)	85
Jackhammer	88
Scraper	88
Dozer	87
Paver	89
Generator	76
Pile driver	101
Rock drill	98
Pump	76
Pneumatic tools	85
Backhoe	85

Table 5.5 Sound levels for various construction sources.

Sound Propagation

In many instances, the noise level of an offending source must be estimated at a specific location. First, the source level at a known distance is obtained through measurements or reference books. The next step is to determine the path that the sound will take to the receiver and what effect the path will have on the level. Factors that affect sound propagation outdoors are distance, terrain, and atmospheric conditions. The effect of each of these factors are discussed next.

Distance. Sound attenuates with distance from the source. For point sources, such as a person speaking, the sound level decreases by 6 dB for each doubling of distance from the source. If a person's voice has a sound level of 65 dBA at 1 m (3 ft.), then it will be 59 dBA at 2 m (6 ft.). For line sources such as steady traffic on roadways, the sound level decreases by 3 dBA per doubling of distance if the intervening terrain is hard like concrete. If the intervening terrain is soft like plant material, the drop-off factor per doubling of distance is 4.5 dBA.

Terrain and Vegetation. Sound traveling over soft ground, such as grass, can be absorbed and attenuated as it travels from source to receiver. This is in addition to the attenuation from distance. If the ground is hard, like concrete, then the attenuation is minimal. Sound traveling through dense woods and shrubbery is attenuated by 5 dBA per 30 m (100 ft.), up to a maximum of 20 dBA.[6] A hill can reduce sound by as much as 30 dBA.

Atmospheric Effects. The air temperature hundreds of meters above the ground is typically cooler than near the ground. This temperature gradient tends to bend sound waves upward to create a shadow zone. Areas in the shadow zone can experience a noticeable reduction in noise (see Figure 5.6). When the temperature above ground is warmer than at ground level, sound is bent downwards. This condition, called temperature inver-

Figure 5.7 ... while wind affects sound propagation by creating a shadow zone upwind.

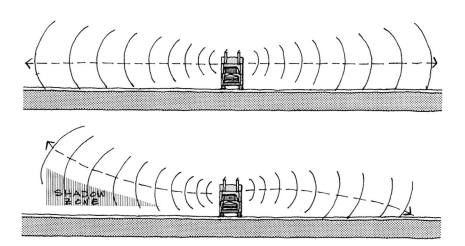

sion, has negligible effects at short distances but may be significant at distances greater than 800 m ($\frac{1}{2}$ mile).[7]

Wind speeds high above the ground are usually greater than near the ground. Therefore, sound traveling against the wind is bent upward to create a "shadow zone," which can cause reductions of 20 to 30 dBA (see Figure 5.7). Sound traveling with the wind is bent downward. This can lower the noise reduction provided by barriers or soft ground. It can also allow sound to propagate over long distances.

Sound energy is also dissipated over distance by "molecular absorption." The amount of molecular absorption depends on temperature and humidity. Sound energy is more readily absorbed at higher frequencies than at lower frequencies.[8] Over short distances of a few hundred feet, atmospheric absorption plays a small role in outdoor noise control problems.

On days of fog or light precipitation, it may appear that sound carries exceptionally well. This is not attributable to acoustical properties of fog or rain but to secondary effects. During light rain, temperature and wind gradients tend to be small so that sound carries farther than on sunny days when "micrometereological inhomogeneities" can create strong gradients.[9]

In summary, atmospheric effects can cause fluctuations in environmental noise over long distances (greater than 400 m). Typically, atmospheric effects will attenuate a noise to levels below those predicted. Rarely will these effects cause environmental noise to be significantly above predicted levels.

Amphitheater Effect. One aspect of propagation that is commonly misunderstood is the "amphitheater effect." For example, if a house is located on a hill above a softball field, voices from the softball field may be heard if ambient noise levels are low and there is line-of-sight between the noise source (people playing softball) and the receiver (people in the house). With low ambient noise levels and line-of-sight, a voice can be heard at great distances.

People sometimes think that the hillside is amplifying the sound and refer to it as the "amphitheater effect." Although the voices may sound as if they were amplified, in actuality, there is no amplification since energy is not added. In at least three projects we have been involved in, neighbors were concerned about the "amphitheater effect" and experiments were performed to analyze the phenomenon. The measurements showed that an echo was present, but it did not increase the noise level.

Noise Impact Assessment

The goal of a noise impact assessment is to determine if a project will cause impacts or whether the project itself will be impacted by the environment. This section discusses criteria for determining impact, how to determine which criteria apply to a project, and preparing noise impact assessments.

Criteria for Determining Environmental Impact. In environmental studies, the focus is on annoyance and disruption of normal activities such as speech communication. The criteria are based on both absolute levels and change in levels.

Absolute Criteria. Absolute criteria are based on levels primarily abstracted from EPA and U.S. Department of Housing and Urban Development (HUD) documents. For outdoor noise, the EPA has determined that a DNL of 55 dB is the level requisite to protect public health and welfare with an adequate margin of safety.[11] Some governmental agencies have adopted this standard. Many urbanized areas have ambient noise levels in excess of a DNL of 55 dB. In these cases, agencies have adopted standards that are 5 to 10 dB higher.

Absolute criteria allow for simple impact assessment. If the measured or calculated noise level exceeds the criterion, there is an impact. If it is less than the criterion, there is no impact. For example, HUD has noise standards that are applied to housing that they subsidize. If the DNL on a development site is below 65 dB, then the project is in an environment considered by HUD to be "normally acceptable." If the DNL is between 65 dB and 75 dB then the site is considered "normally unacceptable," which means that the site must have special noise control measures implemented to allow the project to be approved.[12] Table 5.6 shows the HUD site acceptability standards.

The HUD methodology relies on predictions rather than measurements. This approach tends to overpredict levels and may, therefore, overestimate an impact. According to Theodore Schultz, primary author of the HUD methodology, the screening procedure in the HUD guidelines is deliberately conservative because it recognizes that there will always be some aspect of a particular situation which cannot be taken into account.[13] This concept of conservative design is important in engineering and is sometimes referred to as the "factor of safety." It is always preferable to require additional studies due to a conservative estimate than to allow a project to proceed in a marginal noise environment and later have problems.

	Day/night average sound level (dB)
Acceptable	Not exceeding 65 dB Threshold may be shifted to 70 dB in special circumstances
Normally Unacceptable	Above 65 dB but not exceeding 75 dB
Unacceptable	Above 75 dB

Table 5.6 HUD site acceptability standards.

Criteria Based upon Change in Levels. Studies show that people react to changing noise levels in their community based on the magnitude of the increase.[14] While small changes in environmental noise are imperceptible, larger ones have the potential for causing widespread community response.

To address this issue, some governmental agencies have adopted criteria based upon a change in noise levels. According to FHWA guidelines, a new roadway would cause an impact if it: (1) generates a noise level that approaches or exceeds the noise abatement criteria (NAC) for the particular land use; or (2) causes a substantial increase in noise levels.[15]

The NAC is a "noisiest hour" L_{eq} of 67 dB for residences. Table 5.7 shows the FHWA traffic noise abatement criteria for various locales.

For example, a roadway improvement project is predicted to increase traffic noise from 46 dBA to 62 dBA at residences. The future noise level of 62 dBA does not exceed the absolute criterion, yet the increase of 16 dBA would be considered an impact. Figure 5.8 shows how the addition of a "substantial increase" criterion affects the impact assessment.

State	Descriptor	"Approach or Exceed" Interpretation	"Substantial Increase" Interpretation
California	L_{eq}	Equals NAC	12 dBA increase and ≥65 dBA
Connecticut	L_{10}	Equals NAC	15 dBA or > NAC
Florida	L_{eq}	Within 2 dBA of NAC	≥10–15 dBA
Illinois	L_{eq}	Equals NAC	14 dBA increase or within 2 dBA of NAC and >12 dBA increase
Kentucky	L_{eq}	Equals NAC	≥10 dBA
Maryland	L_{eq}	Equals NAC	≥10 dBA
Massachusetts	L_{eq}	66 dBA	≥15 dBA
Michigan	L_{eq}	Equals NAC	≥10 dBA
Minnesota	L_{eq}	Equals NAC	≥10 dBA
Missouri	L_{eq}	Equals NAC	≥10 dBA
Montana	L_{eq}	Equals NAC	≥10 dBA
New Jersey	L_{eq}	64 dBA	≥10 dBA or >64 dBA
New Mexico	L_{eq}	Equals NAC	≥10 dBA and >57 dBA
New York	L_{eq}	Equals NAC	≥6 dBA
North Carolina	L_{eq}	Within 2 dBA of NAC	≥10 dBA
North Dakota	L_{eq}	Within 3 dBA of NAC	No response
Oklahoma	L_{10}	Equals NAC	No response
Oregon	L_{eq}	Equals NAC	≥10 dBA
Pennsylvania	L_{eq}	66 dBA	≥10–15 dBA
Puerto Rico	L_{eq}	66 dBA	≥10 dBA
Rhode Island	L_{eq}	Equals NAC	No response

Table 5.7 State Department of Transportation interpretations of Federal Highway Administration (FHWA) noise regulations. Noise Abatement Criterion (NAC).

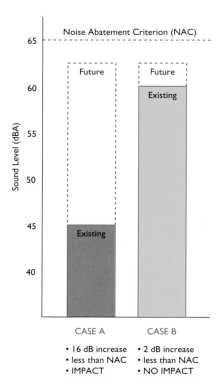

Figure 5.8 How "substantial increase" criterion affects noise impact assessment.

The FHWA "substantial increase" criteria allows individual states to interpret the criterion as they see fit. The information in Table 5.7 shows that some states consider a substantial increase to be 6 dBA, while for others the increase must be at least 15 dBA. On a local level, changes of 3 to 5 dBA may be deemed significant by a city or county.

Schultz performed studies that synthesized the response of people to varying levels of environmental noise.[16] Although he endeavored to create models that accurately estimate human response to noise, one of his reports about aircraft noise states: "even in quiet zones...about 10 percent of the people interviewed complained about aircraft noise; these same people complain about other aspects of their environment in general and may be habitual complainers. We recall, too, that about 25 to 30 percent of the people never complain even in extremely noisy environments."[17] Thus, a wide variety of responses can be expected due to changes in a community's acoustical environment.

How to Determine Which Criteria Apply to a Project. Each project may have one or more applicable criteria. Knowing the criteria in the early phases of a project is vital to formulating a correct approach and minimizing unnecessary measurements or analyses. A good starting point would be the planning code of the governing agency, which usually contains criteria for land use compatibility. Criteria are also contained in the municipal or zoning code as a noise ordinance. Noise ordinances often address mechanical noise, construction noise, or amplified sound systems.

Noise Impact Studies

Noise impact studies typically follow a fairly uniform format. First, applicable acoustic criteria are identified. Second, the noise at the project site is quantified. The next step is to predict future noise levels of the project. The future levels are then compared to the applicable criteria, which are based on: (1) "Land Use Compatibility"—how a project will be affected by the environment at a proposed location. These criteria are typically in terms of absolute levels such as DNL; and (2) "Impacts of the Project on Adjacent Land Uses"—these impacts can be both in terms of absolute levels and change in noise levels.

If the predicted levels exceed the criteria, then there is an impact. If an impact is identified, then feasible ways to reduce or eliminate the impact (mitigation) are specified.

One technique that can help with impact assessment is preparation of noise contours. These contours, usually in terms of DNL, are plotted on a site map. The contours are then compared with various land uses in the study area. Noise contours are shown in Case Study No. 13, Chapter 17.

EIS and EIR. Federal and state environmental laws require noise studies. The National Environmental Policy Act (NEPA) specifies that an Environmental Impact Statement (EIS) be prepared if a federally funded project has the potential to significantly affect the environment. The California Environmental Quality Act (CEQA) requires that an Environmental Impact Report (EIR) be prepared for most new building projects.[18] An EIS is similar to an EIR and tends to follow the typical format for impact studies. In cases where a project has federal funding and is in California, both documents can be combined to form an EIS/EIR.

In general, these impact reports must answer the following two questions: (1) will the project significantly increase existing noise levels? and (2) will the project expose people to severe noise or vibration?

The EIS/EIR must assess future levels with respect to applicable absolute standards as well as determine any significant changes that the project would create. The NEPA and CEQA guidelines, however, along with many governmental planning codes, do not provide specific numerical goals for "significant increase." It leaves room for interpretation.

In the absence of specific governmental criteria, one approach to determine whether an increase is significant can be found in a document from the Urban Mass Transit Administration (UMTA).[19] UMTA adopts the commonly accepted notion that a 3 dBA increase is barely noticeable, a 5 dBA increase is noticeable but not dramatic, and a 10 dBA increase is perceived as a doubling of loudness. Table 5.8 contains the UMTA noise impact criteria.

One aspect of an EIS/EIR that can be confusing is the assessment of alternatives to the project and "no build" scenario. This assessment is accomplished by calculating noise levels for all future scenarios, including alternatives, and then comparing them with the existing conditions. The differences are then compared to the "increase" standards shown in Table 5.8 to determine impact.

Mitigation Measures

If impacts are identified in a noise study, mitigation measures must be presented. Individual mitigation measures are provided so that the project will meet specific acoustic criteria. For example, if an existing roadway generates excessive noise at the site of a proposed project, the developer of the project may be required to provide a noise barrier to control outdoor sound levels to meet acoustic criteria.

Since a project's design is usually at a schematic level during the environmental review phase, it may not be feasible to provide detailed recommendations for all mitigation measures. For environmental studies, it is usually sufficient to determine a range of possible mitigation measures that will allow the project to meet the criterion. For example, a study may recommend that sound-rated windows be specified to meet a criterion. Although the specific rating of each window is not specified, the study should make it clear to decision makers that the project can feasibly meet the criterion with sound-rated windows.

Enough information should be provided to the decision makers so they can determine if the mitigation is commensurate with the community goals. The developer must then decide if the mitigation will still allow the project to be economically viable.

1.	Generally not significant
a.	No noise sensitive sites are located in the project area
b.	Increases in noise levels with implementation of the project are projected to be 3 dBA (L_{eq}) or less at noise sensitive sites and proposed project would not result in violations of noise ordinances or standards.

2.	Possibly significant
	Increases in noise levels with implementation of the project are expected to be no greater than 5dBA (L_{eq}). Determination of significance must consider existing noise levels and the presence of noise-sensitive sites.

3.	Generally significant
a.	Proposed project would cause noise standards / ordinances to be exceeded.
b.	Proposed project would cause an increase in noise levels of 6–10 dBA (L_{eq}) in built up areas.

Table 5.8 Significance of noise impact based upon change in noise levels.

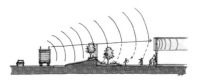

Figure 5.9 Noise barriers work by blocking sound from a noise source to a receiver.

Mitigation can be applied at the source, the path, and the receiver. At the source, measures include quieter equipment or enclosures. The path can be made longer by moving source or receiver, or the sound can be intercepted by a noise barrier. At the receiver, building modifications such as sound-rated windows can be considered.

Outdoor Noise Control

The following section presents methods used to mitigate environmental noise. The most common are the barriers or berms that can be seen along many highways.

Barriers/Berms. Noise barriers include walls, earthen berms, and building structures. Any solid element which blocks noise transfer from a source to a receiver can act as a barrier. The effectiveness depends on its height, density, its distance from the source and receiver, and the relative height of the source and receiver with respect to the barrier. As traffic noise propagates towards a barrier, some of the sound will hit the wall and reflect back while some will diffract, or bend, over the top of the barrier to the other side. Figure 5.9 is a sketch of a barrier and the associated sound paths.

There is an area behind the barrier where a person does not have a view of the traffic. This area is called the shadow zone. Receivers in the shadow zone experience a reduction in traffic noise. A barrier will provide 5 dBA of reduction if it just breaks line-of-sight from the source to the receiver. For every additional 305 mm (1 ft.), the barrier extends above the line-of-sight, the noise will be reduced by an additional 1 dBA. The practical limit for reduction by a freeway wall is about 15 dBA.

Notice how the second story window in Figure 5.9 has a direct line-of-sight to the truck exhaust. Thus, this barrier reduces traffic noise in the

Figure 5.10a Sound levels from roadways can be reduced by using setbacks...

backyard and first floor of the home but does not reduce the intrusion of large diesel truck exhaust noise into the second floor. This is usually handled by sound-rated windows.

Figures 5.10 a and b illustrate how a housing project achieves an outdoor noise goal for two different site plans. The housing in 5.10a requires a significant amount of distance between it and the freeway while the 5.10b achieves the same noise goal with a wall and a small setback.

Barriers can be made of concrete, steel, wood, glass, or stucco. Barriers called *greenwalls* are soil-retaining systems that allow plantings. Regardless of the material, a barrier must have no cracks or gaps and a minimum mass of 10kg/m² (2 psf) to 20 kg/m² (4 psf) depending upon the reduction needed. Alternate barrier designs are shown in Figure 5.11.

Noise reflection off barriers is a common concern. A study by Caltrans and experiments by our firm have concluded that reflections off buildings or barriers typically contribute less than 1 dBA to the overall noise level.[20] Only in rare instances will reflected sound be significant. For example, if there are parallel barriers on both sides of a freeway, reflected sound from one barrier can go over the opposite barrier and cause an increase in noise. This phenomenon is called *barrier insertion loss degradation* and depends on the topography, height of barriers, and distance between barriers.

Earthen berms provide about 3 dBA more reduction than walls of similar heights. Although the additional reduction provided by a berm is advantageous, a berm may not be practical in many circumstances since it takes up significantly more space than a wall.

Buildings. The first row of homes between a roadway and receiver provides about 5 dB of reduction. Each subsequent row provides an additional 3 dBA of reduction. 20 dBA is the upper limit for noise reduction.

Figure 5.10b ...or by incorporating noise barriers.

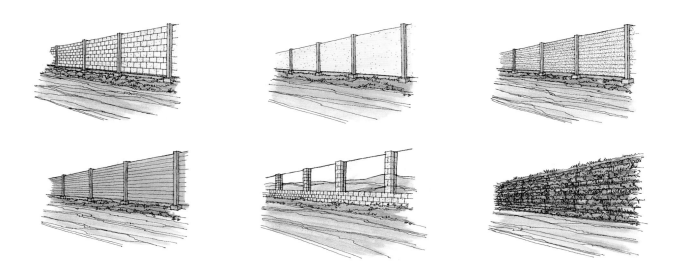

Figure 5.11 Alternate barrier designs: (*top: left to right*) concrete; plaster; steel; wood; glass; earthen wall.

Indoor Noise Control

The reduction provided by a building facade is a composite of the individual components. Indoor noise control is usually accomplished by improving the windows since they are acoustically the weakest link in an exterior wall. Roofs, doors, and attic vents are also important when designing for aircraft flyovers.

Windows. Sound travels both through the glass and around the window frame. A typical facade will reduce noise by approximately 25 dBA with the windows closed; when opened, the facade reduces the sound entering a building by 10 to 15 dBA. The acoustical performance of the window can be improved by increasing the glass thickness and airtightness of the weather-stripping.

Exterior Walls. Many buildings have single stud exterior walls with one layer of gypsum board on the interior and wood siding or stucco on the exterior. This wall reduces noise by about 40 dBA. If more reduction is needed, alternatives such as additional layers of gypsum board, resilient channels, staggered or double-stud construction can be incorporated. Heavier and thicker wall constructions may be required near rail lines or major roadways.

Roof Vents and Other Considerations. Roof vents and air intakes for mechanical ventilation systems can be significant sources of sound leaks. Internally lined ducts at air intakes and discharges may be needed to limit exterior noise intrusion.

Conclusion

As people live closer to noise sources, and each other, control of environmental noise has become increasingly important. With proper planning considerations for environmental noise, our acoustic environment can be improved.

Notes

1. Report of the Administrator of the Environmental Protection Agency to the President and Congress on Noise (Senate Document 92-63, U.S., GPO, Washington, D.C., 1972).

2. Noise Control Act of 1972, Public Law 92-574, 92 Congress, HR 11021, October 27, 1972.

3. Federal Air Regulation (FAR), Part 36.

4. Ibid, page 3-8.

5. Federal Transit Administration, U.S. Department of Transportation, Transit Noise and Vibration Assessment, Final Report, April, 1995.

6. Power Plant Construction Noise Guide, BBN for Empire State Electrical Energy Corporation, May, 1972.

7. Ibid.

8. Ibid.

9. Leo L. Beranek, *Noise and Vibration Control*, (McGraw-Hill Book Company, 1971).

10. Technical Background for Noise Abatement in HUD's Operating Programs. BBN Report 2005R, November 8, 1971.

11. EPA Levels Document, 1974.

12. *The Noise Guidebook*, HUD, 1985.

13. Theodore J. Schultz, personal communication with Charles Salter.

14. Theodore J. Schultz, *Community Noise Rating* (Applied Science Publishers, 1972).

15. Code of Federal Regulations, Title 23, Chapter 772, (23 CFR 772).

16. Theodore J. Schultz, "Synthesis of Social Surveys of Noise Annoyance," *Journal of Acoustical Society of America*, 1978, pp 377-405.

17. Theodore J. Schultz, "Technical Memorandum for Noise Abatement in HUD's Operating Programs," Bolt Beranek and Newman for the U.S. Department of Housing and Urban Development, November 8, 1971.

18. CEQA: California Environmental Quality Act, Public Resources Codes, Sections 21000-21178.1

19. UMTA Circular C 5620.1, U.S Department of Transportation, Urban Mass Transit Administration, October, 1979.

20. Interstate 680 Livorna-Stonecastle Soundwall Study, California Department of Transportation, July, 1994.

6

Room Acoustics

David R. Schwind, FAES

Planning • The Growth of Multipurpose Spaces • Design Criteria • Architectural Design Parameters—Small Rooms • Architectural Design Parameters—Large Rooms • Seating • Terraced Seating • Reflected Sound • Specular Reflection • Diffusion • Echoes • Flutter Echoes • Focusing • Reverberation • Standing Waves • Sound Levels in Rooms • Sound Transparent Surfaces • Sound Absorbing Materials • Porous Absorbers • Panel Absorbers • Resonators • Air Absorption • Music Hall Evaluation and Active Acoustics • Objective Parameters • Subjective Impressions • Active Acoustics • Appendices

Is room acoustics design an art or a science? This frequently asked question highlights the complex and esoteric nature of acoustical design. Many have posed this question, particularly as it relates to concert halls. Room acoustics is in fact a combination of both art and science. Scientific theory plays an important role in defining acoustical measurements and analysis techniques. Still, the best acoustical theory must be combined with creativity, intuition, and experience to be implemented effectively. In other words, while applied acoustics is heavily based in theory, it is also much improved by the empirical judgment of an experienced acoustician. This chapter explores the science and the engineering principles of room acoustics design, while also discussing the connection between acoustical measurements, calculations, and human preference.

Room acoustics as a discipline involves the study and analysis of direct and reflected sound. Appropriate room acoustics are essential in all spaces where sound is to be transmitted to a listener; this includes both

speech and music. Room acoustics design criteria are determined according to the room's intended use. Music, for example, is best appreciated in
spaces that are "warm" and reverberant. Speech, by contrast, is more intelligible in rooms that are less reverberant and more absorptive. This means
the criteria that create good speech intelligibility are very different from
the criteria that create a space suitable for listening to and appreciating
music. It is possible to create suitable acoustics for both speech and music
in the same space, although this is rarely accomplished without some
degree of compromise.

The term room acoustics typically brings to mind spaces where
music is performed and recorded: concert halls, recording studios, and scoring stages, for example. While acoustics are especially important to the success of these spaces, a much wider variety of facilities benefits from well-
designed acoustics. Lecture and convention halls, classrooms, board rooms,
council chambers, courtrooms, places of worship, theaters, cinemas, and
broadcast studios all depend on their acoustical quality. Speech intelligibility is essential in all of these spaces. Different acoustic design criteria are
required for rooms where music is to be played, where "natural" acoustics
help support unamplified musical instruments. Appropriate room acoustics
are also required to reduce reflected noise in such spaces as restaurants, lobbies, offices, libraries, and factories.

This chapter is organized to highlight good acoustical design principles. For this reason, examples of spaces that benefit from a particular acoustical approach are presented to the reader throughout the chapter. To make this
chapter more readable, it is divided into five sections. Part I discusses the
growth of multipurpose spaces (theaters), acoustical planning, and architectural design criteria. Part II explains the principles of reflected sound in
rooms. Part III is devoted to sound pressure levels in rooms and also to
acoustically transparent and sound-absorbing materials. Part IV is concerned
primarily with music and room acoustics. Objective room acoustics measurements are compared here with subjective acoustical impressions. A brief
exploration of the developing concept of active (electro-enhanced) acoustics
rounds out part IV, which is followed by the chapter's conclusion. Part V of
the chapter makes up the appendices, which are for those readers who are
interested in the more technical aspects of acoustical room design. Using
three different music hall shapes as examples, this information illustrates the
closing gap between acoustical calculations and measurements.

I. Planning and Design Criteria

Planning

In order to achieve the best results, the acoustical requirements for a new
facility should be considered early in the design phase. These considerations include: (a) *Room shape*—a room's shape, in part, determines the "signature sound" of a space; (b) *Space allowances*—audience and performer
capacities, wall thickness, storage requirements; (c) *Space adjacencies*—functional proximity requirements and separation between noisy and quiet

functions; (d) *Materials*—a room's finish materials and their mountings determine the amount of sound absorbed and reflected.

These considerations are often individually established by an acoustical consultant working in conjunction with architects, interior designers, and other specialists such as theater consultants.

The Growth of Multipurpose Spaces

Since 1980, there has been an explosion of multipurpose and performing arts theaters in North America. In addition to the major urban centers that have historically devoted halls and theaters to concerts, theatrical performances, dance, films, and lectures, many suburban and rural towns have built or are planning to build such facilities. Often, these projects are part of redevelopment, urban renewal, historic preservation, or adaptive reuse programs. It is not uncommon for these new facilities to have capacities in the range of 400 to 700 seats. However, unlike the major cities, which can afford to construct separate symphony, opera, and dramatic venues, smaller cities and towns often require facilities that are functionally multipurpose in order to accommodate a variety of artistic pursuits. While many of the theaters built today are multipurpose, a great number are also constructed for commercial film exposition. These cinemas typically range in size from 80 to 400 seats, and are usually built in multiplexes of at least four, and as many as twenty, separate theaters.

The acoustical trend in urban theaters has primarily focused on renovation and adaptive reuse. Many architecturally ornate vaudeville theaters became cinemas during the 1960s and 1970s. More recently, these theaters, as well as the cinema palaces built in the twenties and thirties have become too expensive to operate due to their large, typically more than 1,000 seats, capacities and to the advent of more profitable multiplexes. At the same time, with the increasing number of performing arts groups dedicated to drama, musical theater, dance, and symphony, as well as the historical preservation efforts of community and national groups, many of these one-time vaudeville and film palaces have become performing arts centers. Adaptive reuse has also extended to the conversion of places of worship into music and recording halls. Suitable acoustics for these new theaters play an important role in their overall success, community acceptance, utility, and profitability.

Design Criteria

The design criteria for any room should be based on its estimated percentage of use for a particular function. This is particularly important for multipurpose spaces that may need to serve, for example, both as a lecture facility as well as for music recitals. Often such different requirements pose a design conflict that is difficult to resolve, especially if the room is large. While higher levels of reverberation are often suitable for listening to music, the same levels often reduce speech intelligibility. As a general rule, speech is intelligible in rooms having a reverberation time of one second or less. Conversely, music is composed of a wide variety of repertoire and genres, each of which has its own desirable range of reverberation or "live-ness" provided by the room.

If the room is to function as a multipurpose space, then it is essential that the room's varying functions be prioritized. Although it is possible to create multipurpose rooms that can accommodate several functions, the acoustical design for these multipurpose spaces, particularly in larger rooms, usually demand some degree of compromise. For example, if a community center is designing an auditorium that will be used primarily as a lecture facility throughout the year, but will also be used for the presentation of musicals during the summer, then the center might decide to create a space that is optimal for lectures and also suitable (but not ideal) for appreciating music. It is possible to reduce or even eliminate this compromise using active acoustics as discussed in Part IV of this chapter.

In cases where speech intelligibility must be optimized, such as in the dialogue of a film sound track, more specific acoustical criteria are required. Clear dialogue intelligibility is, after all, essential to the story-telling art of film making. Reverberation time criteria are developed as a function of room volume and frequency. Establishing acceptable reverberation times that vary with respect to frequency has proven to be successful in numerous film studio sound-mixing and screening applications. Such criteria are also useful in other situations where the intelligibility of dialogue is paramount.

In addition to criteria for reverberation time, spaces used for critical listening should be designed with concern for the audio signal's imaging and echoes. Imaging includes the apparent size and location of sounds that are part of audio reproduction. Recent psychoacoustic research has defined the early sound field thresholds for perception of reflections, changes to the audio image, and echoes based on the sound level of reflections, and their delay after the direct sound.[1]

If a critical listening space is to be designed to be neutral, that is without added "coloration," then the early reflection levels should lie at or below the threshold for image shift, as shown in Figure 6.1. The threshold for image shift is the level at which a sonic image appears to move from its actual location. Achieving these relatively low reflection levels in a studio control room requires treatment of all surfaces involved in providing first order sound reflections to the listener. One surface which is involved that cannot be treated by the studio design consultant is the upper surface of the mixing console. Future mixing console designs should consider using control surfaces made of porous material, such as sintered aluminum.

Another consideration in the design of critical listening spaces is eliminating "rattles and resonances" often associated with metal fixtures, such as lighting, ducts, diffusers, and furniture. Difficulties are often resolved by applying visco-elastic damping material. Damping is normally available as sheet material with a self-adhesive backing or in liquid form. The sound intensity produced by a vibrating surface is normally proportional to the velocity of the panel vibration. Damping reduces the panel velocity and hence the sound.

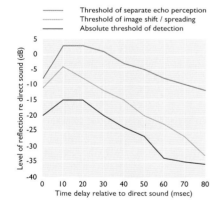

Figure 6.1 Audibility of discrete reflections v. time of arrival, with speech as the test signal.

Numerous sources state that locating sound-reflecting surfaces near a speaking person is essential to achieving adequate loudness and intelligibility for an audience of more than 50 people. If the room is sufficiently quiet and absorbent, however, this may not be the case. In fact, early sound reflections from nearby adjacent surfaces can result in listeners localizing the talker (auditorally) to another location than where he or she is actually standing. Roman and Greek amphitheaters produced good speech intelligibility in part because of the low ambient noise levels in the surrounding environment, and in part because these outside theaters did not produce significant reverberation. It has also been observed that substantial additional vocal effort is not necessary in relatively absorptive rooms seating up to 500 people, provided that the ambient background noise level is very low (NC 20 or less).

Many speech intelligibility analyses culminate in a signal-to-noise ratio calculation. The *signal* radiates directly from the talker or loudspeaker. Reflected room sound may be considered "noise," because the earliest reflections affect timbre and localization phenomena, while later reverberation is known to mask speech. Therefore, all reflections can be considered a form of noise correlated to the signal. Reverberation is especially devastating to the intelligibility of a signal such as speech heard by a hearing-impaired person or someone listening *monaurally,* "with only one channel of audio information." A common example of the latter is the speakerphone through which distant talkers are difficult to understand due to their location in a reverberant field and to the monaural listening condition. Figure 6.2 depicts reverberation time criteria for good speech intelligibility in cinemas, a venue type devoted to storytelling and the intelligibility of dialogue

Figure 6.2 Reverberation time criteria for cinemas and screening theaters. (*left*) Desirable reverberation time envelope as a function of frequency. (*right*) Maximum reverberation time in the 500 Hz octave band for optimum dialogue intelligibility.

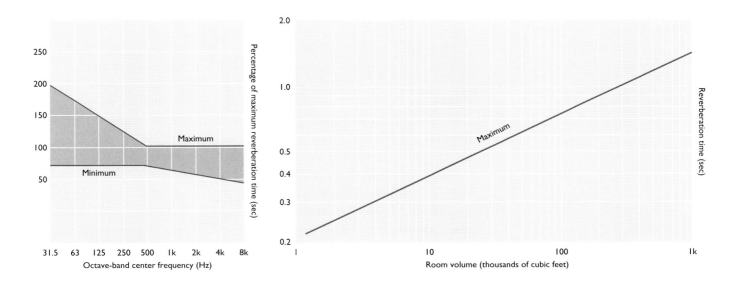

In order to optimize speech intelligibility, the direct-to-reflected sound (signal-to-noise) ratio should be 15 dB or greater. Early sound reflections can aid in increasing the early-to-late sound ratios in reverberant spaces to help overcome the interfering late or reverberant sound. In the case of music, the human hearing response integrates the early sound. This *early* sound comprises direct sound and early reflections occurring within the first 80 msec after the arrival of the direct sound at the listener. The *late* sound occurs after 80 msec. Acceptable early-to-late sound ratio criteria for classical music generally ranges between 0 and 3 dB.

Architectural Design Parameters—Small Rooms

In rectangular rooms with volumes of less than 283 m^3 (10,000 cu. ft.), room dimensions that are equal or exact multiples of one another should be avoided to prevent axial standing-wave frequencies from coinciding. When standing waves coincide, the sound pressures become additive, over-emphasizing a particular low frequency bass note. A design objective is to pick a room aspect ratio of length, to width, to height (i.e., proportions), that evenly distributes standing waves through the low frequency range. Two of the most even distributions are theoretically obtained with aspect ratios 1.9:1.4:1 and 1.44:1.19:1.[2]

Architectural Design Parameters—Large Rooms

When planning new theaters and music halls, a check of total room volume is desirable. Table 6.1 depicts typical volume requirements for different types of performance spaces.

Ceiling height also plays an important role in delivering early reflections in music halls. Typical ceiling heights yielding an adequate density of early sound are: (1) 500 seats: 9–11 m (28–34 ft.); (2)1,500 seats: 13.5–16 m (42–50 ft.); (3) 2,500 seats: 18–21 m (55–65 ft.)

Typically, the maximum ceiling height (h) can be determined from:

EQUATION 6.1

$$h = 0.85 \left[\frac{V}{T_{500}} \right]^{1/3}$$

Where V is the room volume m^3 (cu. ft.), and T$_{500}$ is the desired reverberation time at 500 Hz.

Ceiling heights can vary from these guidelines, particularly when the room shape or reflectors address the required early reflections. Ceiling reflectors positioned over an orchestra should be 5 to 10 m (16 to 32 ft.) above the platform in order to be considered a benefit.[3] Generally, the smaller the ensemble, the lower the reflectors may be. Desirable maximum stage width is approximately 18 m (55 ft.) and stage depth is 12 m (36 ft.).

The acoustical literature highlights the benefits of early lateral reflections, which originate from side walls. It is important, however, to recognize that not all lateral reflections are delivered exclusively by side walls; many acousticians recognize that the ceiling and the ceiling to side-wall reflections are also critical.[4] The shape and angle of the ceiling and of the

Room type	Volume/seat m^3/seat	(cu. ft./seat)
Lecture hall & cinema	4.3–5.7	(150–200)
Multi-purpose hall	5.7–7.1	(200–250)
Opera	8–10	(280–350)
Music and recital halls	10–11.4	(350–400)

Table 6.1 Volume requirements for various performance spaces.

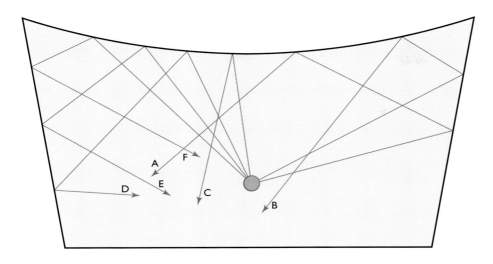

Reflections	Path
A & B	Wall, ceiling, listener
C	Ceiling, listener
D, E & F	Ceiling, wall, listener

Figure 6.3 Vertical cross-section of music hall looking at the stage. The room shaping is used to deliver early reflections evenly over the audience and to maximize "enveloping sound."

side walls can aid the delivery of early reflections to every seat—even in cases where the hall is relatively wide. Figure 6.3 illustrates the importance the ceiling has in delivering early sound to the audience. The ceiling reflects sound directly; it propagates to and from the walls and down to the audience. Concert halls relying solely on natural acoustics are generally recommended to have capacities below 2,500 seats and hall volumes less than 28,300 m³ (1,000,000 cu. ft.). In greater sizes, it is nearly impossible to supply adequate early reflections and perceived loudness levels, unless large suspended sound-reflecting panels and/or active electro-acoustics are employed to enhance the reflected sound.

Seating

Intimacy, so vital to live performances, is related in part to the audience being close to the performance. Historically, deep balconies and wide fan seating plans have been used to bring audiences closer to the stage. Deep balconies reduce the quality of musical acoustics by shading the audience underneath from ceiling reflections. Often the top balcony offers the best sound due to the reception of reverberation and intense early reflections from the ceiling and little possibility of echoes due to the near proximity of the rear wall. Wide fan seating plans, while generally workable for drama and other speech-oriented events, become a detriment for musical performances in theaters seating over 500 occupants because the angled side walls send most of the acoustic reflections to the rear wall, rather than to the audience or to the ceiling (see Figure 6.4). Those with the best sight lines in the center suffer the most from a lack of early reflections. This lack of early reflections has proven to be the nemesis of many concert facilities.

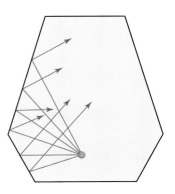

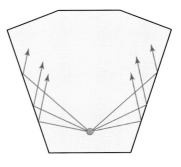

Figure 6.4 (*top*) Typical reverse-fan shaped room wall reflections. (*bottom*) Typical fan-shaped room wall reflections. (*Note:* most reflections are directed toward the rear wall—not the center portion of the audience.)

Concert hall scale model with terraced seating (see text at right).

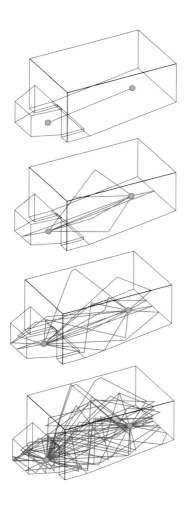

Figure 6.5 Paths of sound propagation: (*from top*) direct sound; first order reflections; second order reflections; third order reflections.

Terraced Seating

Visual intimacy can also be accomplished by creating sections of audience (approximately 300 viewers seated in 10 rows of seats) that are terraced separately from other sections. By elevating some sections, the further sections obtain unobstructed sight and sound lines. This can help to reduce bass loss due to *seat-dip* caused by sound grazing over many rows of seats. The architect Hans Scharoun pioneered this approach to seating in Berlin's Philharmonie. Other terraced halls include the Leipzig Gewandhaus, the Tokyo Suntory Hall and the proposed Los Angeles Walt Disney Concert Hall. (See photo at left.) Even a small performance hall such as Aspen's Harris Concert Hall benefits by stepping up the rear section of seats. Acoustics benefit further in larger concert halls when additional sound–reflecting wall surfaces are provided at the edges of the seating sections. These surfaces can help increase the density of early reflected sound, particularly for the center seats, which are farthest from the room boundaries and could have the greatest gap in time between the direct sound and the early reflections, or Initial Time Gap (ITG). In a concert hall, the ITG should not exceed 20 to 25 msec.[5]

While aisle seating is the most convenient for egress, often the aisles occupy the space where some of the best seats would be. Continental seating, which has side aisles, has a distinct acoustical advantage: no seats are next to the walls. The acoustic response is always poorer next to walls. In this zone, the listeners' ears often receive dramatically different signals! In theaters and music halls containing a large audience that covers a majority of the floor, it should be kept in mind that the audience can be the most significant absorptive surface in the room.

II. The Basic Principles of Room Acoustics

Reflected Sound

The main difference between indoor and outdoor sound propagation is in the level of reflected sound. Indoor environments naturally create more reflected sound than do outdoor environments. Reflected sound can be divided into three distinct categories: (1) early and middle–reflected sound; (2) reverberation (late–reflected sound); (3) standing waves. The corresponding behavior of reflected sound is analyzed in three domains: (1) geometric; (2) statistical; (3) modal (the wave nature of sound).

Early reflections contribute more to the subjective perception of reverberance, or "liveness" of a space. Early and middle reflections occur within the first quarter of a second after arrival of the direct sound. Early sound is considered to be 40 msec after arrival of the direct sound for speech while for music 80 msec is more appropriate. The number and quantity of early and middle reflections delivered to any particular listening location depends largely on the room's shape. For this reason, geometric analysis, which involves the study of reflected sound propagation paths modeled as rays radiating from the source of sound, is particularly useful for tracing echo paths and for studying the uniformity of early reflected sound in medium-

and large-sized spaces. It is also particularly important in spaces larger than 300 seats and 2,830 m³ (100,000 cu. ft.) in volume (see Figure 6.5). It is these early reflections that are so critical for spectacular unamplified music acoustics. To a large degree, they form the signature sound of the space.

Once sound reflections have built up to a point where they are not discernible as discrete events, the late reverberation process takes over. In most well-designed spaces, reverberation is a statistical phenomenon, no longer relying on specific room shape and sound propagation paths. For this reason, the statistical study of room acoustics, which ignores the path of specific reflections but considers reflected sound as an aggregate probability, is employed with respect to reverberation. Statistical analysis methods are applicable to rooms with relatively uniform sound absorbing material distribution and reasonable aspect ratios.

In spaces having a diffuse sound field, where sound is uniformly distributed throughout the space, reverberation decays logarithmically, although the decay sounds even and consistent to the human listener. The reverberation time is defined as "the time for reflected sound to decay 60 dB." Generally, it is necessary to avoid assessing early-sound reflections as part of reverberation since the reflections contribute to sound build-up, rather than to sound decay. The first 10 dB of decaying sound reflections are generally not used to determine the reverberation time, which is determined from the remaining decay.

The wave nature of acoustics is best illustrated by standing waves (room modes), which occur between wall surfaces (see Figure 6.6). This phenomenon is particularly important to bass response in relatively small rooms less than 283 m³ (10,000 cu. ft.) in which there are fewer standing waves per frequency band width, thereby exaggerating those standing waves that do exist.

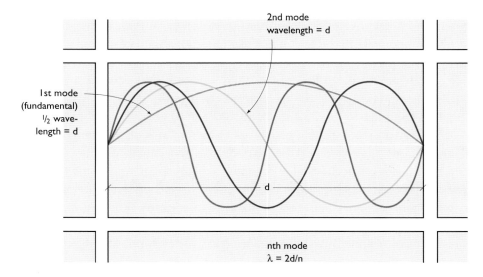

Figure 6.6 Vertical section through room depicting four axial room modes (standing waves).

Specular Reflections

The manner in which sound reflects depends on the shape, texture, and material of the room boundary. Specular reflections, those reflections con-

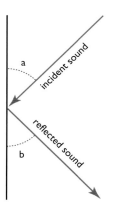

Figure 6.7 Lambert's Law: the angle of reflection (b) is equal to the angle of incidence (a).

forming to Lambert's law of reflection, where the angle of incidence equals the angle of reflection (see Figure 6.7), typically occur at smooth, hard, and relatively flat surfaces. For a surface to be a good reflector of sound, its dimensions should be at least one wavelength or larger than the lowest frequency being reflected. For instance, the wavelength of the musical note middle C (256 Hz) is approximately 1.35 m (4½ ft.) long. Two octaves higher, a little above 1 kHz, the wavelength measures just over 0.345 m (13 in.). In order to adequately reflect low-frequency sounds, which have larger wavelengths, the reflectors must be relatively large.

Diffusion

Sound can also reflect in a diffuse manner. The reflection is fragmented into many reflections having less intensity, which are scattered over a wide angle creating a uniform sound field. Diffusion can be created in a variety of ways, most often by introducing surfaces having irregularities in the form of angled planes or convex surfaces sized at least as large as the wavelength being diffused. Three-dimensional surface ornamentations, columns and statuary serve as diffusing elements and were integral to the acoustics of 17th, 18th, and 19th century performance spaces. The depth of the diffusing undulations must be at least one-tenth the wavelength being diffused (Figure 6.8). However, it is possible, if attempting to create a relatively low-frequency diffuser, for example the octave below middle C, which has a wavelength of 2.7 m (9 ft.), to have specular reflections at higher frequencies. For this reason, in some concert halls, there are macro as well as micro diffusive elements to accommodate diffusion in different frequency (and therefore wavelength) ranges. Fractal mathematics could help create surfaces that diffuse sound over a greater frequency bandwidth. This is accomplished by duplicating the shape of the macro element at micro-scale on the surface of the macro element. Most common diffusers work well between 800 Hz and 4 kHz. Reflections are usually comprised of both diffuse and specular components. In the case of specular reflections, most of the acoustic energy travels in the specific direction dictated by Lambert's law. However, some energy is diffused. Similarly, many diffusers have a strong lobe of directional energy directed along the specular reflection path.

It is common to think of room acoustics geometrically in terms of a reflected ray diagram analysis, but this assumes total specularity, which is an idealization. After three or four consecutive reflections of a ray, the analysis is no longer accurate due to diffusion. Ray diagram analysis has proven to be most useful for time and sound-level distribution of the early reflections and echoes, which are most likely to be specular. Where numerous listening locations are involved (such as at 1,000 seats), it can become a laborious task to calculate the geometric path for each reflection and is seldom done. However, room acoustics software packages are available that can calculate sound paths for a large number of listening locations. Using such a software program, early reflected sound response for three typical room shapes were analyzed and compared. The shapes are the shoe box with sloped floor, the fan for improved sight lines, and the modified reverse fan with improved early sound response. Refer to the Appendix following this chapter for the diagrams which illustrate these common shapes of performance spaces.

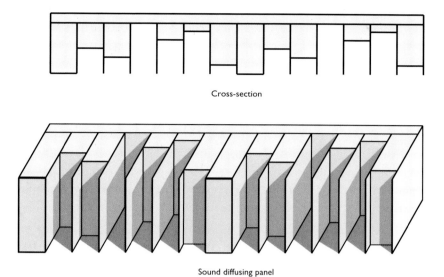

Cross-section

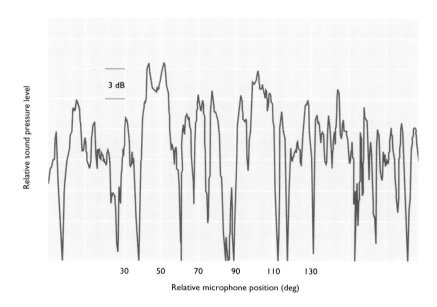

Sound diffusing panel

Figure 6.8 (*top*) Prime root diffusing (PRD) panel and (*bottom*) its measured response indicating that reflected sound is dispersed at a fairly uniform level through a wide range of angles.

Relative sound pressure level

3 dB

30   50   70   90   110   130

Relative microphone position (deg)

Echoes

Echoes are reflections that can be heard distinctly and separately from the early reflected and reverberant sound. Design criteria for echo detection depend on the type of space being constructed. For most general purposes involving speech communication, echoes are normally heard due to intense reflections arriving 40 msec and later after the direct sound signal has reached the listener. In other words, the difference in path length between the direct sound and the reflected sound is at least 13.8 m (46 ft.), corresponding to a propagation time of 40 msec or greater. The reflection forming the echo must be intense enough to dominate those surrounding it. Ironically, echoes are most commonly detected in the front rows of an auditorium and on stage. This results from the front row being farthest from the rear wall, thus generating the largest path length difference between the direct sound and the sound reflected from the rear wall or the combination of the ceiling and the rear wall (see Figure 6.9). Sometimes, only a performer or lecturer is able to perceive an echo. A common example of this phenomenon is a hotel ballroom where a directional sound system radiates sound onto the wall behind the audience and returns a strong echo to the lectern. The echo is not apparent to the audience because of the strong direct sound signal delivered by the loudspeakers. The lecturer, who is not in the direct sound field produced by the loudspeakers (because of the need to control feedback into the microphone), perceives the echo. Typically, such echoes can be suppressed using sound-absorbing or sound-diffusing materials. Even surfaces as small as 10 m² (100 sq. ft.) can require treatment to suppress an echo. Generally, very absorptive rooms must be designed with extreme care in regard to the placement of reflective materials.

Figure 6.9 Cross-section of auditorium (*top*) and sound pressure level response at listener (*bottom*): t_1 is the travel time for the direct sound d; t_2 is the travel time for the reflected sound r. Reflection r is more intense than other reflections arriving at the same time and is perceived as an echo. Sound absorbing material on the rear wall would reduce r's intensity and eliminate the echo.

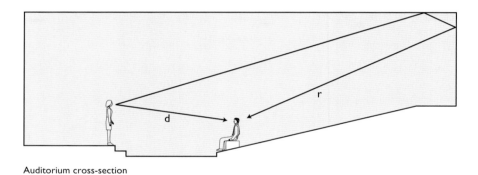

Auditorium cross-section

Flutter Echoes

A flutter echo results when sound travels back and forth between two parallel surfaces and is attenuated more slowly than reflections from other surfaces. Thus, relatively more intense reflections are radiated from the reflective and parallel surfaces (see Figure 6.10). Flutter echoes, which are usually perceivable at frequencies of 250 Hz and greater, largely rely on parallel room boundaries to be sustained. Angling room boundaries, therefore, can help eliminate high frequency flutter echoes.

Focusing

The cardinal rule in the design of rooms is to avoid sound reflectors that focus in the plane of listening. A focusing surface concentrates sound energy, which may then be intense enough to be perceived as an echo. Surfaces such as domes, barrel-vaulted ceilings, and concave rear walls can cause sound focusing and are notorious for generating strong echoes. Such architectural elements should be designed with extreme care to avoid acoustical defects.

It is necessary to design the radius of the curved surface so that the focal point is well above the listening plane or well outside of the room boundaries (Figure 6.11). It is often possible to create the visual effect of a dome or barrel vault using lighting and a much larger radius, which places the focal point well outside the room. The exact focal point depends on the location of the source. In fact, if the focal point is kept high enough above the listening plane and closer to the concave surface than to the audience, curved surfaces can help diffuse sound. After the sound "rays" pass through the focal point, the rays diverge, diffusing sound.

Reverberation

Reverberation is directly proportional to room volume, inversely proportional to the surface area, and inversely proportional to the amount of sound-absorbing material. Excess reverberation results in a blurring of sounds and can reduce speech intelligibility. It is possible to reduce reverberation by the following means: (1) adding sound-absorbing material; (2) reducing room volume; (3) increasing surface area.

Reverberation time is the measure used to quantify reverberation and is the time required for sound reflections to decay 60 dB, one millionth of their original amplitude. A 3-dimensional "waterfall" reverberation decay is shown in Figure 6.12. The Sabine reverberation formula presented below, named for the physicist who first recognized this relationship, applies to rooms that have a relatively diffuse (uniform) sound field.

EQUATION 6.2

$$T = \frac{KV}{S\overline{\alpha}}$$

Where T is the reverberation time (sec.); K is a constant equal to 0.16 for metric units and 0.05 for English units at room temperature; V is the room volume m^3 (cu. ft.); S is the room surface area m^2 (cu. ft.); $\overline{\alpha}$ is the average absorption coefficient.

(Note that S$\overline{\alpha}$ is the absorption in units of square meters or square feet, hence the descriptors, metric sabins and English sabins.) Most acoustical measurements and theory are based on the diffuse sound field, which assumes spaces of reasonable proportions (length, width, and height aspect

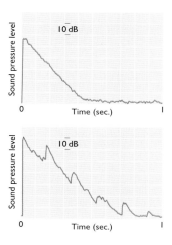

Figure 6.10 (*top*) Reverberation decay of sound pressure level indicative of adequate diffusion. (*bottom*) Reverberation decay of sound pressure level indicative of flutter echo.

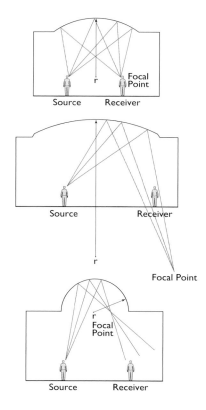

Figure 6.11 (*top*) Undesirable focusing in the listening plane; (*middle*) a larger radius (r) of curvature places the focal point well outside of the room; (*bottom*) a small radius of curvature places the focal point well above the listening plane, resulting in a diffuse sound at the listener.

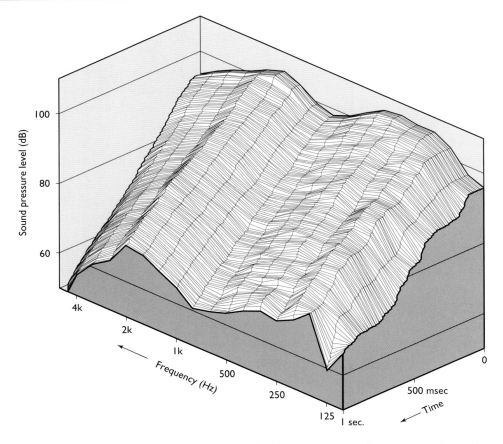

Figure 6.12 Reverberation decay plotted as a function of frequency and time.

ratio) and evenly distributed sound–absorbing material. For example, typical open plan offices have an unusual aspect ratio with very low ceilings compared to the horizontal dimensions. A diffuse sound field cannot be expected in the open office space. A diffuse sound field implies uniform sound levels throughout the listening area. If sound-absorbing material is concentrated on a single surface such as a gymnasium ceiling, then, using Sabine's formula, the reverberation time can depart very significantly from that calculated. An alternative equation was developed by Fitzroy to handle the special case of disparate room dimensions or sound-absorbing material concentrated on a single pair of surfaces, as follows.

EQUATION 6.3

$$T_f = K \frac{V}{S^2} \left[\frac{x^2}{A_x} + \frac{y^2}{A_y} + \frac{z^2}{A_z} \right]$$

Where T_f is the reverberation time (secs.); K is a constant equal to 0.16 for metric units and 0.05 for English units; V is the room volume m^3 (cu. ft.); S is the total room surface area m^2 (sq. ft.); x, y, and z are the areas of parallel rectangular room surface pairs m^2 (sq. ft.); A_x, A_y, and A_z are the total absorption in sabins corresponding to x, y, and z.

While there are other reverberation time equations, such as those described by Norris–Eyring and Fitzroy, for example, the Sabine equation was the first developed, and it remains valid in most cases. In order to determine the reverberation time in a diffuse room, it is necessary to sum up all of the room's sound absorption due to each surface material's contribution. This can be accomplished in each frequency range by multiplying the surface area by the sound-absorption coefficient for a particular frequency range, and then summing in that frequency range for all materials located within the space (see Appendix 2 for more information). Just as reflections are not entirely specular or diffuse, no material is entirely sound-absorbing or sound-reflecting. Rather, a material can have both absorptive and reflective properties, sometimes being reflective in the mid- and high-frequency ranges and absorptive in the low. (A more complete discussion of sound-absorbing material occurs in Part III of this chapter.)

As a general guide, it is not advisable to concentrate large amounts of sound-absorbing material on one surface only, particularly where that surface is distant from a group of listeners. For example, it is not advisable in larger spaces (such as a gymnasium) to concentrate sound-absorbing material only on the ceiling; this concentration would dramatically reduce the efficiency of the sound absorption. In order for a diffuse sound field to exist, sound-absorbing material needs to be distributed over both the wall and ceiling surfaces. In a rectangular space, for example, it is not good design practice to concentrate sound-absorbing material on two parallel surfaces or on two pairs of parallel surfaces. This simply reduces reflections coming from the absorptive surfaces and may result in an echo by enhancing the audibility of the reflected sound from the remaining pair (or pairs) of room surfaces. The reflections from the absorptive surfaces are decreased in amplitude, resulting in a relative increase in the amplitude of the remaining reflections.

Standing Waves

Standing waves are also known as *room modes*. Room modes are most easily perceived when listening to low-frequency tones in small rooms having hard surfaces, for example, a transformer in a concrete vault, which often produces the maximum sound level at a frequency of 120 Hz. If the room dimensions are coincident with the wavelength of 120 Hz (approximately 2.9 m), the sound pressure level of the 120 Hz tone is a maximum at the wall and integral multiples of a $\frac{1}{2}$ wavelength away from the wall. Minimum sound pressures are also at integral multiples of a $\frac{1}{2}$ wavelength, starting $\frac{1}{4}$ wavelength from the wall. This is heard as a variation in loudness, with differences of up to 20 dB (a perceived quadrupling in loudness) being observable to a listener moving about the room. This acoustic phenomenon can be observed in a shower stall, in which one or two mid-bass frequencies are amplified causing strong resonance, which results from single standing waves. While it is fun to sing in the shower for this reason, it adds coloration, often called "boominess." Professional recording artists and engineers usually consider this "boominess" to be undesirable.

Standing waves usually occur between hard parallel wall surfaces and are of particular concern in relatively small rooms, such as music practice rooms, voice recording booths, small audio-control rooms, and other spaces used for recording or for monitoring recordings. In an idealized case, it can be assumed that walls are infinitely rigid and stiff, so that minimum sound absorption occurs, and there is little phase difference between the incident sound and the reflected sound at the point of reflection. Rooms in which two or more major dimensions (for example, length, width, and height) are equivalent to multiples of half wavelengths are notorious for causing additive standing waves and undesirable resonances. The frequency of resonance is higher in small rooms due to the smaller dimensions and shorter wavelengths. For this reason, standing waves are a more important consideration in small rooms where the frequency of interest lies within the normal speech range of 100 Hz to 5 kHz. It is noteworthy that standards require acoustical laboratories to have the lowest useful one-third octave frequency band contain at least ten modes (standing waves) to assure reasonably accurate measurements. This requirement results in a smoother frequency response (less amplification of a single frequency), due to overlapping modes. The lower-limiting frequency is usually 100 Hz. For this reason, laboratories do not usually measure below this frequency, in spite of the fact that there is a growing need for data below 100 Hz. In order to have statistically reliable data at low frequencies, there needs to be a sufficient number of overlapping modes (standing waves); otherwise, severe variations in the sound pressure level result from one location to another, depending upon where the wave maximum or minimum occurs and where the measurement is taken. Using longer measurement periods with a moving microphone can help to average these differences. In order to decrease the lower-limiting frequency (allowing lower frequencies to be tested) and reduce these variations, it is necessary to increase the room volume. The Schroeder frequency (f_s) is the transition from single modal resonances, where standing waves are a concern, to overlapping normal modes. Above this frequency, single room modes are normally not a concern.

EQUATION 6.4

$$f_s = 2000 \sqrt{T/V}$$

Where T is the reverberation time (sec.); V is the room volume m^3.

In studios used for the production or reproduction of audio material, sufficient low-frequency absorption is important. The sound absorption in this case acts as damping, reducing the amplitude and broadening the frequency range of the resonance.

In the idealized case of massive and rigid room boundaries, the sound pressure maximum of the sound wave occurs at the surface boundary. A pressure minimum occurs ¼ wavelength away from the boundary. Conversely, the acoustic velocity (of vibrating air molecules) must be zero at the rigid boundary (i.e., no motion). Therefore, the acoustic velocity is a maximum ¼ wavelength away from the wall (Figure 6.13). This is generally a good location for an absorber, to damp dominant room modes (i.e., resonances). At room corners all room modes achieve a pressure maximum and a velocity minimum (assuming massive rigid boundaries). Room modes exist whether or not the walls of the space are angled or are parallel.

The criteria for acoustical design with respect to modal resonances is to provide adequate low frequency damping in the form of absorption and to have as many room modes per one-third octave band as possible.

Usually, those room modes that propagate axially and are parallel to the room's length, width, and height are dominant. However, the mode field is three dimensional! Usually, the axial modes are the most poorly damped, due to the perpendicular incidence with the room boundaries, effectively making the sound-absorbing material appear to be the thinnest. When sound strikes the wall at an angle, the sound-absorbing material is effectively thicker. Two other types of room modes exist (Figure 6.14): (1) those involving two pairs of surfaces, such as all four walls or a pair of walls and the floor/ceiling, known as *tangential room modes*; (2) those involving all three pairs of surfaces, known as *oblique room modes.*

If reverberation is examined as a statistical decay of room modes, the oblique modes generally decay more rapidly, followed by the tangential modes. The axial modes, being most poorly damped, are normally the slowest to decay. Where there is not an adequate density of modes, it is extremely important for the axial modes to be adequately damped. Otherwise, long extensions of reverberation time can occur in a narrow-frequency range, resulting in a perceptible resonance. Angling room boundaries does not eliminate standing waves, but does redistribute them by slightly altering their frequency. This, however, is not a substitute for adequate damping.

III. Sound Levels & Sound Absorption

Sound Levels in Rooms

There are four categories of sound fields that contribute to the sound level. They are the near field, the far field, the direct field, and the reverberant field. From the listener's perspective, there are the direct and the reflected sound levels. A sound source has a near and a far field (refer to Figure 6.15). The near field distance usually relates to the source dimensions, as discussed in Chapter 20. The larger the source, therefore, the larger the near sound field. Then, when moving away from the source, direct field attenuation takes over until the listener has moved far enough back to enter the reverberant field. There are two primary ways to have a larger direct field: provide a more directional sound source or add more sound-absorbing material (thereby reducing reverberation).

Direct sound radiated from the source in the far field attenuates proportionally to the inverse square of the distance away from the source

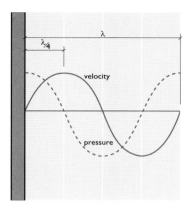

Figure 6.13 Room mode sound pressure maximum at the wall. Acoustic molecular velocity is a maximum ¼ wavelength from wall.

Axial – one dimensional

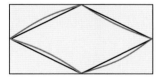

Tangential – two dimensional

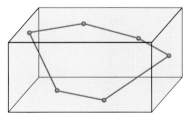

Oblique – three dimensional

Figure 6.14 Types of room modes (standing waves).

(6 dB per doubling of distance). Sound usually attenuates very little in the near field. Sound levels can be predicted using the following equation:

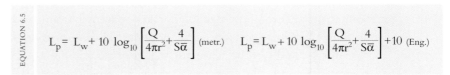

$$L_p = L_w + 10 \, \log_{10} \left[\frac{Q}{4\pi r^2} + \frac{4}{S\overline{\alpha}} \right] \text{ (metr.)} \qquad L_p = L_w + 10 \, \log_{10} \left[\frac{Q}{4\pi r^2} + \frac{4}{S\overline{\alpha}} \right] + 10 \text{ (Eng.)}$$

EQUATION 6.5

Where L_p is the sound pressure level (dB re: 20 μPa) at distance r (m, ft.) L_w is the sound power level of the sound source (dB re: 1 pW); Q is the directivity of the source; S is the surface area of the room (m², sq. ft.); $\overline{\alpha}$ is the average sound absorption coefficient.

The first term of the equation containing r^2 relates to the attenuation of the direct sound field which is 6 dB per doubling of distance, and the second term containing $S\overline{\alpha}$ to the reverberant field. The above equation is applicable to empty rooms having diffuse sound fields with uniformly distributed sound-absorbing material.

The directivity (Q) is equal to 1 for a spherical omni-directional source. Q equals 2 for a hemispherical source. Q of the human voice approximately equals 4, and a loudspeaker typically used in cinema sound reproduction with a dispersion pattern of 90 degrees wide by 40 degrees high has a Q of slightly greater than 10. (Dispersion patterns are normally quoted at the maximum included angle 6 dB below the on-axis response.)

Recently, a new theory has been proposed for theaters and concert halls, in which sound propagates over a diffuse sound-absorbing plane, namely the seats and audience.[6] The reverberant sound field does not result in virtually constant sound levels as in the case of an empty room. This phenomenon becomes important when evaluating direct-to-reverberant sound level ratios for speech and music in larger spaces. (Refer to this chapter's Appendix, and the early-to-late sound index discussed later.)

A parameter often used to evaluate the "liveness" of a room is the *critical distance,* also known as the *reverberation* or *room radius.* Reverberation radius is "the distance from the source of sound where the direct sound field and reverberant sound field have the same sound level." This is not only a property of the room, but also of the sound source's directivity. In concert halls, this distance is approximately 5 m (15 ft.), whereas in a film-screening theater, the room radius may be 20 m (62 ft.). The following equation may be used to calculate the reverberation radius.

$$r = 0.057 \left[\frac{QV}{T} \right]^{1/2}$$

EQUATION 6.6

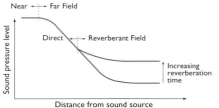

Near ←→ Far Field

Direct ←→ Reverberant Field

Increasing reverberation time

Sound pressure level

Distance from sound source

Figure 6.15 Sound pressure level reduction as a function of distance from the source.

Where r is reverberation radius (m); Q is the directivity of the source; V the room volume (m³); T is the reverberation time (sec.).

EQUATION 6.7

$$r = 0.141 \left[QA \right]^{1/2}$$

This is an alternate form of Equation 6.6, where A is the total room absorption (metric sabins).

In concert halls, almost all of the listeners are farther from the orchestra than the room radius; the audience is largely in the reverberant field. In a cinema, however, where reverberation would impair speech intelligibility, most of the listeners should be in the direct sound field. It is noteworthy that in the case of film sound reproduction, all of the sound effects, including reverberation, are recorded as part of the film sound track and do not generally require assistance from the room reflections.

In a factory where there are numerous noise-producing machines and hard floor, wall, and ceiling surfaces, the reverberant sound field can add significantly to the overall sound level. A similar situation occurs in a crowded restaurant or at a party where numerous conversations occur simultaneously. This is particularly true where the background noise is not controlled by noise emissions from a single source, but rather by an ensemble. In this case, the addition of sound-absorbing material is beneficial to the reduction of noise. The noise levels can typically be reduced in the reverberant field up to a maximum of 10 dB (perceived as a halving of loudness). More typically, however, the noise levels are reduced by about 6 dB.

The following equation can be used to estimate the reduction in noise levels in the reverberant far field.

EQUATION 6.8

$$LR = 10 \, \log_{10} \left[\frac{A_2}{A_1} \right]$$

Where LR is the reduction in sound level (dB); A_1 is the initial total room absorption (sabins); A_2 is the final total room absorption (sabins).

In many cases, sound-absorbing material is added to the facility's ceiling. The ceiling is commonly closest to the sources of sound, which are distributed evenly in a room with a low ceiling. In this way, each sound source has equal access to the absorption, so no individual sound becomes dominant by not being absorbed. Where the ceiling is high, sound-absorbing material should be evenly distributed on the ceilings and walls.

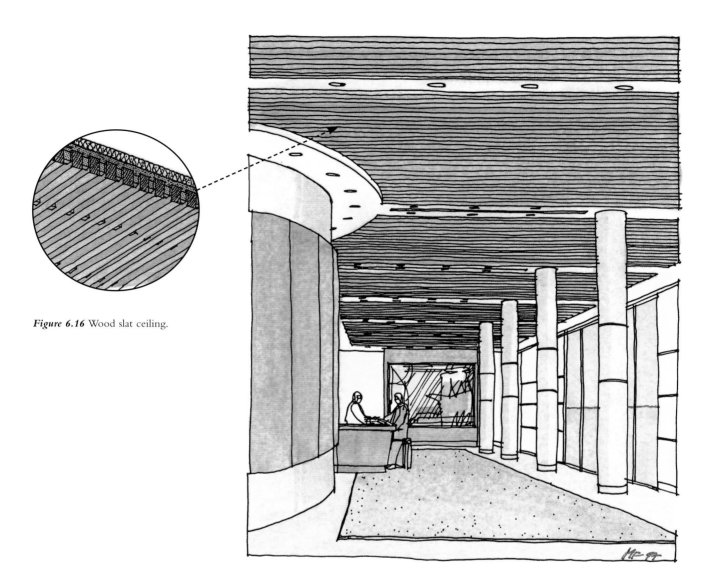

Figure 6.16 Wood slat ceiling.

Sound Transparent Surfaces

In cases where visual and acoustic aesthetics cannot be accommodated by a common room shape or by the distribution of materials, acoustically transparent surfaces can play a helpful role. Sound transparent materials are used to conceal acoustical treatments and to avoid what could otherwise be considered a conflicting design appearance. With proper selection, a wide range of materials can work as acoustically transparent finishes. These include fabric, woven wire mesh, perforated metal, and spaced wood slats (Figure 6.16).

The most common acoustically transparent material is loosely woven fabric, such as a light-weight muslin or other fire-retardant cloth, often seen covering fibrous, sound-absorbent materials such as acoustical wall panels (Figure 6.17). Most fabrics used as transparent materials also have a sound-absorbing quality, which can become an issue in spaces designed to be reverberant. The excess absorption comes from restricting the air flow through the fabric, resulting in vibration of the fabric fibers and therefore dissipating acoustic energy as heat.

In the case of rigid perforated material, sheet metal is the most commonly used (Figure 6.18). In general, rigid, perforated materials are relatively transparent except at high frequencies where increasingly more sound is reflected. Another drawback of perforated metal is that it is relatively expensive and prone to rattles and resonances, if not properly attached. The most common mistake when using perforated materials is to examine only the percentage of open area and assume that transmissivity (acoustical transparency) is directly proportional to the open area alone. In fact, the transmissivity of perforated material is governed by three parameters: perforation hole size, material thickness, and percentage of open area.[7] The high-frequency transmissivity can be increased by using the thinnest possible material, the smallest hole diameter, and the greatest open area (the greatest number of holes). Generally, the most acoustically transparent sheet metal materials have the smallest hole diameter and the greatest open area. The smallest available hole diameter is approximately equal to the material thickness, due to manufacturing constraints. One common transmissive element is the motion picture screen, which is perforated to allow the main channel loudspeakers behind the screen to be heard through the screen. In 1980, it was recognized that a large percentage of high-frequency sound was being re-reflected behind the screen due to the relatively poorer high-

Figure 6.17 Acoustical wall panels.

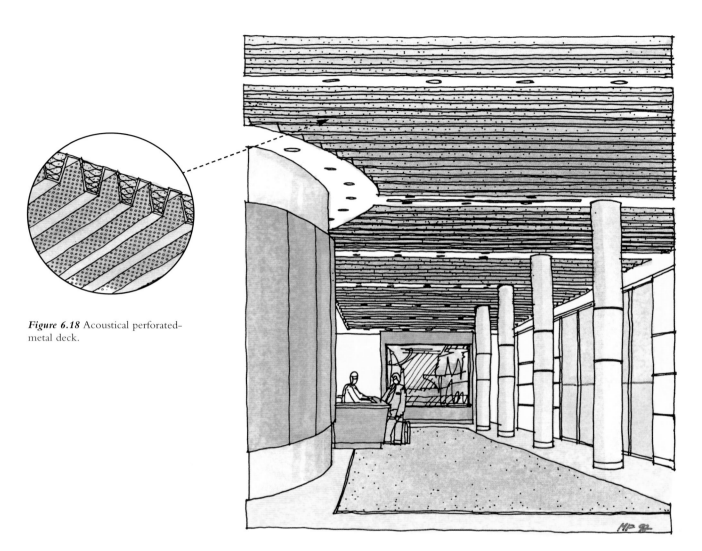

Figure 6.18 Acoustical perforated-metal deck.

frequency transmissivity of this perforated material. In order to prevent sounds from leaking back into the theater and reducing clarity, the practice of placing sound-absorbing material behind the screen was developed.

Sound Absorbing Materials

All materials have some sound-absorbing properties. Incident sound energy that is not absorbed must be reflected, transmitted, or dissipated. A material's sound-absorbing properties can be described as a sound absorption coefficient in a particular frequency range. The coefficient can be viewed as a percentage of sound being absorbed, where 1.00 is complete absorption (100%), and 0.01 is minimal (1%). To determine the amount of sound absorption a room surface has, the surface area is multiplied by the absorption coefficient to yield a result in sabins (units of area m^2 or sq. ft.). The total sound absorption within a room can be obtained by adding the amount of absorption in sabins attributed to all surfaces in each frequency range. (Refer to Appendix 2.)

Incident sound striking a room surface yields sound energy comprising reflected sound, absorbed sound, and transmitted sound. Most good sound reflectors prevent sound transmission by forming a solid impervious barrier. Conversely, most good sound absorbers readily transmit sound. Sound reflectors tend to be impervious and massive, while sound absorbers are generally porous, light-weight material. It is for this reason that sound transmitted between rooms is little affected by adding sound absorption to the wall surface.

There are three basic categories of sound absorbers: porous materials commonly formed of matted or spun fibers, panel (membrane) absorbers having an impervious surface mounted over an airspace, and resonators created by holes or slots connected to an enclosed volume of trapped air. In some cases, the absorptivity of each type of sound absorber can be dramatically influenced by the mounting method employed.

Porous Absorbers

Common porous absorbers include carpet, draperies, spray-applied cellulose, aerated plaster, fibrous mineral wool and glass fiber, open-cell foam, and felted or cast porous ceiling tile. Generally, all of these materials allow

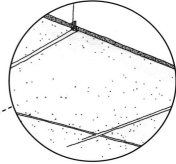

Figure 6.19 Lay-in acoustical ceiling tile.

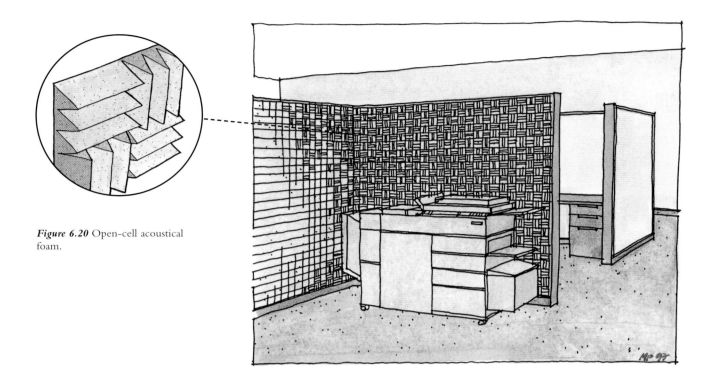

Figure 6.20 Open-cell acoustical foam.

air to flow into a cellular structure where sound energy is converted to heat. Porous absorbers are the most commonly used sound-absorbing materials. Thickness plays an important role in sound absorption by porous materials. Fabric applied directly to a hard massive substrate such as plaster or gypsum board does not make an efficient sound absorber due to the very thin layer of fiber. Generally, better sound absorption is provided by thicker porous materials. Thicker materials generally provide more bass sound absorption, or damping. Sound-absorbing materials are rated using the noise reduction coefficient (NRC), a single figure-of-merit rating averaging four-octave band sound absorption coefficients from 250 Hz to 2 kHz, the primary speech frequency range. Thick cut pile carpeting on a hair jute pad can have an NRC of 0.35, while a glue-down carpet tile may only have an NRC of 0.10. Typical 15 mm (⅝ in.) thick ceiling tiles have an NRC rating of 0.50 when mounted in a lay-in grid ceiling. A 25 mm (1 in.) thick glass-fiber-based ceiling tile typically has an NRC rating of 0.80 or greater (Figure 6.19). Providing an air cavity behind porous absorbers, such as the air plenum above ceiling tile, normally increases the low-frequency sound absorption.

Open-cell foam panels are effective sound absorbers because they offer increased surface area due to the contoured surface of the foam (Figure 6.20). Like foam panels, suspended baffles yield more sabins per unit area of material. Baffles are absorbent because both the front and back sides of each baffle is exposed to the sound field (Figure 6.21). Baffles are typically used in spaces that require high levels of sound absorption such as railway stations and industrial manufacturing plants. Where thicker, more abuse-resistant finishes are required, porous sound-absorbing materials are concealed behind protective material that is perforated or slotted. Wall car-

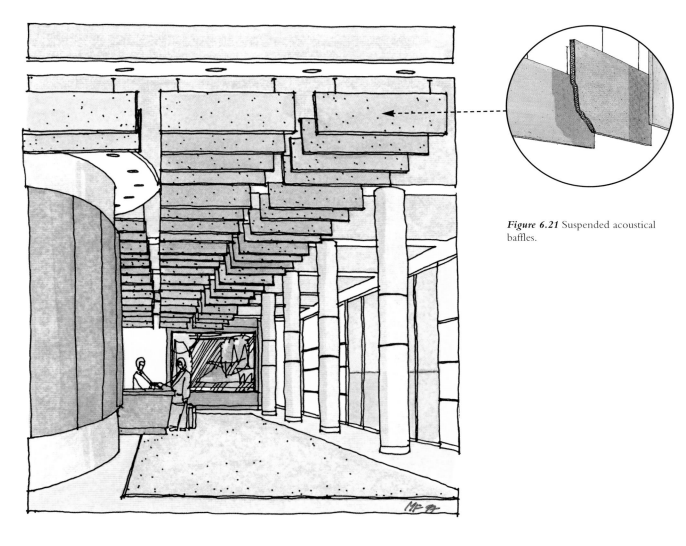

Figure 6.21 Suspended acoustical baffles.

pet that is furred over sound-absorbing fibrous board (Figure 6.22) is also used. This type of carpet is commonly applied to the rear and side walls of movie cinemas. Quilted sound-absorbing material is also very abuse-resistant (Figure 6.23). It is available with a massive septum to help reduce sound transmission. This material is used in a wide range of situations. For example, baffles are frequently hung around noisy machinery and larger arrays of baffles are suspended on the walls of sound stages and recording studios.

Typically, a substance that prevents air flow into the porous material will result in its decreased sound absorption. For example, applying bridging paint or fabric backed with a non-porous (impervious) material reduces sound absorption. Frequently, acoustical plasters, when repainted, lose their sound-absorbing properties. In any case, acoustic plasters and other spray-applied sound-absorbing treatments rely on proper application in multiple coats in order to create porous cavities (Figure 6.24). When these treatments are improperly applied, their absorptive properties are greatly reduced.

Laboratory test reports providing sound-absorption coefficients are based on measuring the sound-absorbing materials in a diffuse field with

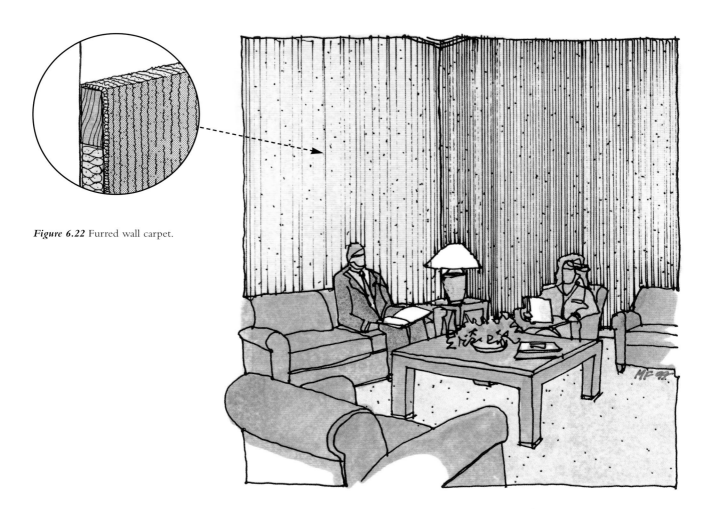

Figure 6.22 Furred wall carpet.

sound striking the material being tested at all angles of incidence (random incidence). This is rarely the case in actual field conditions, since normally there is a primary direction for sound striking a surface radiating from a stationary source. Where sound incidence is perpendicular to the material, a sound-absorption coefficient less than that reported may often be observed. The sound-absorption coefficient at perpendicular incidence is best measured using an impedance tube. Sound-absorbing materials can be deficient in absorbing power at perpendicular incidence in a narrow frequency range, resulting in a flutter echo.

Occasionally, laboratories report sound-absorption coefficients exceeding 1.00. This anomaly is created by edge defraction and the fact that only the horizontal surface area of the material is considered, and the increase of surface area (due to the edges associated with the thickness) is ignored.

Panel Absorbers

Typically, panel absorbers are non-rigid, non-porous materials which are placed over an airspace that vibrates in a flexural mode in response to sound pressure exerted by adjacent air molecules. Common panel (membrane) absorbers include thin wood paneling over framing, lightweight impervious ceilings and floors, glazing, and other large surfaces capable of resonating in response to sound. Panel absorbers are usually most efficient at absorbing low frequencies. This fact has been learned repeatedly on orchestra platforms where thin wood paneling absorbs most of the bass sound, robbing the room of a quality described as "warmth." Typically, the maximum sound-absorption coefficient that can be provided by panel absorbers is 0.5 in a relatively narrow-frequency range. Membrane panel absorbers can be tuned to a particular frequency. It is possible to broaden the frequency range being absorbed by adding porous material in the air-space behind the panel absorber, also called *cavity damping*. The resonant frequency of a panel absorber is typically the frequency at which maxi-mum absorption will occur. The resonant frequency can be calculated using the equation below.

EQUATION 6.9

$$f = \frac{483}{[md]^{1/2}} \quad \text{(metric)} \qquad f = \frac{170}{[md]^{1/2}} \quad \text{(English)}$$

Where f is the resonant frequency (Hz); m is the surface density, kg/m^2 (psf); d is the depth of the airspace behind the panel absorber, cm (in.).

Typically, the resonant frequency of panel absorbers lies below 400 Hz. The greater the airspace behind the membrane and the higher its surface den-sity, the lower the resonant frequency will be. Membrane absorbers can be useful for helping damp standing waves between 100 and 200 Hz. A typi-

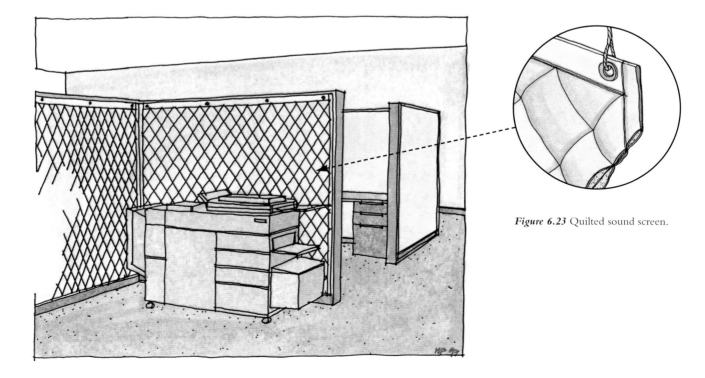

Figure 6.23 Quilted sound screen.

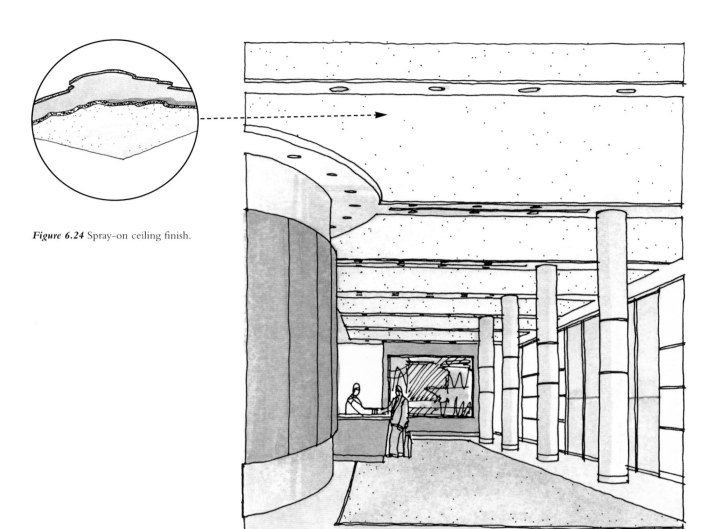

Figure 6.24 Spray-on ceiling finish.

cal membrane absorber is 6 mm (¼ in.) tempered hardboard over a 37 mm (1½ in.) airspace containing 25 mm (1 in.) thick glass fiber.

It is possible to combine the attributes of a porous absorber with those of a membrane absorber. For instance, a porous absorptive layer could be surface-applied over a membrane absorber. Specialized porous absorbers have been developed for use in locations where cleaning is necessary, such as in kitchens and operating rooms. In these cases a thin plastic film is applied over the glass fiber. Usually, the low-frequency absorption is enhanced by the membranous characteristics of the film. However, when the two absorbers are combined, the high-frequency absorption is somewhat compromised.

Resonators

Resonators typically act to absorb sound in a narrow-frequency range; they include some perforated materials and materials that have openings (holes and slots). The classic example of a resonator is the Helmholtz resonator, which has the shape of a bottle. The resonant frequency (f_c) is governed by the size of the opening, the length of the neck, and the volume of air trapped in the chamber.

EQUATION 6.10

$$f_c = 2\pi c \left[\frac{A}{V(l + nd)} \right]^{1/2}$$

Where c is the speed of sound; A is the neck cross-sectional area; l is the length of neck; d is the diameter of the neck opening; n typically varies between 0.85 for a small neck diameter and 0.60 for a resonator without a neck; V is the chamber volume.

Slotted concrete masonry units (CMU) are building materials that are based on the use of resonators. Slots open the hollow core (or chamber) to the room (Figure 6.25). Often, these resonators are tuned to relatively low frequencies. For such a resonator to be an efficient absorber, it is necessary

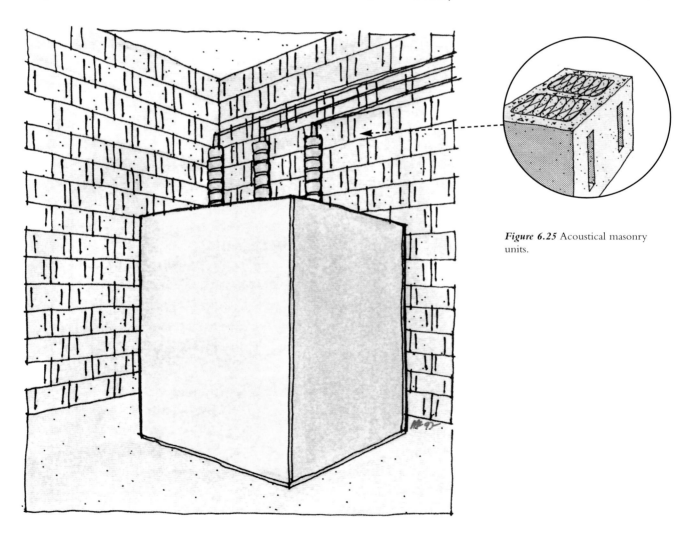

Figure 6.25 Acoustical masonry units.

to locate the mouth of the opening near an area of high–acoustic velocity. Usually, such resonators have fairly narrow tunings and are ideal for damping standing waves or absorbing sound emitted by a tonal source, such as a transformer.

Typically, perforated materials only absorb the mid-frequency range unless special care is taken in designing the facing to be as acoustically transparent as possible. (See the discussion of transparent surfaces earlier in this chapter.) Slats usually have a similar acoustic response. The amount of high-frequency absorption is determined by the dimension of the slat and its ability to reflect a particular wavelength, and the space between slats exposing a higher percentage of absorptive material (Figure 6.26). Normally, wood grilles with very narrow slats come closest to having the same absorption characteristic as the sound-absorbing material being covered.

Long, narrow slots can be used to absorb low frequencies. For this reason, long, narrow air distribution slots in rooms for music production should be viewed with suspicion since the slots may absorb valuable low-frequency energy. For example, in a music recording stage specifically

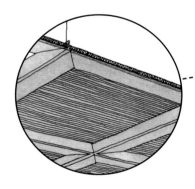

Figure 6.26 Lay–in slatted wood tile.

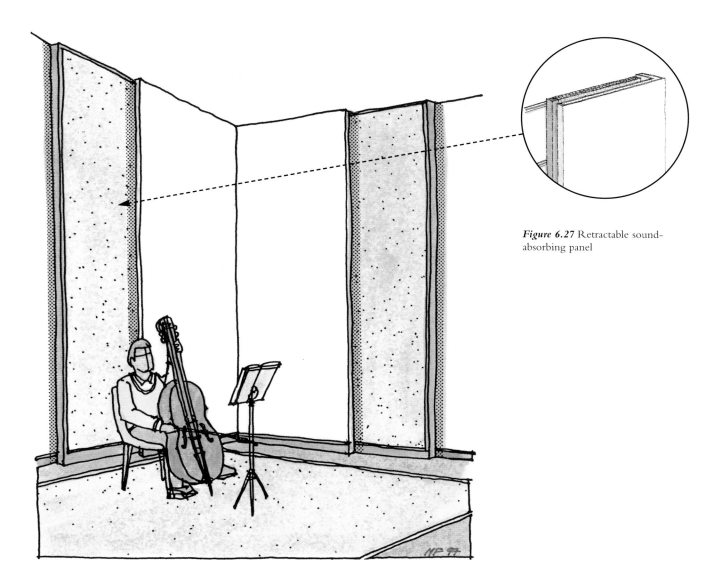

Figure 6.27 Retractable sound-absorbing panel

designed to have variable reverberation times by using sliding sound ab-sorptive panels, an interesting sound-absorbing phenomenon occurred (Figure 6.27). When the sound-absorbing panels were pulled out of the slots and exposed in the room, the reverberation time was reduced, except at frequencies below 80 Hz. When the absorptive panels were in the slots, the room reverberation increased except at frequencies below 80 Hz. Why did the low frequencies behave differently? A 19 mm ($\frac{3}{4}$ in.) gap, 10 m (31 ft.) long, was created when the panels were in the slots. The slots were filled with the concealed 100 mm (4 in.) thick sound-absorbing material. Normally, with sound-absorbing material concealed in the slots, a longer reverberation time is expected. This reversal can be attributed to the absorption sitting in the unsealed slots, which absorb more sound below 80 Hz than when the absorptive panels are out of the slots and exposed in the room. To eliminate the low-frequency paradox, the slots must be sealed when the sound-absorbing material is concealed in the slot.

Air Absorption

Sound absorption in air can be significant in rooms larger than 2,800 m^3 (100,000 cu. ft.) in a frequency range above 1 kHz. Air absorption becomes much more significant at relative humidities below 30 percent. In music halls having large interior volumes, air absorption is often responsible for the reduced reverberation time at high frequencies when compared to the mid-frequency range. Air absorption in a room is directly proportional to the distance traveled by the reflected sound prior to reaching the listener; absorption is therefore greater in larger rooms.

IV. Music Hall Evaluation and Active Acoustics

THE final assessment of the suitability of a room's acoustics is subjective and entirely dependent upon human perceptions and preferences. While acousticians have had the ability to measure sound pressure level and reverberation time for decades, the subjective impression of reverberance has not always correlated well with the objective measurement. In addition, the terminology describing the subjective impressions has been poorly standardized until recently. In order to have a meaningful discussion, it is necessary to standardize the terminology, as indicated in Table 6. 2.

Table 6.2 Standard terminology.

Subjective term	Description of precept	Proposed objective measure
Loudness	Strength or loudness of a sound	Total sound (pressure) level (Source strength)— G (A-weighted)
Clarity	Articulation—the ability to hear definition and detail, often relating to speech or faster tempo music	Early-to-late sound ratio—C_{40}, C_{80} (level adjusted)
Intimacy	Apparent closeness of sound	ITG or "Initial time gap"
Reverberance	Perception of reflected sound and liveness	Early sound reflections, EDT (125 Hz to 4 kHz SEG ratio, ISE-T_5)
Envelopment	Immersion in a sound field, the sense of being surrounded	Late lateral sound level (after 80 msec)
Brightness	Relative loudness of treble or high frequency sounds compared to mid-frequency sound	2 to 4 kHz sound level and reverberation time
Bass warmth	The relative loudness of bass or low frequency sounds compared to the mid-frequency sounds	Early low frequency sound level—125 to 500 Hz values of G in the first 50 msec

The next section contains definitions for the objectively measurable parameters that are then related to subjective impressions in the section that follows. Following these two sections, the field of active acoustics is briefly discussed.

Objective Parameters

Reverberation time (T) is the time it takes the reflected sound to decay 60 dB subsequent to the abrupt cessation of the sound source.

Source strength (G) is the ratio of sound pressure level compared to the reference level of the direct sound field without reflections measured at 10 m (33 ft.) from the same source. In concert halls, the values typically range between 0 and +10 dB, demonstrating the amplification provided by room reflections.

The early-to-late sound ratio (Clarity (C_{40}, C_{80})) is the sound pressure level including the direct- and early-reflected sound (up to 40 msec for speech and up to 80 msec for music), divided by the total sound energy arriving after 40 and 80 msec respectively. C_t can be estimated from the following equation:

EQUATION 6.11

$$C_t = 10 \, \log_{10} \left[\frac{Ve^{(0.04r+13.82t)/T}}{312Tr^2} + e^{(13.82t/T)} - 1 \right]$$

Where C_t is clarity; t is the time (sec.), defining the extent of the early sound field (usually 0.04 or 0.08); r is the distance from the sound source m (ft.); T is the reverberation time. (See Appendix 1 for predicted values of C_t for a 12,000 m³ (39,372 cu. ft.) hall at varying distances and reverberation times.)

C_{40} values of +5 dB are considered good for speech intelligibility, while those less than 0 dB are poor. C_{80} values considered desirable are +5 dB for electronically amplified music, 0 dB for classical music, and –2 dB for romantic classical music.

Lateral energy fraction compares the sound energy arriving laterally with the sound energy arriving from all directions. It has been shown that this measurement does not account for all spatial effects heard by listeners, and that no parameter yet measures the subjective diffuseness of a reverberant sound.[8] The acceptable range of variation in early (0–80 msec) lateral energy fraction is 0.1 to 0.35 for unoccupied music halls with the larger value being preferred.

Interaural cross-correlation (IACC) is the degree of correlation between signals arriving at a listener's ears. IACC rates on a scale of 0 to 1.

Early decay time (EDT) is the initial reflected sound decay defined as "the slope of the early sound decay occurring in the first 10 dB of decay normalized to 60 dB, making it comparable to reverberation time." In a perfectly diffuse sound field, the early-decay time would be equal to the reverberation time. This is the case in many concert halls. However, in some halls, where reverberance is less apparent, the early-decay time is less than the reverberation time due to a lack of early-reflected sound.

Sound energy growth (SEG) curve is a new parameter depicting the growth in sound energy during the first 200 to 300 msec (Early and Middle).

Instantaneous sound envelope (ISE) is a new parameter yielding the sound amplitude and arrival time of individual reflections.[9] ISE is especially useful for comparison with calculated reflectograms generated in computer models. ISE shows individual reflections contribution to the SEG (See Figure 6.36 in Appendix 3). The author's intention in introducing these two new terms is to provide a better comparison between measurements (ISE), calculations (from ray or image computer models), and subjective impressions (SEG) based on energy parameters.

Initial time gap (ITG) is the time between the direct sound arrival at a listener and the first reflections. For good concert acoustics, the ITG should not exceed 20 msec.

Subjective Impressions

It is well known that the subjective impression of reverberance is not solely a function of the reverberation time, but rather a relationship between sound-energy growth, early-decay time, early-to-late sound ratio, and the specific type of program material being transmitted. Reverberance is much more apparent when listening to speech than to classical music, due to the need for greater articulation. Similarly, fast tempo percussive music, such as jazz, is more easily "muddied" by reverberation than is classical music. Classical music of the romantic period can tolerate still longer reverberation times. Baroque organ music can thrive on long reverberation times of 3 to 6 seconds. It is possible for a listener to perceive the same level of reverberance in very different spaces. For instance, similar impressions of reverberance can be had in a 5,700 m³ (200,000 cu. ft.) room having a reverberation time of 2 seconds as in a 57 m³ (2,000 cu. ft.) space with a reverberation time of 0.5 seconds![10]

One interesting study, which is based on subjective impressions, indicates that concert listeners may fall into two preference groups: those who prefer reverberance and those who prefer intimacy. This same study also revealed parameters strongly correlated to overall acoustic impression as indicated graphically in Figure 6.28.

The link between the subjective impression parameters and the objective measurement parameters is currently an area of intense study.[11] Recently, the valid frequency ranges for certain parameters were identified as follows: (1) early-decay time: 125 Hz to 4 kHz; (2) early-to-late sound ratio: 500 Hz to 2 kHz; (3) sound strength: 125 Hz to 4 kHz; (4) lateral energy fraction: 125 Hz to 1 kHz.

Other research results indicate high levels of correlation between loudness and A-weighted source strength. Reverberance was found to be highly correlated with early-decay time. Clarity was found to highly correlate with the early-to-late sound ratio averaged over the 500 Hz and 1 kHz bands. Brightness (treble) was found to highly correlate with late sound strength (after 80 msec) subtracting the 1 and 2 kHz average value from the 4 kHz value.

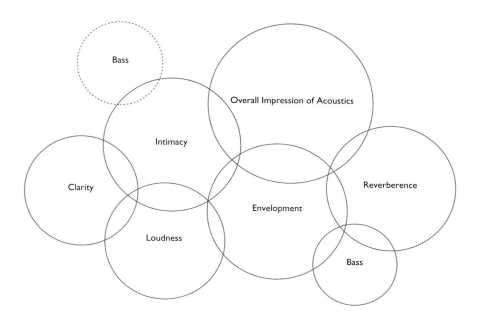

Figure 6.28 Relationship of acoustical terms based on subjective impressions.

Active Acoustics

Active acoustics is the emerging field of enhancing natural acoustics with an electroacoustic system employing a digital reverberation processor. While the concept of adding electronically generated reflections to those provided by the room is not new, recently, the ability to synthesize electronic reflections, which cannot be distinguished from those occurring naturally, has become easier. This field has a bright future; its applications include both new construction and retrofit in existing buildings. This form of active acoustics adds the electronically generated sound field to that provided by the room, unlike active noise control, which seeks to reduce the sound level through sound field cancellation. Active acoustics potentially allows for the design of multiuse facilities based on speech. For music, the space can be made more reverberant with more early reflections by using an active sound field. Active acoustics can potentially be implemented using as few as two appropriately placed microphones and a digital reverberation processor having two inputs and four outputs. Recently, such systems have been implemented successfully in the Elgin Theater in Toronto, the Luther Burbank Community Arts Center in Santa Rosa, and the Tsai Center at Boston University.[12]

Until now, the implementation of such systems has been slow due to the lack of enthusiasm for electronics and electroacoustics in the application where their benefits would be most felt—classical music. This, however, is changing, in large part due to the active role electronics are playing in our daily lives. If electronically enhanced reverberance cannot be distinguished from natural reverberance produced in the best halls, then wide acceptance for electronically assisted reverberance should follow. One major advantage of synthesized reverberation is that it can be adjusted to suit a particular performance or performance-venue configuration, particularly those where seating plans are variable. From the architectural standpoint, provisions need to be made for an array of loudspeakers concealed within the walls and ceiling of a reverberance-enhanced space. Concealing loudspeakers remains a challenge where smooth wall and ceiling surfaces occur.

Conclusion

Is room acoustics an art or a science? This chapter has explored that question by illustrating the connection between newer sound-measurement techniques and aural impressions formed by listeners' preferences. Recent technology has refined the acoustician's ability to predict a room's acoustical requirements. It is now possible, for example, to provide active acoustical enhancement by introducing synthesized sound reflections through an array of loudspeakers, thus improving the quality of the transmitted sound dramatically. More specific design criteria are also evolving to suit different uses. Acknowledging the uniqueness of the design criteria required for each space is vital to the success of the facility, especially if it is multipurpose.

Art implies intuition and mastery. Science can aid in the development of both. But what role does luck play? Were the grand masters simply lucky? Is it luck or skill that allows an artist to appeal to a broad audience? It is in fact a combination of both. Today's room acoustics, like many arts, is an opinion-dominated field, one that is influenced as much by history as it is by technology.

References

1. Floyd Tool, "Speakers and Rooms for Stereophonic Sound Reproduction." (proceeding of Audio Engineering Society Eighth International Conference, 1990)

2. Joseph Milner and Robert Bernhard, "An Investigation of the Modal Characteristics of Nonrectangular Reverberation Rooms." *Journal of the Acoustical Society of America,* vol. 85, no. 2 (February 1989).

3. T.J. Shultz, "Room Acoustics in the Design and Use of Large Contemporary Concert Halls." (proceedings of the 12th International Congress on Acoustics, 1986).

4. Y. Toyota and M. Nagata et al., "A Study of Room Shape of Concert Halls" (technical paper GAA presented at 120th Acoustical Society of America, November 1996).

5. Leo L. Beranek, *Concert and Opera Halls: How They Sound* (Acoustical Society of America, 1996).

6. M. Baron and L. Lee, "Energy Relations in Concert Auditoriums." *Journal of the Acoustical Society of America,* vol. 84, no. 2 (August 1988) pp. 618–628.

7. T.J. Shultz, "Acoustical Uses for Perforated Materials: Principles and Applications" (Industrial Perforators Association, 1986)

8 M. Baron and L. Lee, pp. 618–628.

9 D. Schwind and A. Nash, et al., "The Early Sound Field in Performance Halls" (Audio Engineering Society Preprint 4108. 99th Convention, October 1995).

10. D. Griesinger, "Subjective Loudness of Running Reverberation in Halls and Stages" (proceedings of W.C. Sabine Centennial Symposium, June 1994), pp. 89–92.

11. G. Soulodve and J. Bradley, "Subjective Evaluation of New Room Acoustic Measures." *Journal of Acoustical Society of America,* vol. 98, no. 1 (July 1995) pp. 294–301.

12. D. Griesinger, pp. 89-92.

V. Appendices

This section is for the reader who seeks more technical information about detailed calculations and measurements of room acoustics. The techniques outlined in this appendix demonstrate that the gap is closing between acoustical calculations and results measured in real spaces. For example, the reader is invited to compare the calculated data of Appendix 1 with the measured results presented in Appendix 3. Although it should be kept in mind that the hall dimensions vary, and, as indicated in Table 6.5, trends in the data are readily apparent.

Appendix 1: Early Sound Field Calculations

Comparison of Three Music Performance Hall Shapes

Three different concert hall shapes are compared using a computer program based on a ray-tracing algorithm called "Odeon" Version 2.6D. The three hall shapes compared are the shoebox, the fan, and the reverse fan. None of the halls have balconies. The sound source (s1) has been placed in comparable positions and receivers (listeners) are located 10, 20, and 30 meters from the stage lip. The 30-meter position only applies to the reverse-fan shape, due to its greater length. The hall volumes are all 12,000 cubic meters corresponding to audience capacities of approximately 1,200. The stage sizes are all similar, corresponding to 200 square meters. Three dimensional views of the halls, shown as wire frame models, are depicted in Figures 6.29, a, b, and c, and show receiver positions near the center and sides.

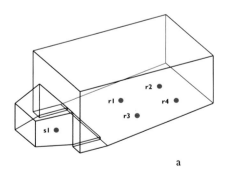

a

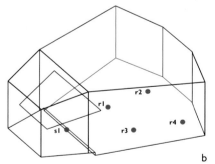

b

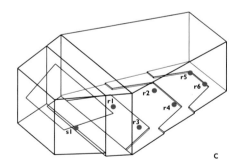

c

Figure 6.29 Concert Hall Wire frame models: (a) shoebox; (b) wide fan; (c) reverse fan.

Figures 6.30 a, b, and c all depict the halls with the source (s1) on the left and receiver (r2) on the right, located 20 meters from the stage lip. Paths of sound propagation shown as rays emanating from the source are only depicted for those reaching the receiver. Only first and second-order reflections are shown. Figures 6.30 a, b, and c also depict polar plots of incident sound at the receiver in the horizontal-plan view and polar plots of the incident sound in the vertical view. In both cases, the listener is facing the stage on the left side of the page. These polar plots correspond to the three hall shapes and the relative source-receiver positions indicated. Polar plots show the direction and intensity of sound received by the listener. In comparing the horizontal polar responses, note that the incident sound from the front direction is narrower for the fan and wider for the reverse fan than for the shoebox. These data indicate the reverse fan should create a greater sense of spaciousness and envelopment.

Figures 6.31 a, b, and c are cross-sectional end views which face the stage and show both source and receiver (r4) locations. Note that the receivers are at a higher elevation, due to the floor slope. Figure 6.31a for

Figures 6.30 a, b, and c (*top to bottom*) (a) shoebox reflections; (b) fan reflections; (c) reverse fan reflections. Reflections received by the listener are indicated in the horizontal and vertical planes.

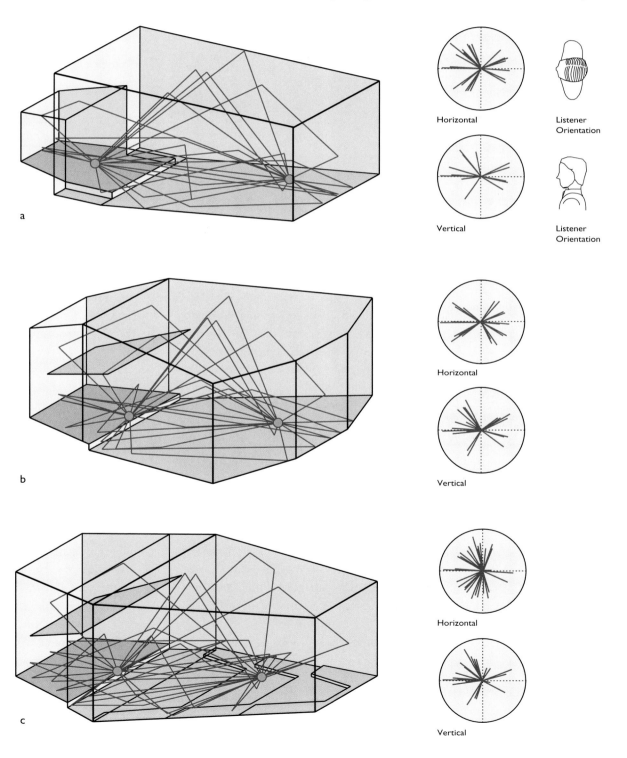

the shoebox clearly shows the receiver obtaining reflections from both sidewalls via the ceiling. Figure 6.31c indicates numerous such reflections. The fan shape in Figure 6.31b shows the sound-reflecting canopy over the stage aiding the reception of only one reflection from the house right wall. Note the relative width of the three proposed hall shapes, the fan being widest and the shoebox narrowest.

Figures 6.32 a, b, and c correspond to the three hall shapes and illustrate the temporal distribution of early reflections and their relative amplitude along with the initial time delay gap. Reflections up to the third order are shown.

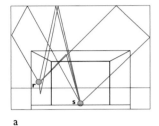 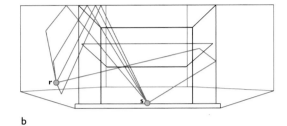 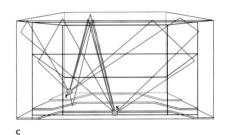

a   b   c

Figure 6.31 a, b, c Comparison of sidewall and ceiling reflections: (a) shoebox; (b) fan; (c) reverse fan.

The shoe box has an ITG (Initial Time Gap) of approximately 20 milliseconds, whereas the other two shapes have ITGs of approximately 40 milliseconds for a listener at r2, who is located approximately 20 meters from the stage. The ITG of the fan and reverse fan-shaped halls could be reduced by lowering or stepping the ceiling down at the stage end of the hall. This modification, however, could have an adverse impact on room volume and overall reverberation time. An alternative to lowering the ceiling, is to suspend large sound-reflecting panels, which would not affect room volume.

Note the relative density and number of early reflections shown on the left side of Figures 6.32 a, b, and c. The reverse fan of Figure 6.32c is superior in this regard.

The right sides of Figures 6.32 a, b, and c show estimates of the reverberant decay. The upper line is smooth and represents the reverse-integrated decay and the lower line is typical of a single decay due to an impulse without averaging. Large positive peaks well above the adjacent data are indicative of potential echoes. This is of particular concern for the fan-shaped hall. Also note the longer period the sound level is sustained equivalent to the direct sound for the reverse fan; the decay does not begin immediately, but instead allows the reverberance to "bloom." This accounts for a longer early-decay time, which corresponds to hall "liveness."

Table 6.3 compares the three hall shapes by listener location. Note that the number of early reflections increases near the walls and toward the rear of the halls in all cases. Also note that in all cases, the reverse fan has a higher total number of reflections followed by the shoebox and the fan. Note that the early decay times at 500 Hz for all listener locations in the reverse fan hall are typically longer than the other shapes. Table 6.3 also

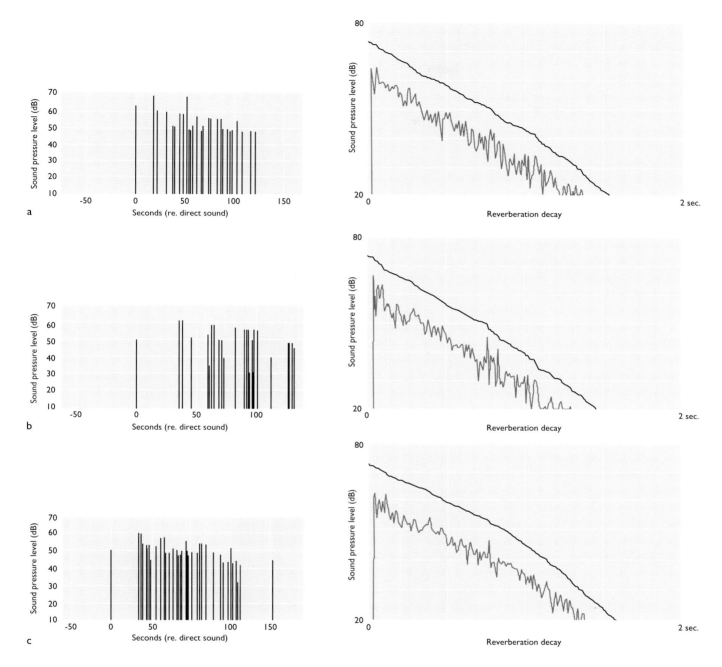

compares the LEF (Lateral Energy Fraction), which is most consistent in the shoebox but rises to higher levels in the reverse fan.

Figure 6.33 compares the listener locations 10 and 30 meters (r1 and r5) from the stage. Note the higher density of early reflections in the more distant seat. This trend can also be seen when comparing receiver location r2 to r5 in Table 6.3.

Figure 6.32 a,b, and c Calculated early (*left*) and late (*right*) sound reflections for the situations depicted in Figure 6.31: (a) shoebox; (b) fan; (c) reverse fan.

Narrow Fan	Listener locations			
Reflection Orders	R1	R2	R3	R4
1st	7	7	7	7
2nd	12	19	14	19
3rd	16	28	24	28
Total Reflections	35	54	45	54
LEF @ 500Hz	0.03	0.11	0.26	0.16
EDT (T=1.7)	1.7	1.8	1.6	1.7

Shoe Box	Listener locations			
Reflection Orders	R1	R2	R3	R4
1st	5	7	5	7
2nd	13	19	15	20
3rd	27	33	29	25
Total Reflections	45	59	49	63
LEF @ 500 Hz	0.16	0.17	0.27	0.19
EDT (T=1.8)	1.9	1.8	1.8	1.9

Reverse Fan	Listener locations					
Reflection Orders	R1	R2	R3	R4	R5	R6
1st	8	8	8	8	8	8
2nd	21	22	20	24	29	29
3rd	24	36	28	33	46	49
Total Reflections	51	66	56	65	83	86
LEF @ 500Hz	0.06	0.26	0.18	0.34	0.28	0.31
EDT (T=1.7)	1.9	2.4	2.2	2.3	1.9	2.0

Table 6.3 Comparison of hall shapes at various listener locations. The number of reflections are listed by reflection order. Calculated Lateral-Energy fractions (LEF) and Early-Decay Times (EDT) are listed by listener location.

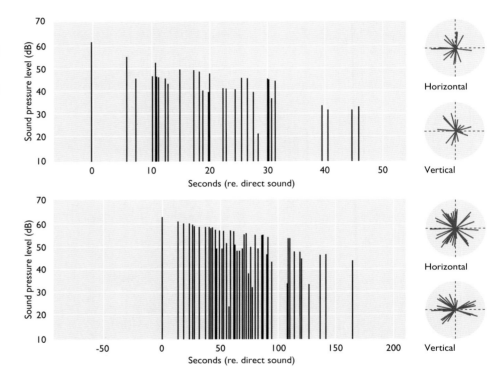

Figure 6.33 Reverse fan data from listener locations 10 and 30 meters from the stage.

Early-to-Late Sound Index Calculations

The results of early-to-late sound index (clarity) calculations using Equation 6.11 are presented for 40 millisecond (Figure 6.34) and 80 millisecond (Figure 6.35) integration times in the halls having a volume of 12,000 cubic meters. The 40 milliseconds integration time is normally used to assess speech and 80 milliseconds is used for classical music. The results are displayed in a three dimension plot as a function of listener distance from the sound source and the hall's reverberation time. In both cases, note the higher levels of clarity at reduced reverberation times and near the source.

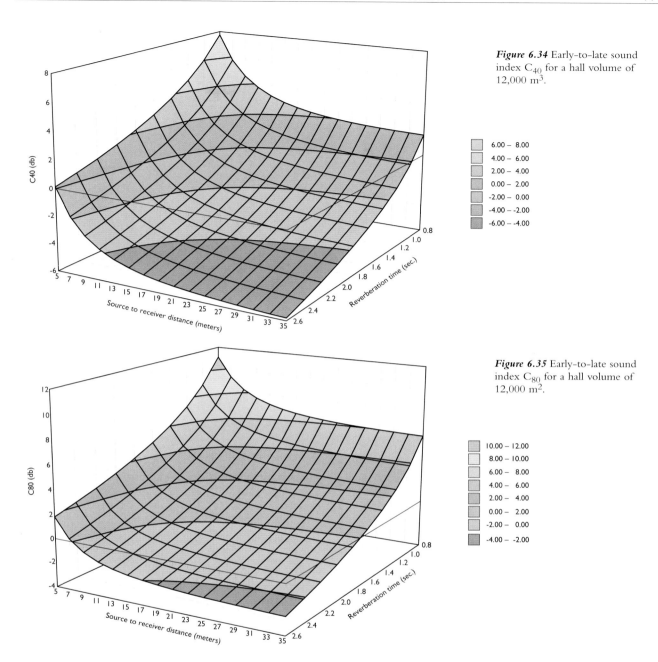

Figure 6.34 Early-to-late sound index C_{40} for a hall volume of 12,000 m^3.

	6.00 – 8.00
	4.00 – 6.00
	2.00 – 4.00
	0.00 – 2.00
	-2.00 – 0.00
	-4.00 – -2.00
	-6.00 – -4.00

Figure 6.35 Early-to-late sound index C_{80} for a hall volume of 12,000 m^2.

	10.00 – 12.00
	8.00 – 10.00
	6.00 – 8.00
	4.00 – 6.00
	2.00 – 4.00
	0.00 – 2.00
	-2.00 – 0.00
	-4.00 – -2.00

Appendix 2: Reverberation Time Calculation Using Sabine's Equations

Table 6.4 is an example of a typical reverberation time calculation most commonly used by acoustical engineers for the last ninety years. Absorption coefficients are multiplied by the surface area to obtain a total number of sabins for each surface and material. The total absorption is summed in each octave band for use with Sabine's equation (Equation 6.2). These reverberation times are based on statistical probability and may differ from the results obtained using ray-tracing algorithms. This difference is more evident for the case of unevenly distributed sound-absorbing material in spaces having large aspect ratios such as very long and narrow spaces. In this example, a reverberation time of 2.2 seconds at 500 Hz agrees reasonably with the ray-tracing results.

Absorption Coefficients

Material	Surface	125	250	500	1,000	2,000	4,000
60 mm plaster	Ceiling	0.10	0.05	0.04	0.03	0.03	0.03
2 layers of gypsum board	Walls	0.20	0.14	0.12	0.11	0.10	0.09
25 mm sound absorbing panel	Rear wall	0.09	0.32	0.76	0.95	0.99	0.99
Wood over space	Stage	0.15	0.11	0.10	0.07	0.06	0.07
Audience	Main floor	0.72	0.79	0.86	0.88	0.88	0.88

Sabins (m²)

Surface	Area (m²)	125	250	500	1000	2000	4000
Plaster Ceiling	812	81	41	32	24	24	24
Gypsum board walls	1409	282	197	169	155	141	127
Sound absorbing panel rear wall	265	24	85	201	252	262	262
Wood stage	300	45	33	30	21	18	21
Audience	514	370	406	442	453	453	453
Air absorption (sabins)		0	0	0	0	24	96
Totals	3300	802	762	874	905	922	983
Octave band reverberation time (sec.)		2.4	2.5	2.2	2.1	2.1	2.0

Table 6.4 Reverberation time calculation.

Appendix 3: Early Sound Field Measurements

Introduction

In a performance hall, the early sound field establishes its "signature sound" and is responsible for many of the hall's subjective attributes. The temporal distribution of sound arriving within the first quarter second seems to be largely responsible for the perceived "liveness" of the hall. Useful techniques for examining the early sound field are the measured sound energy growth (SEG) curve and instantaneous sound envelope (ISE). These techniques examine sound growth rather than decay. In addition to providing a clear, easy-to-interpret data presentation, traditional parameters such as early-to-late sound index (C_{80}), rise time, and the initial-time-delay gap can be extracted or read directly from these curves (Figure 6.36). This alternative data presentation is intended to help correlate integrated sound energy parameters to the overall response of the hall as well as to subjective response. The reflection energy cumulative curve (RECC) is another descriptor involving the growth of early sound reflections. The RECC, however, does not include the direct sound. Most importantly, the contribution of individual reflections (ISE) to the growth of sound energy (SEG) can readily be seen.

Measurements

A number of recorded impulses in performance spaces have been collected. The acoustical excitation for these measurements is generated using either a balloon burst or by firing a starter's pistol. Both methods provide suitable test spectra. A balloon burst results in a relatively "pink" spectrum (equal sound energy in any constant percentage bandwidth). The pistol produces more of a "white" spectrum (equal sound energy in any constant bandwidth). The pistol also produces a higher output level, which improves the signal-to-noise ratio in larger spaces. The bandwidth of these test signals spans the four octave bands from 250 Hz to 2 kHz. Since these sources have relatively constant sound power from impulse to impulse, it is possible to make meaningful inter- and intra-hall comparisons for the purpose of evaluating loudness.

These recorded signals were processed with the Hilbert Transform to obtain a magnitude, or "envelope" function referred to as the instantaneous sound envelope or ISE (see Figure 6.36). The ISE depicts the arrival of the direct and reflected sound at the measurement microphone subsequent to a test impulse. The advantages of this presentation are: (1) the magnitude can be displayed on a logarithmic amplitude scale, and (2) the delay between arrivals is easier to discern than on a traditional oscilloscope display. The first peak in the ISE is sound arriving directly from the source. Later peaks represent room reflections. ISE is similar in data presentation format to the energy time curve (ETC); the difference is that the ETC is derived from an electric (stimulus-response) measurement, whereas the ISE is the result of a measured acoustical impulse.

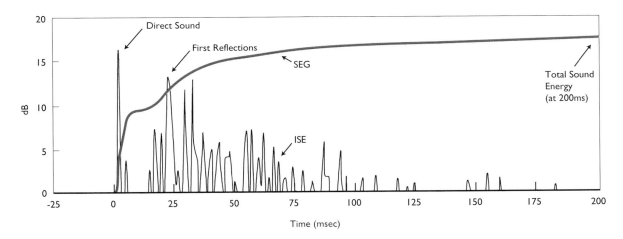

In addition to the ISE, the recorded signal was squared and integrated over time. This function is referred to as sound energy growth, or SEG, also shown in Figure 6.36. The SEG depicts the build-up of sound energy during the first 200 msec.

When the SEG and ISE are plotted on the same time axis, the contribution of individual reflections to the growth of sound energy can be seen. For example, significant time delays between arrivals in the ISE result in plateaus in the SEG.

Figure 6.36 Instantaneous Sound Envelope (ISE) and Sound Energy Growth (SEG). ISE represents the magnitude of the sound pressure at the microphone as a result of an acoustical impulse. The SEG depicts the build-up of early sound energy.

Discussion

Figure 6.37 a through d contains comparative data for two music performance halls, a scoring stage, and a film screening room. All data were analyzed using the ISE and SEG techniques. These figures all appear on the same relative amplitude scale for comparison. Several of these facilities have received high critical acclaim from musicians, audiences, and the media. Three of these facilities are discussed as case studies in Chapter 17. Table 6.5 presents the physical characteristics and reverberation times for these facilities.

The Screening Room is designed for film and lectures. The relatively small size, short reverberation time (and resulting high speech intelligibility) in this room make it unique in this study. Here, the sound energy growth reaches its final value within 60 msec.

The Music Recital Hall has a reverse-fan-shape plan with a stepped ceiling and is used primarily for music recitals. It has adjustable 50 mm (2 in.) thick sound-absorbing panels that slide on rails that can cover up to half of each side wall. With the sound-absorbing panels withdrawn, reflections between 20 msec and 80 msec have greater amplitudes than the direct sound (see Figure 6.37b). In this highly acclaimed space, it is noteworthy that, for the first 190 msec, reflection amplitudes are within 5 dB of the direct sound.

The Symphony Hall measurements were performed in the Loge prior to its recent renovation (1993). This hall has retractable fabric banners extending along the junction between the wall and ceiling. With the banners retracted, there are several reflections within 5 dB of the direct sound level.

Scoring stages are used to make orchestral recordings for motion pictures. Based on limited comparative studies, there does not seem to be a consensus among recording engineers as to optimal perceived "reverberance." However, the reduced intensity of the early reflections when compared to the direct sound is likely a significant factor.

One indicator of perceived liveness is the density and duration of significant early reflections (i.e., reflections within 5 dB of the direct sound level, T_5). Based on the investigations conducted thus far, the spaces with the most perceived liveness appear to have significant reflections sustained for at least 50 msec (T_5 in Table 6.6). The best listening spaces are also relatively free of long plateaus during the initial rise of the sound energy growth.

Table 6.5 Characteristics of spaces studied.

Facility	Volume (m³)	No. of Seats	Floor Area (m²)	T [s]
Dolby Screening Room San Francisco, CA	800	49	–	0.3
Harris Music Recital Hall Aspen, CO	4800	500	–	1.8
Symphony Hall San Francisco, CA	28000	3200	–	2.1
Todd-AO Scoring Stage Studio City, CA	5600	–	630	2.2

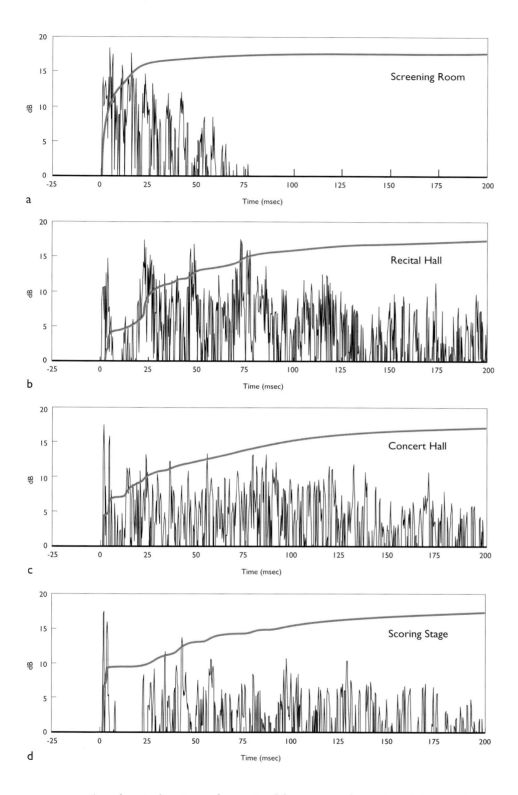

Figures 6.37 a–d *(from top):* (a) ISE & SEG in a 49-seat screening room where T=0.3s; (b) ISE & SEG in a 500-seat musical recital hall where T = 1.8s; (c) ISE & SEG in a 3,000-seat concert hall where T = 2.1s; (d) ISE & SEG in a 5,600 m^3 scoring stage where T = 2.2s.

Another indication of perceived liveness is the ratio of the SEG at 200 msec to the SEG at 10 msec. The halls with the most perceived liveness in this study have a ratio between 8 and 12 dB (see Table 6.6).

Performance Hall	T$_5$ (msec)	Ratio SEG$_{200\ msec}$ / SEG$_{10\ msec}$ (dB)
Screening Room	30	5.2
Recital Hall		
no absorption	190	12.5
with absorption	130	9.8
Symphony Hall		
banners retracted	90	9.8
banners extended	50	6.2
Scoring Stage	50	7.8

Table 6.6 Duration of significant reflections (T$_5$) and sound energy growth ratio.

References

J. S. Bradley, "A Comparison of Three Classical Concert Halls." *Journal of the Acoustical Society of America*, vol. 89 (1991), pp. 1176-91.

M. Baron, *Auditorium Acoustics* (Chapman and Hall, 1993).

V. L. Jordan, *Acoustical Design of Concert Halls and Theaters* (Applied Science Publishers, Ltd., 1980).

L. L. Beranek, *Music, Acoustics, and Architecture* (J. Wiley & Sons, 1962).

Y. Toyota, K. Oguchi and M. Nagata, "A Study of Characteristics of Early Reflections in Concert Halls," (paper presented at 2nd Joint Meeting of the ASA and ASJ, Hawaii, November, 1988).

M. R. Schoeder, "Complimentarity of Build-up and Decay Processes," *Journal of the Acoustical Society of America*, vol. 40 (1966).

L. Cremer and H. A. Muller, *Principles and Applications of Room Acoustics*, vols. 1 & 2 (Applied Science Publishers, 1982).

D. Griesinger, *Further Investigations into the Loudness of Running Reverberation* (Lexicon, Waltham, 1995).

D. Griesinger, "How Loud is My Reverberation?" (paper presented at the AES 98th Convention, Paris, February, 1995). pp. 25-28.

J. S. Bendat, *The Hilbert Transform and Applications to Correlation Measurements* (Bruel & Kjaer-Naerum, 1987).

Jeffrey Borish (Ph.D. thesis, CCRMA, Department of Music, Stanford University, 1983).

Nicholas A. Edwards, "An Investigation Employing an Images Model of Room Acoustics," (PP5 paper presented at 106th ASA meeting, November, 1983).

Mathematical Definitions of ISE and SEG

The Instantaneous Sound Envelope, $S_E(t)$, is defined as:

EQUATION 6.12

$$S_E(t) = 20 \, \log_{10} \sqrt{p^2(t) + \tilde{p}^2(t)} \ [\text{dB}]$$

The time signal, $p(t)$, is the measured acoustic signal at the receiver microphone as result of an acoustic impulse excitation (e.g., balloon pop, pistol shot). The imaginary part of the time signal, $p(t)$ is the Hilbert Transform of the time signal, $\tilde{p}(t)$.

EQUATION 6.13

$$\tilde{p}(t) = \mathcal{H}[p(t)]$$

The Hilbert Transform of a time signal is defined as:

EQUATION 6.14

$$\mathcal{H}[p(t)] = \frac{1}{\pi} \int_{-\infty}^{\infty} p(t) \, \frac{1}{t-\tau} \, d\tau$$

This corresponds to a -90 degree phase shift, or a shift of $\frac{1}{4}$ wavelength. The ISE therefore represents the magnitude, or envelope, of the received sound level versus time.

The "Sound Energy Growth" (SEG) curve, is defined as:

EQUATION 6.15

$$G_{SE}(t) = 10 \log_{10} \int_{0} p^2(t) \, dt \quad [\text{dB}]$$

The SEG shows the accumulation of sound energy during the first 200 msec.

7

Sound Insulation

Anthony P. Nash, P.E.
& Charles M. Salter, P.E.

Principles of Sound Insulation • The Effect of Frequency and Mass • Dual Panel Partitions • Limitations of Dual Panel Partitions • Structurally Isolated Constructions • Air Leaks • Coincidence Dips • Sound Transmission Terminology and Calculations • TL is a Power Ratio • Composite TL • Measuring TL • Sound Isolation • Noise Reduction • Flanking Sound • TL v. Frequency • Sound Transmission Class (STC) • Impact Insulation Terminology and Calculations • Sound and Impact Insulation Ratings • Data for Walls • Data for Wood Framed Floor/Ceilings • Data for Concrete Floating Floors • Data for Impact Insulation

EVERYONE has experienced unwanted sound intrusion—a television in the next room, a loud neighbor walking on the floor above, or a jet flying over. Often, measures are required to reduce intrusive noise. This chapter addresses one of the most essential techniques in acoustics: reducing the transmission of sound through solid barriers in buildings. This form of sound reduction is referred to as *sound insulation*. Part I of this chapter describes the principal concepts involved in the transmission of airborne and impact sound. Part II contains examples of sound-rated constructions along with rules governing their acoustical performance.

I. Principles of Sound Insulation

SOUND insulation depends on the interaction between sound in air and vibration in solids. A sound wave impinging on a solid panel causes the surface to vibrate. The vibrating panel then induces air on the other side to be set into motion, re-radiating some of the original sound, which causes the opposite side of the panel to act as a new, but weaker, sound source. The reduction of sound energy in this manner is called *sound transmission loss* (TL). The sound energy that is not transmitted through the panel is either dissipated or reflected.

The Effect of Frequency and Mass

As the frequency of sound increases, more force is required to vibrate the panel in order to maintain the same sound level on the opposite side. For every doubling of frequency, the TL increases by about 6 dB. Thus a panel has greater TL at higher frequencies than at lower frequencies. Similarly, as the mass of a panel is increased, more force is required to make it vibrate. For this reason, a massive panel has greater TL at all frequencies than a lighter panel. According to the "mass law" theory, doubling the mass of a panel will also increase its TL by 6 dB. Most common panels in buildings conform to the mass law. In architectural acoustics, we call such panels *partitions,* whether they are walls, floors, or ceilings.

Dual Panel Partitions

Two panels separated by a hollow air cavity are called a *dual panel partition*. When sound impinges on this type of partition, further energy-conversion losses occur due to the air space between the panels. If the total mass of the system is held constant, the TL of a dual panel partition will always be greater than that of a single panel. This TL improvement is called an *increase over mass law.* The depth of air space separating the panels and their combined masses determine the lowest frequency where the TL of the dual panel will begin to exceed that of the single panel.

Doubling the air space of a dual panel partition increases the TL by about 5 dB. Figure 7.1 compares the TL of single panel partitions to dual panel partitions and shows that a deep air space is especially helpful for improving high-frequency TL.

This principle of the dual panel partition has profound implications when high levels of sound insulation are needed. For example, dual panel partitions constructed with gypsum board can provide TL performance equivalent to a solid concrete wall weighing eight times as much. Minimizing the weight of sound-insulating systems can be important to the economics of building construction; for this reason, lightweight dual panel partitions are used extensively.

Limitations of Dual Panel Partitions. The dual panel partition is quite sensitive to structural coupling that enables mechanical energy to bypass the air space between the two panels. For example, the sound insulation of a dual

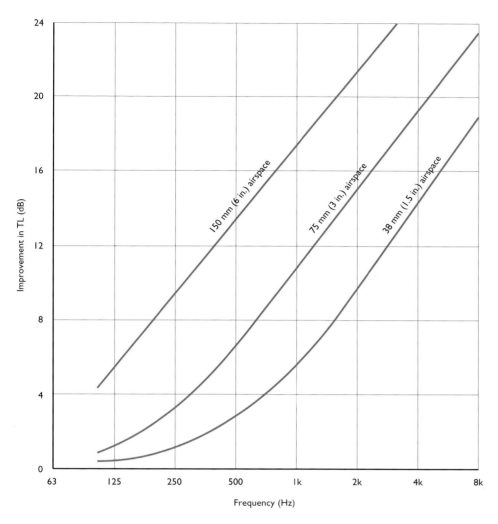

Figure 7.1 TL comparison between dual panel partitions and single panel partition of equal mass.

panel partition connected together with wood studs is less than the same assembly without such connections. Proven techniques for minimizing structural coupling in dual panel partitions include use of light gauge metal studs and double studs as discussed in Part II of this chapter.

Two distinct coupling mechanisms are created by the air cavity between the two panels. At low frequencies, the trapped air is relatively stiff, and the two panels are mechanically linked by a spring-like element. At higher frequencies, the short sound wavelengths can resonate within the cavity itself, locally creating high sound pressures at the inner surfaces of the panels. These resonances primarily affect the TL of a dual panel partition in the upper speech frequencies (around 2000 Hz).

Placing sound-absorbing material between the two panels makes the trapped air behave as if it were more resilient. Since a resilient spring transmits less force from one panel to the other, adding sound-absorbing material is equivalent to increasing the separation between the two panels. As a consequence, the low-frequency TL is improved. The sound-absorbing material also tends to suppress cavity resonances in the high-frequency range; thus, high-frequency TL is also improved. In practice, adding sound-absorbing material within a dual panel partition can increase the TL by 5 to 10 dB. The TL improvement is influenced by the thickness of this sound-absorbing material, however, its density is usually irrelevant. Inexpensive, lightweight glass-fiber building insulation has nearly the same effect on TL as more costly, heavier mineral wool.

Structurally Isolated Constructions

To attain high levels of sound insulation, both additional mass and isolation are required. A specially isolated floor is the system of choice when attempting to achieve high TL between two rooms situated one above the other. The isolated floor is called a *floating floor* and is often used in mechanical equipment rooms, recording studios, and acoustical testing facilities. The typical floating floor construction is a 102 mm (4 in.) thick concrete slab that "floats" above the supporting floor on vibration isolators, as shown in Figure 7.9. Double masonry or combined masonry-stud constructions are examples of walls comparable to "floating floors."

Air Leaks. It is vital that any partition intended to serve as a sound-insulating barrier not have air or sound leaks. Long cracks are especially detrimental since all sound frequencies can easily find a path through these openings. For this reason, the perimeter of a partition should be sealed airtight with a flexible sealant.

Coincidence Dips. Sometimes a TL plot shows a "dip" or depression like that for 6 mm (¼ in.) monolithic glass shown in Figure 7.2. This depression is called a *coincidence dip* and is caused by interaction of sound and vibration in a panel. This coincidence dip can cause flaws in the quality of sound transmitting through partitions. For example, with 13 mm (½ in.) gypsum board, the coincidence dip occurs at 3,150 Hz. In 6 mm (¼ in.) glass, the coincidence dip is at 2,500 Hz. The physical explanation for this dip is discussed in numerous books on acoustics such as *Noise and Vibration Control* by Leo Beranek.

The effects of a coincidence dip do not normally appear in partition installations; however, there are situations where it can lead to problems. For example, if a 6 mm (¼ in.) single-glazed window is installed in a building facade that is parallel to a roadway, individual vehicles will generate sound waves that arrive at the window over a range of angles. Depending on the instantaneous angle of each vehicle with respect to the window, the coincidence frequency will vary as the vehicles pass by. The sound transmitted into the building will have a "swishing" character as the coincidence dip moves up and down in the frequency range. This characteristic can be quite disconcerting in a quiet office next to the window. The amount of TL degradation at the coincidence frequency depends inversely on the energy dissipated within the panel. For instance, if the 6 mm (¼ in.) glass window described above were constructed using two 3 mm (⅛ in.) glass sheets laminated together with clear plastic, mechanical energy would be dissipated by the interlayer. The red line in Figure 7.2 shows the TL improvement provided by laminated glass consisting of two pieces

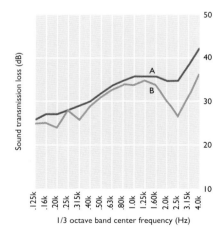

Figure 7.2 TL comparison of 6 mm (¼ in.) glass to 6 mm (¼ in.) laminated glass. A is laminated glass; B is monolithic glass.

of 3 mm (⅛ in.) thick glass with 0.8 mm (0.03 in.) plastic inner layer. The use of laminated glass is a simple modification that virtually eliminates the "swishing" character of traffic noise heard within a building.

Sound Transmission Terminology and Calculations

TL is a Power Ratio. This chapter began by describing TL as a reduction of acoustical energy when sound encounters a solid panel. It is helpful to think of TL as a ratio of the arriving- to transmitted-sound power per unit area of a panel. Whether the area is large or small, a given panel type is presumed to have the same TL. Thus, TL is an acoustical property independent of the panel size and is usually quantified in decibels. Table 7.1 lists the TL values of some common constructions.

		Frequency (Hz)					
		125	250	500	1000	2000	4000
Walls							
1	38 x 89 mm (2 x 4 in.) studs with 13 mm (½ in.) gypsum board both sides with 89 mm (3½ in.) insulation.	15	31	40	46	50	42
2	38 x 89 mm (2 x 4 in.) staggered studs with 13 mm (½ in.) gypsum board both sides with 89 mm (3½ in.) insulation.	31	37	47	52	56	50
3	38 x 89 mm (2 x 4 in.) double stud with 13 mm (½ in.) gypsum board both sides with 89 mm (3½ in.) insulation.	37	48	57	65	70	70
4	102 mm (4 in.) double masonry brick with 57 mm (2¼ in.) airspace.	42	47	56	63	67	71
Doors							
5	45 mm (1¾ in.) hollow wood door with brass weather strip.	16	15	17	18	26	31
6	45 mm (1¾ in.) solid core wood door with perimeter gaskets.	32	35	37	40	42	39
Windows							
7	3 mm (⅛ in.) monolithic glass.	21	22	26	30	33	28
8	6 mm (¼ in.) laminated glass.	26	29	32	35	37	42
9	13 mm (½ in.) laminated glass.	29	32	36	37	41	51
10	5 mm (³⁄₁₆ in.) double glass with 102 mm (4 in.) air space.	28	35	42	50	47	48

Table 7.1 TL of common building constructions.

Composite TL. The transmission of sound energy through a panel is analogous to heat loss from a room. Each element comprising the room's surfaces has a unique characteristic for conducting heat at a given temperature differential. For example, a window conducts heat more readily than does an insulated wall. The total heat loss from a room can be calculated by considering the "transmissibility" of each element multiplied by its surface area.

The total sound transmitted through several different panel elements is calculated in a similar manner. This calculation results in the net sound-insulation performance called composite transmission loss.

The nomogram in Figure 7.3 was developed to minimize the mathematics involved in computing composite transmission loss.

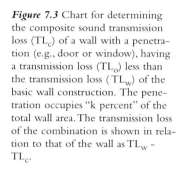

Figure 7.3 Chart for determining the composite sound transmission loss (TL$_c$) of a wall with a penetration (e.g., door or window), having a transmission loss (TL$_o$) less than the transmission loss (TL$_w$) of the basic wall construction. The penetration occupies "k percent" of the total wall area. The transmission loss of the combination is shown in relation to that of the wall as TL$_w$ - TL$_c$.

TL$_c$ = Composite Transmission Loss
TL$_w$ = Transmission Loss of Wall
TL$_o$ = Transmission Loss of window, door, or opening

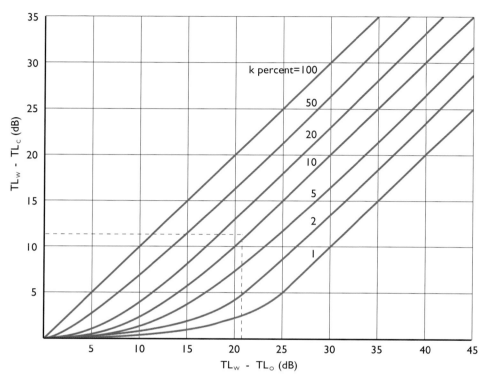

Comment about TL predictions. With any predictive methodology that has simplifications, some field conditions do not always correspond with the assumptions in the method. For instance, an opening in the partition may have various shapes or aspect ratios. The composite TL calculating method in Figure 7.3 is more accurate when the open area is a crack rather than a circular hole.

Example problem. For a wall containing a window, the composite TL$_c$ would be somewhere between the TL$_w$ of the wall and the TL$_o$ of the window. Assume a total wall/window area equals 2.7 m (9 ft.) high by 4.3 m (14 ft.) long. The wall construction equals 38 x 89 mm (2 x 4) staggered studs (construction No. 2 in Table 7.1). The window construction equals 1 m (3 ft.) by 1.5 m (5 ft.) with 3 mm (⅛ in.) thick glass (Construction No. 7 in Table 7.1).

The calculation is done as follows: First, find the difference between TL$_w$ and TL$_o$ at the frequency of interest. TL$_w$ of the wall at 500 Hz is 47 dB; TL$_o$ of the window at 500 Hz is 26 dB.

The difference of TL$_w$ - TL$_o$ is 21 dB; enter Figure 7.3 in the horizontal axis at 21 dB and draw a vertical line. The window represents 12 percent of the total wall window area—1.4 m² (15 sq. ft.) ÷ 11.7 m² (126 sq. ft.) = 12%. Locate the diagonal 12 percent line (between 10 percent and 20 percent), and find the intersection between the 12 percent line and the vertical line at 21 dB. The intersection occurs at 12 dB on the vertical axis. Therefore, 12 dB is subtracted from TL$_w$ to determine TL$_c$. The composite TL$_c$ of the wall/window construction is 35 dB at 500 Hz.

Figure 7.4 compares the measured and calculated TL of a wall and window element separately and in combination. The composite TL is about 4 dB lower than the wall TL at 1000 Hz as a result of the window.

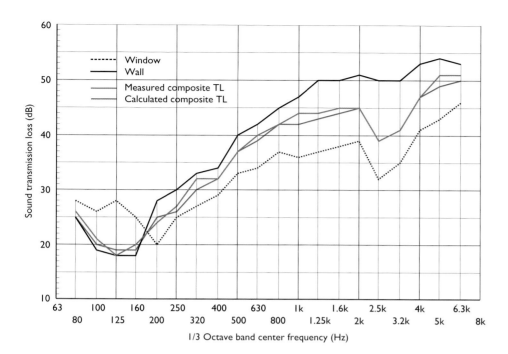

Figure 7.4 Comparison of measured and calculated transmission loss for a composite construction.

Note how well the measured and calculated values agree. The National Bureau of Standards (NBS) also found excellent correspondence between measured and calculated TL values for composite constructions.[1]

When a partition has three or more elements in it, the calculation of the composite TL is done in several steps. First the composite TL due to elements 1 and 2 is calculated. Then these results are combined with element 3 to calculate the resulting composite transmission loss of the first three elements. This process continues until all the elements are combined into a final value representing the TL performance of the composite system.

Gaps or sound leaks in a construction can significantly limit the resulting transmission loss. For an opening in a wall, the TL should be taken as 0 in all frequencies. An opening in a wall representing 10 percent of the wall area would limit its composite TL to a maximum of 10 dB, 1 percent to 20 dB, 0.1 percent to 30 dB, etc. An opening in a wall representing a 0.1 percent of crack in a wall represents a 3 mm (⅛ in.) gap at the base of a wall that is 3 m (10 ft.) high.

The sound transmission loss of composite construction can be calculated using Equation 7.1:

$$\text{TL}_{composite} = 10 \log_{10} \left[\frac{S_1 + S_2 + S_3 + \cdots + S_n}{S_1\tau_1 + S_2\tau_2 + S_3\tau_3 + \cdots + S_n\tau_n} \right]$$

EQUATION 7.1

S_1 to S_n represent the various surface areas of the components that make up the composite construction. τ_1 to τ_n relate to the "transmissibility" of the individual components as defined in Equation 7.2:

EQUATION 7.2

$$TL = 10 \log \frac{1}{\tau} \text{ or } \tau = \frac{1}{10^{\frac{TL}{10}}}$$

An example of how to use Equations 7.1 and 7.2 to calculate composite transmission loss is shown on the next page.

Measuring TL. TL is determined in a laboratory environment by methods that have proven to be both convenient and accurate. In the laboratory, amplified electronic noise is used to generate a sound field in one of a pair of rooms separated by the partition being tested. Both rooms are very reverberant so the sound field is diffuse; angles of sound arriving on one side of the partition (and leaving the other side) are completely random.

Sound Isolation. Sound isolation is defined as the ratio of the sound pressures between two spaces. Unlike TL, sound isolation is affected by factors that are controlled in the acoustical laboratory such as sound absorption in the receiving room and the area of the partition common to both rooms. To improve acoustical privacy in buildings, it is possible to either (1) increase the TL of the partition or (2) alter the non-TL factors such as sound absorption in the receiving room and the common partition area.

Noise Reduction. The sound isolation between two spaces is quantified in a manner similar to TL except that the values of sound isolation are called noise reduction (NR). Mathematically, NR equals the source room sound pressure level (in decibels) minus the receiving room sound pressure level (in decibels). NR equals the TL of a partition plus a correction term. The correction term is 10 log A/S, where A is the sound absorption in the receiving room and the S is the surface area of the intervening partition.

For example, assume that two rooms are separated by a 4.6 m (15 ft.) long and 3 m (10 ft.) high partition having a TL of 40 dB at 500 Hz. Both rooms have 19 metric sabins (200 sabins) of absorption at 500 Hz. The NR between the two rooms is:
NR = TL + 10 log (A/S)

$$NR = TL + 10 \log \left[\frac{A}{S}\right]$$

$$= 40 + 10 \log \left[\frac{19}{4.6 \times 3}\right] \text{ metric}$$

$$= 40 + 10 \log \left[\frac{200}{15 \times 10}\right] \text{ English}$$

$$= 41 \text{ dB at } 500 \text{ Hz}$$

As the common partition area increases, more sound is transmitted and the NR value decreases. Adding sound absorption to the receiving room means that the NR will increase since less sound energy exists. If the receiving room is fully furnished and has sound–absorbing walls and ceiling, the NR could exceed the TL by about 5 dB. If the receiving rooms is highly reverberant, the NR could be 3 dB less than the TL. For two adjoining furnished rooms, the typical NR value is equal to the TL of the partition.

Flanking Sound. Sound isolation between two rooms can be degraded by building elements other than the intervening partition. For example, there could be a sound path through an air duct that connects the two rooms. Clearly, the air duct is not a partition; rather, it is a separate sound-transmitting element. In architectural acoustics, these secondary means of sound transmission are called *flanking paths.*

For these and other reasons, consultants who perform field tests of acoustical privacy in buildings may be initially asked to measure the sound insulation (TL) of a partition but end up reporting the sound isolation (NR) of an architectural system. Thus, sound insulation (TL) is useful for comparing laboratory data for various constructions and sound isolation (NR) is indicative of the acoustical privacy experienced by a building occupant.

TL v. Frequency. Chapter 4 explains that sound can be divided into a number of frequency bands. Each frequency band may have a different sound pressure, thereby giving the spectrum its unique character. When evaluating sound insulation, these individual bands are treated as separate entities much like separate cells are treated in a spreadsheet. There is an important reason for this approach—most constructions that control, absorb, or reflect sound have different values at different frequencies. This frequency-dependent characteristic certainly applies to sound transmission loss. Consequently, it is standard practice for acoustical laboratories in North America to measure TL in 16 one-third octave bands from 125 Hz to 4 kHz.

Sound Transmission Class (STC). Using 16 discrete TL values to describe a partition's performance is cumbersome. To simplify communications within the building industry, a single-number rating scheme called sound transmission class (STC) was developed by ASTM in the 1960s. The scheme is based on the premise that sound-rated partitions in buildings are mainly intended to control the audibility of speech. The original ASTM committee decided that a graphical plot of TL versus frequency for a 230 mm (9 in.) thick concrete wall should correlate well with people's acoustical privacy expectations. The decision was based on social surveys performed in Europe, where noise codes had existed for many years.[2] The shape of this plot was adopted as a standard contour to determine the STC rating for all types of sound-rated partitions.

The STC rating is determined from the TL values as follows: (1) the STC contour is fitted over a plot of the measured TL data; (2) the single-number STC rating is found where the TL of the standard contour intersects a frequency band which was arbitrarily selected to be 500 Hz. Figure 7.5 illustrates the standard contour overlaid on a plot of the TL data for the 6 mm (¼ in.) glass TL from Figure 7.2. There are regions where the STC contour arches above the TL data; these regions are called *deficiencies*. The contour is raised as high as possible on the plot of TL data until the accumulated deficiencies begin to exceed an arbitrary value established for the STC protocol.

Sound transmission class (STC) has proven to be a useful measure for ranking the acoustical performance of various constructions. Some single panel and double panel partitions are compared in Figure 7.6. This figure reinforces the fact that it is possible to achieve high STC ratings by increasing the air space between two lightweight panels.

The STC scheme is also used to assign a single-number rating to measured noise reduction data. In this case the rating is called *noise isolation class* (NIC). NIC values can be less than or greater than the STC of the partition, depending on the arrangement of the spaces and other factors described previously.

A 4 m high by 10 m long concrete wall contains a 1.5 m high by 1.5 long window, a 1 m wide by a 2.5 m high door, and a 0.5 m high by 1.5 m long louver. Calculate the composite transmission loss at the 500 Hz using the following data.

Glass window
TL_g @ 500 Hz = 24 dB
Surface area S_g = 2.25 m^2

Door
TL_d @ 500 Hz = 20 dB
Surface area S_d = 2.5 m^2

Louver
TL_l @ 500 Hz = 0 dB
Surface area S_l = 0.75 m^2

Concrete wall
TL_w @ 500 Hz = 40 dB
Surface area S_w = 34.5 m^2

$$\tau_g = \frac{1}{10^{24/10}} = 0.004$$

$$\tau_d = \frac{1}{10^{20/10}} = 0.01$$

$$\tau_l = \frac{1}{10^{0/10}} = 1$$

$$\tau_w = \frac{1}{10^{40/10}} = 0.0001$$

$$TL_{Composite} = 10 \log x$$

$$\left[\frac{S_g + S_d + S_l + S_w}{S_g \tau_g + S_d \tau_d + S_l \tau_l + S_w \tau_w} \right]$$

$$= 10 \log x$$

$$\left[\frac{2.25 + 2.5 + 0.75 + 34.5}{2.25(0.004) + 2.5(0.01) + 0.75(1) + 34.5(0.0001)} \right]$$

$$= 10 \log \left[\frac{40}{0.009 + 0.025 + 0.75 + 0.00345} \right]$$

$$= 10 \log \left[\frac{40}{0.78745} \right] = 10 \log(50.796)$$

TL composite = 17 dB

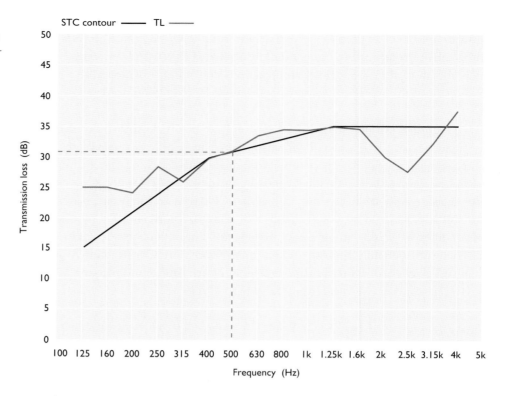

Figure 7.5 STC contour. (*Note:* the STC of a particular contour is indicated by its TL value at 500 Hz).

Impact Insulation Terminology and Calculations

Although the preceding sections discussed airborne sound insulation, controlling noise from footsteps is also important in multi-story buildings. A test method and sound-rating scheme for footstep noise has been developed that is comparable to the STC rating. The resistance of a floor/ceiling to footstep noise transmission is called *impact sound insulation.*

Testing Floor/Ceilings. The test is performed by placing a special mechanical hammering device on the floor called a *tapping machine.* First developed in Germany before World War II, the machine name was presumably inspired by its five lightweight hammers that are intended to simulate the "tapping" of high-heeled shoes on hard floor surfaces. The steel hammers are lifted and dropped onto the floor in rapid succession, generating high sound pressures in the room below. In the lower room, the resulting sound pressure levels are measured in frequency bands and the data are then fitted to a type of standard contour akin to the STC method. In this case, the single figure rating is called *impact insulation class* (IIC). The IIC rating scheme has been devised so a better floor is assigned a higher IIC value. (refer to Figure 7.7).

The range of impact insulation values is arbitrary; in order to simplify communication, IIC values have been designed so they correspond numerically with STC values. In California, for example, the State building code requires that floor/ceiling constructions in multi-family dwelling units achieve minimum laboratory ratings of STC 50 and IIC 50.

IIC Performance. The impact insulation performance depends on both the floor surface material and the construction details of the floor/ceiling

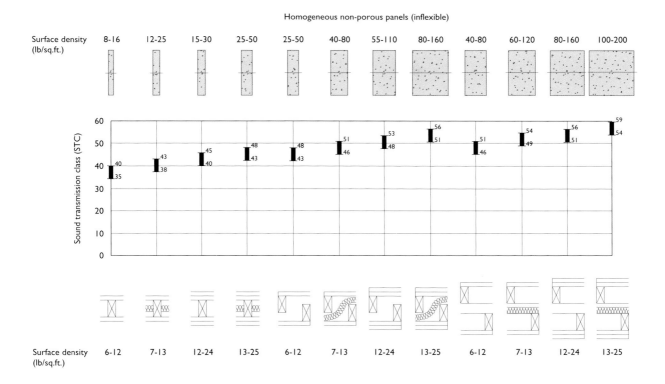

itself. Soft floor coverings such as thick carpet and pad significantly increase IIC values since the carpet fibers cushion the fall of the tapping machine hammers. A framed-floor construction can also provide excellent performance if the principles of a double panel partition are utilized.

Limitations. The tapping machine does not replicate human walking; therefore, the resulting IIC rating may not correspond with a person's impressions of various floor/ceiling systems. For example, a wood-frame floor/ceiling can have a satisfactory IIC rating, yet footsteps can still be heard in the room below as a "booming" sound. The tapping machine and the IIC rating have both been discussed and debated by experts for decades. Nevertheless, this scheme for evaluating impact insulation is still with us.

II. Sound and Impact Insulation Ratings

This part of the chapter lists STC and IIC ratings for various wall and floor/ceiling constructions to illustrate the sound-insulation principles explained in Part I. In some cases, however, it is necessary to include a discussion of TL in frequency bands.

These data are taken from the following acoustical testing laboratories: NRCC (National Research Council of Canada); Riverbank Acoustical Laboratories; Geiger & Hamme Laboratories; OCF (Owens/Corning Fiberglas); Kodaras Acoustical Laboratories.

Various constructions are compared and grouped into eight categories: (1) concrete; (2) gypsum board; (3) single stud; (4) resilient channel; (5) staggered stud; (6) double stud; (7) metal stud; (8) masonry block.

Figure 7.6 STC values for various wall constructions.

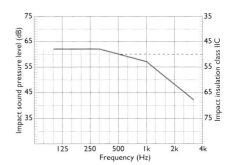

Figure 7.7 IIC contour.

Data for Walls[3]

Compare 1a to 1b: A 33% increase in mass results in a 3 point increase in STC. The theoretical TL improvement due to increase in mass is 20 log (weight ratio) = 20 log (1.33) = 2.5 dB

Compare 2b to 2a: A 25% increase in mass results in a one point increase in STC.

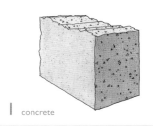

| concrete

2 gypsum board

1a

150 mm (6 in.) concrete
366 kg/m² (75 psf)
STC 55 (RIVERBANK)

1b

200mm (8 in.) concrete
464 kg/m² (95 psf)
STC 58 (RIVERBANK)

2a

13 mm (½ in.) gypsum board
STC 28 (NRCC)

Compare 2c to 2a: A doubling of mass results in a 3 point increase, not 5 points as anticipated by the General Rule. A 5 dB increase does occur at frequencies between 125 Hz and 500 Hz. STC increase does not follow the mass law due to other effects occurring in high frequency range.

Compare 2d to 2a: Increase is only 2 STC points even though both mass and air space is doubled. Reverberant sound in air cavity causes lower than expected STC value. Resonances degrade TL throughout speech frequency range.

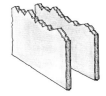

2 gypsum board

2b

16 mm (⅝ in.) gypsum board
STC 29 (NRCC)

2c

Two layers 13 mm (½ in.) gypsum board, laminated
STC 31 (NRCC)

2d

Two layers 13 mm (½ in.) gypsum board, 65 mm (2½ in.) air space, no studs
STC 30 (NRCC)

Compare 2e to 2d: Sound absorption in cavity causes a 14 point increase in STC by decreasing the stiffness of the trapped air and controlling resonances.

Compare 3a, b, and c: "Identical" constructions tested in different laboratories result in a 5 point STC range. STC rating for this construction is very sensitive to grip of fasteners between gypsum board and stud. Variation could also be caused by difficulties in measuring TL at low frequencies.

2 gypsum board

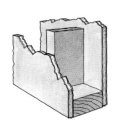

3 single-stud

2e

Two layers 13 mm (½ in.) gypsum board, 65 mm (2½ in.) air space, 50 mm (2 in.) batt insulation, no studs
STC 44 (NRCC)

3a

13 mm (½ in.) gypsum board each side; 38 x 89 mm (2 x 4) studs, 400 mm (16 in.) o.c.
STC 34 (NRCC)

3b

13 mm (½ in.) gypsum board each side; 38 x 89 mm (2 x 4) studs, 400 mm (16 in.) o.c.
STC 30 (Geiger & Hamme)

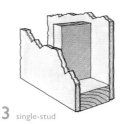

3 single-stud

Compare 3d to 2d: The 6 STC point increase is not expected since the studs are bridging the cavity. It is possible that studs are reducing cavity resonances, thereby improving TL.

Compare 3e to 3d: Sound absorption improves TL above 500 Hz. Four point improvement in STC is partly caused by subtle differences in connections of gypsum board to studs.

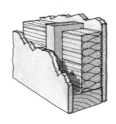

3c

13 mm (½ in.) gypsum board each side; 38 × 89 mm (2 × 4) studs, 400 mm (16 in.) o.c.
STC 35 (OCF)

3d

38 × 89 mm (2 × 4) studs, 600 mm (24 in.) o.c.; 13 mm (½ in.) gypsum board each side
STC 36 (NRCC)

3e

38 × 89 mm (2 × 4) studs, 600 mm (24 in.) o.c.; 13 mm (½ in.) gypsum board each side, 50 mm (2 in.) batt insulation
STC 40 (NRCC)

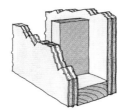

3 single-stud

Compare 3f to 3e: STC rating is 3 points lower due to closer stud spacing.

Compare 3g to 3c: Doubling of mass results in a 4 point increase in STC.

Compare 3h to 3g: Adding insulation results in a 6 point increase in STC.

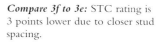

3f

38 × 89 mm (2 × 4) studs, 400 mm (16 in.) o.c. 13 mm (½ in.) gypsum board each side; 50 mm (2 in.) batt insulation
STC 37 (NRCC)

3g

38 × 89 mm (2 × 4) studs, 400 mm (16 in.) o.c.; Two layers 13 mm (½ in.) gypsum board each side
STC 39 (OCF)

3h

Same as #3g, with 89 mm (3½ in.) batt insulation
STC 45 (OCF)

3 single-stud

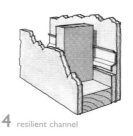

4 resilient channel

Compare 3i to 3c: A 50 percent increase in mass results in a 3 point increase in STC.

Compare 3j to 3i: Adding insulation only increases STC rating by 2 points. Structural coupling controls the STC.

3i

38 × 89 (2 × 4) studs, 400 mm (16 in.) o.c.; Two layers 13 mm (½ in.) gypsum board one side; one layer 13 mm (½ in.) gypsum board other side
STC 38 (OCF)

3j

Same as #3i, with 89 mm (3½ in.) batt insulation
STC 40 (OCF)

4a

38 × 89 mm (2 × 4) wood studs, 400 mm (16 in.) o.c.; 13 mm (½ in.) gypsum board; resilient channels 600 mm (24 in.) o.c.; 13 × 75 mm (½ in. × 3 in.) gypsum filler strip along base plate
STC 39 (OCF)

Compare 4b to 4a: A 7 point improvement by adding insulation to cavity with isolated construction.

Compare 4b to 4c: Bridging a resilient channel construction reduced STC by 4 points.

Compare 4e to 4b: Doubling of mass results in a 10 point increase in STC.

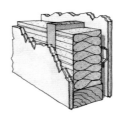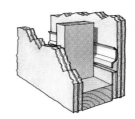

4 resilient channel

4b

Same as #4a, with 89 mm
(3½ in.) batt insulation
STC 46 (OCF)

4c

Same as #4b, without gypsum
board filler strip
STC 50 (OCF)

4d

38 x 89 mm (2 x 4) studs,
400 mm (16 in.) o.c.; two layers 13 mm (½ in.) gypsum
board each side; resilient channels, 600 mm (24 in.) o.c.; 13 x
75 mm (½ x 3 in.) gypsum
filler strip along base plate
STC 52 (OCF)

Compare 5b to 5a: Insulation improves STC by 7 points.

Compare 5b to 4c: Staggered stud wall has about the same perfor-mance as a resilient channel single stud wall.

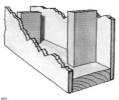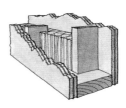

4 resilient channel   5 staggered stud

4e

Same as #4d, with 89 mm
(3½ in.) batt insulation
STC 56 (OCF)

5a

38 x 89 mm (2 x 4) studs,
600 mm (24 in.) o.c., staggered
300 mm (12 in.) o.c. on 38 x
140 mm (2 x 6) plates; 13 mm
(½ in.) gypsum board
STC 42 (OCF)

5b

Same as #5a, with 89 mm
(3½ in.) batt insulation
STC 49 (OCF)

Compare 5c to 5b: Doubling mass results in 6 STC point improvement.

Compare 5e to 5c: Spacing of stag-gered studs does not significantly affect STC rating.

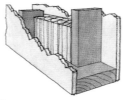

5 staggered stud

5c

38 x 89 mm (2 x 4) studs,
600 mm (24 in.) o.c. staggered
300 mm (12 in.) o.c. on 38 x
140 mm (2 x 6) plates; two
layers 13 mm (½ in.) gypsum
board each side; 89 mm
(3½ in.) batt insulation
STC 55 (OCF)

5d

Same as #5c, minus one layer
gypsum board on one side
STC 53 (OCF)

5e

Same as #5c, with stud spacing
at 400 mm (16 in.) o.c. and
staggered every 200 mm
(8 in.) with 50 mm (2 in.) batt
insulation
STC 54 (OCF)

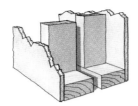

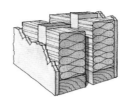

6 double stud

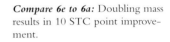

*Compare **6b** to **6a**:* Insulation results in a 9 STC point improvement.

*Compare **6c** to **6b**:* Doubling insulation results in a 3 point increase in STC.

6a

Double row of 38 x 89 mm (2 x 4 in.) studs, 400 mm (16 in.) o.c. on separate plates spaced 25 mm (1 in.) apart; 13 mm (½ in.) gypsum board
STC 47 (OCF)

6b

Same as #6a, with 89 mm (3½ in.) batt insulation on one side
STC 56 (OCF)

6c

Double row of 38 x 89 mm (2 x 4) studs, 400 mm (16 in.) o.c. on separate plates, spaced 25 mm (1 in.) apart; 13 mm (½ in.) gypsum board; 89 mm (3½ in.) batt insulation in both stud cavities
STC 59 (OCF)

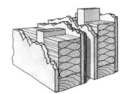

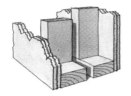

6 double stud

*Compare **6e** to **6a**:* Doubling mass results in 10 STC point improvement.

6d

Double row of 38 x 89 mm (2 x 4) studs, 400 mm (16 in.) o.c. on separate plates, spaced 65 mm apart. 16 mm gypsum board on each end of 4 stud faces + 6 mm gysum. board on one interior face, 89 mm batt insulation, both stud cavities
STC 45 (OCF)

6e

Double row of 38 x 89 mm (2 x 4) studs, 400 mm (16 in.) o.c. on separate plates, spaced 25 mm (1 in.) apart; two layers, 13 mm (½ in.) gypsum board each side
STC 57 (OCF)

6f

Same as #6e, with 38 mm (1½ in.) batt insulation
STC 62 (OCF)

6 double stud

6g

Same as #6e, with 89 mm (3½ in.) batt insulation in both stud cavities
STC 63 (OCF)

6h

Same as #6e, minus one layer 13 mm (½ in.) gypsum board
STC 52 (OCF)

6i

Same as #6f, minus one layer 13 mm (½ in.) gypsum board
STC 59 (OCF)

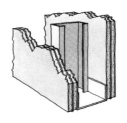

7 metal stud

7a

92 mm (3⅝ in.) metal studs, 600 mm (24 in.) o.c.; 16 mm (⅝ in.) gypsum board
STC 39 (OCF)

7b

Same as #7a, with 75 mm (3 in.) batt insulation
STC 44 (OCF)

7c-d

92 mm (3 ⅝ in.) metal studs, 600 mm (24 in.) o.c.; two layers 13 mm (½ in.) gypsum board each side • STC 50 (OCF)
#7d: Same as #7c, with 75 mm (3 in.) batt insulation
STC 56 (OCF)

Compare 7d, 5c, and 4e: Tested in same laboratory, two layers of gypsum board on each side, insulated and isolated, achieved about the same STC rating.

Compare 8b to 8a: 33% more mass results in a 2 point increase in STC.

Compare 8c to 8b: Filling cells and sealing the surfaces results in a 7 point increase in STC.

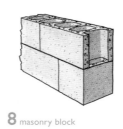
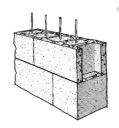

8 masonry block

8a

200 x 200 x 450 mm (8 x 8 x 18 in.) hollow concrete masonry units —11.4 kg/block (25 lbs/block). Both sides sealed with paint.
STC 46 (KODARAS)

8b

Same as #8a, except 16.4 kg/block (36 lbs/block)
STC 48 (KODARAS)

8c

200 x 200 x 450 mm (8 x 8 x 18 in.) hollow concrete masonry units —160 kg/m² (33 psf). Hollow units filled with cement, both sides sealed with paint
STC 55 (KODARAS)

Data for Wood Framed Floor/Ceilings

Wood framed floor/ceilings are commonly used in the construction of low-rise housing. Table 7.2 compares laboratory TL values due to changes in floor toppings, ceiling attachment, and insulation.

Table 7.2 STC ratings for wood framed floor/ceiling constructions (diagrammed in Figure 7.8). *Notes:* All constructions are topped with 38 mm (1½ in.) cellular concrete except B and F; Construction B has 13 mm (½ in.) plywood in lieu of cellular concrete; Construction F has 19 mm (¾ in.) gypsum cement in lieu of 38 mm (1½ in.) cellular concrete. All tests conducted at GEIGER & HAMME in 1970 and 1971.

Insulation batts		Gypsum board ceiling	STC rating
A.	None	16 mm (⅝ in.) nailed directly	48
B.	76 mm (3 in.)	16 mm (⅝ in.) on resilient channels	49
C.	165 mm (6½ in.)	16 mm (⅝ in.) nailed directly	50
D.	None	16 mm (⅝ in.) on resilient channels	55
E.	89 mm (3½ in.)	16 mm (⅝ in.) on resilient channels	58
F.	76 mm (3 in.) mineral wool	13 mm (½ in.) on resilient channels	58

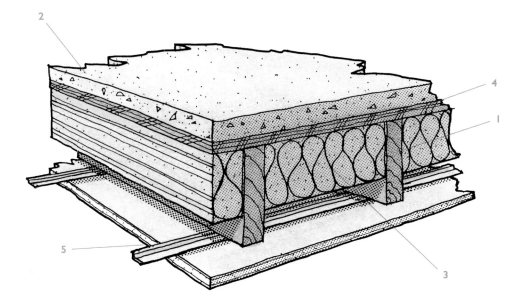

The following can be gleaned from the measured data for floor/ceiling constructions: (1) The combination of resilient channels and batt insulation increases the STC by 10 points; (2) 19 mm ($^3/_4$ in.) gypsum cement or 38 mm (1$^1/_2$ in.) cellular concrete adds about 9 points to the STC rating when the ceiling is isolated; (3) 19 mm ($^3/_4$ in.) gypsum cement provides about the same sound insulation performance as 38 mm (1$^1/_2$ in.) cellular concrete; and (4) Mineral wool is acoustically equivalent to glass fiber batt for a given thickness.

Figure 7.8 Wood-frame floor/ceiling construction. *Key:* (1) 38 x 235 mm (2 x 10) joists; (2) 38 mm (1$^1/_2$ in.) cellular concrete or 19 mm ($^3/_4$ in.) gypsum cement topping; (3) batt insulation or mineral wool; (4) 16 mm ($^5/_8$ in.) plywood subflooring, (5) 13 mm ($^1/_2$ in.) resilient channels.

Data for Concrete Floating Floors

Figure 7.9 shows a floating floor construction on structural concrete slab with a resiliently suspended ceiling below. The separate gypsum board walls help control the flanking of sound around the floor/ceiling construction. This construction can attain STC 80 with the isolated wall constructions but would be limited to about STC 65 without the "anti-flanking" isolated wall constructions. This construction is called a *room within a room* technique since the walls, floors, and ceilings are double and isolated from one another.

Data for Impact Insulation

Table 7.3 lists the IIC values for laboratory tests conducted on (1) wood frame and (2) concrete floor/ceiling constructions. Commercial flooring products are available that allow hard-surfaced floors such as ceramic tile and hardwood to achieve IIC values that meet the building code requirements.

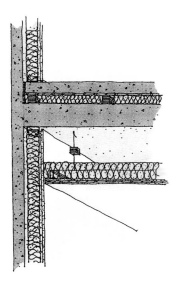

Figure 7.9 Floating floor construction with isolated ceiling and walls.

2. Concrete slab without ceiling

Floor finish	IIC
None	35
Parquet	50
Sheet vinyl	51
Cushioned vinyl	58
Carpet/pad	80

Table 7.3 IIC ratings of floor/ceiling constructions. Ceilings are 16 mm (⅝ in.) gypsum board. Batts are 76 mm (3 in.) thick.

1. Wood frame

Lightweight concrete	Insulation batts	Resilient channels	Floor finish	IIC
Yes	None	Yes	Sheet vinyl	36
None	None	None	Parquet	39
Yes	None	None	Parquet	40
Yes	None	None	Cushioned vinyl	41
Yes	None	Yes	Parquet	50
Yes	Yes	Yes	Parquet	52
Yes	None	Yes	Cushioned vinyl	55
Yes	None	None	Carpet/pad	61
None	None	None	Carpet/pad	61
Yes	None	Yes	Carpet/pad	76
Yes	Yes	Yes	Carpet/pad	79

Notes

1. NBS results were published in *NBS Building Science Series 77: Acoustical and Thermal Performance of Exterior Residential Walls, Doors and Windows,* November 1975.

2. M. David Egan, *Architectural Acoustics* (McGraw-Hill Book Company, 1988).

3. Data compiled in *Catalog of STC and IIC Ratings for Walls and Floor/Ceiling Assemblies*, California Office of Noise Control, 1981.

10 General Rules about Sound Insulation

1. A doubling in mass is expected to cause a 5 STC point increase.
2. For air spaces greater than 25 mm (1 in.) between panels, every doubling is expected to increase STC by about 5 points if reverberant sound in cavity is controlled.
3. Adding insulation in a construction with direct framing attachment is of limited value—typically about a 2 point improvement.
4. Increasing the number of direct framing attachments results in lower STC.
5. In constructions with rigid framing attachments, the TL values can vary significantly due to subtle differences in connections.
6. STC increases by 5 to 10 points when insulation is added to isolated constructions.
7. Staggered stud construction is acoustically comparable to resilient channel construction.
8. The small air space between the two single stud walls significantly reduces the STC rating as compared to a double stud partition (compare 6d to 6f).
9. Light gauge 0.6 mm (25 ga.) metal studs are acoustically equivalent to wood studs and resilient channels or staggered wood stud construction.
10. It is important to seal both faces of a concrete masonry wall in order to control sound leaks. (Sealing can be achieved by painting, plastering, or gypsum board furring.)

Conclusion

The principles of sound insulation are important when addressing noise control in buildings. They are helpful in understanding the information discussed in Chapter 17, 18, 19, and 21.

8

Building Vibration

Anthony P. Nash, P.E.

Fundamentals • Vibration Considerations in Buildings • Sources of Vibration • Effects of Vibration • Models for Building Design • Measurements of Floor Performance

THE role of structure-borne vibration as a transmitter of unwanted sound is discussed in Chapters 7 and 9. Another undesirable characteristic of vibration is that it can be sensed directly by people in buildings. This chapter discusses how low-frequency building vibration can annoy occupants and disrupt sensitive equipment.

Fundamentals

The human response to building vibration includes both audible and "feelable" sensations. At frequencies above 100 Hz, vibration usually manifests itself as sound. At frequencies below 20 Hz, vibration is physically sensed by the human body. It is rare for vibration below 20 Hz to be heard as sound or for vibration above 100 Hz to be felt; however, between 20 and 100 Hz, some combination of audible and tactile sensations are possible.

Certain types of scientific or industrial processes are also affected by vibration. Such effects can be observed as a cause of "errors" in the operation of sensitive equipment used in research laboratories and semi-conductor fabrication plants.

People sense building vibration with their entire body, thus we call their reaction a "whole body" response. Human beings are surprisingly sensitive to whole body vibration between 5 and 10 Hz. Over the past 70 years, experts have come to agree on the amplitudes and frequencies that define "feelability"—the onset of human perception for continuous vibration.

Although the human response to steady-state vibration is well-defined, there is less certainty when the vibration is intermittent. One example of intermittent vibration is human footsteps. In addition to objective factors such as frequency, amplitude, and duration, transient vibration can involve psychological factors that relate to context. In other words, individuals tend to assign different emotional values to the same physical sensation. For instance, a person anxious about structural integrity in a building will be more sensitive to a momentary vibration than someone who is less concerned about the building's structural integrity. The corollary is that an anxious person can be "de-sensitized" by being told that a slight vibration transient does not mean one's safety is jeopardized.

Terminology

There are several descriptors for quantifying vibration amplitudes in solids: *acceleration, velocity,* and *displacement. Acceleration* is defined as the rate of change in *velocity,* and velocity is the rate of change in *displacement.* In the field of engineering vibration, these terms are used where instantaneous motion is the principal concern. For instance, we disregard the steady acceleration of the earth's gravitational field when discussing the fluctuating acceleration imposed upon an apple falling towards the ground. Similarly, the constant speed of a car is ignored when quantifying its ride quality as a vibration velocity. The explicit definitions for these descriptors imply that we are only interested in the rapid dynamic fluctuation of an object about its stable position.

In the propagation of sound, the term particle velocity is used to describe the dynamic motion of air molecules influenced by a sound pressure wave front. The usage of *vibration velocity* in this chapter should not be confused with the particle velocity of sound waves discussed in Chapter 6.

Vibration Considerations in Buildings

Although vibration is considered early in the design phase of certain laboratory facilities, for the majority of buildings, disturbances from vibration can still come as a surprise. Even for buildings that meet code requirements, some components of the structural system can vibrate unexpectedly when people go about their ordinary activities. In the early stages of space planning, the design professional should anticipate the possibility of vibration affecting sensitive occupancies. In this sense, "occupancies" can include both people and equipment.

Factors to consider when addressing building vibration include: (1) Who or what is vibration-sensitive? (2) What is the context of the vibration—is it in a laboratory or office building? and (3) What sets the structure into motion: machinery, footsteps, or road traffic?

Floor Systems. All portions of a building's structure vibrate to a lesser or greater degree. In practice, however, the floor system is the most common component of a building capable of disrupting both people and equipment. There are two reasons for the floor being of primary interest: (1) people and vibration-sensitive equipment are supported by the floor; and (2) people and equipment are usually the principal *generators* of vibration.

There is a duality implied in these two points—both people and equipment can assume alternate roles, being either generators or sensitive receivers of vibration. This duality is discussed again in this chapter.

One common floor system is the so-called *slab-on-grade* where the concrete is wholly in contact with the earth. This floor tends to have minimal vibration response in comparison with floors supported above grade. That is, a slab-on-grade tends to be non-resonant, whereas a supported floor system contains resonances that amplify its response to the same input force.

Other flooring systems include those supported by walls and columns. The principal categories of these supported floor systems are *monolithic* and *framed*. A monolithic floor typically comprises a thick concrete slab, which can be either cast in place or assembled from pre-cast elements. A framed floor system usually contains a thin slab, metal deck, or plywood sheet supported by a grillage of beams, girders, or joists.

The "waffle slab" is one example of a monolithic floor system. The underside of this type floor contains hollow coffers—it looks like a gigantic waffle iron. The deep, hollow coffers of the waffle slab provide a very high stiffness-to-mass ratio; as a result, its resonances occur at frequencies above 20 Hz. This characteristic is desirable because the floor's response to vibration sources is well-controlled, making it a popular choice for vibration-sensitive industrial facilities such as computer chip manufacturing. The disadvantage of the waffle slab is that it is time consuming and costly to construct.

Framed floor systems for general-purpose multistory buildings range from massive (steel beams supporting a concrete slab) to light (wood joists supporting a plywood sheet). The former tends to be used for long spans, and the latter for short spans.

The likelihood of annoying floor vibration often increases with the span of the framing. Although it is more common for long-span floors to have vibration problems, it is not always the case. The reasons for one floor performing better than another is not yet well understood by design professionals.

One factor contributing to this incomplete understanding is that most floor systems are not analyzed for their dynamic response. Predicting the dynamic response of a floor system requires specialized skills and a detailed knowledge of the floor's construction including: (1) its span and width; (2) the material properties of the supporting beams or joists; (3) the floor diaphragm itself; (4) the behavior of joints in the floor diaphragm; and (5) the location of full-height partitions.

Some of this information can never be precisely known, even for floors in existing buildings. The amount of data and time required to analyze a floor system are two of the reasons that combining economical floor design with low-vibration performance is still an imperfect art.

The three vibration descriptors (acceleration, velocity, and displacement) are redundant—that is, if one knows the vibration velocity of an object at some frequency, then, through a mathematical relationship, the acceleration and displacement are also known. The sole benefit of three separate descriptors occurs when discussing the effects of random vibration upon people and equipment. By definition, random vibration has no single frequency; rather, it contains a broad range of frequencies. The sound of rain is an audible example of random vibration. Without knowing the frequency content of a vibration spectrum, it is not possible to convert vibration amplitudes of one descriptor into amplitudes of another—we must choose one of the three descriptors a priori.

The choice of a vibration descriptor is, to a certain extent, arbitrary. If one is measuring vibration with an accelerometer, then it is reasonable to quantify the amplitude using *acceleration*. In this case, the vibration might be expressed as meters per second per second or inches per second per second. Partly for historical reasons, vibration *velocity* is commonly used to specify vibration limits for rotating machines. In this case, the specification might be stated in meters per second or inches per second. Sometimes vibration *displacement* criteria are given by optical equipment manufacturers in order to reduce the likelihood of blurred images. These criteria could be stated in micrometers or microinches.

The practical distinction among the three descriptors is the emphasis of the low-frequency versus the high-frequency regions of the vibration spectrum. For example, a high acceleration at high frequency is generally harmless to a machine, however, even moderate accelerations at low frequency can be damaging.

Sources of Vibration

Walking. Both steel and wood-frame floors can respond in an annoying manner when excited by dynamic forces such as people walking, running, or jumping. In some instances, vibration from machinery can also be disruptive to the occupants or operations within a building. When people walk on a suspended floor system, they generate a transient force with each step. For an adult, the footstep rate is about 1.7 per second and the peak transient force is 700 newtons (150 pounds). Thus, ordinary walking represents a significant mechanical energy source and causes some floors to respond with noticeable vibration.

Equipment. In buildings, mechanical equipment can disrupt people both by generating tactile vibration as well as structure-borne noise. In order to reduce vibration transmission, mechanical equipment is fitted with vibration isolation devices (as described in Chapter 9). The converse of isolating vibrating equipment is to isolate sensitive devices within the building. Floating floors and special isolation tables are two means by which sensitive receivers can be protected from sources of building vibration.

Effects of Vibration

Human Response. For steady-state vibration, people respond with equal sensation to a constant vibration velocity from 8 Hz to 80 Hz (see Figure 8.1). Around 10 Hz, a person can barely sense a vibration velocity of 0.1 millime-

Figure 8.1 Threshold of human perception for vibration. From 8 to 80 Hz, the steady state threshold is 100 microns/sec. (i.e., 4000 micro inches/sec.).

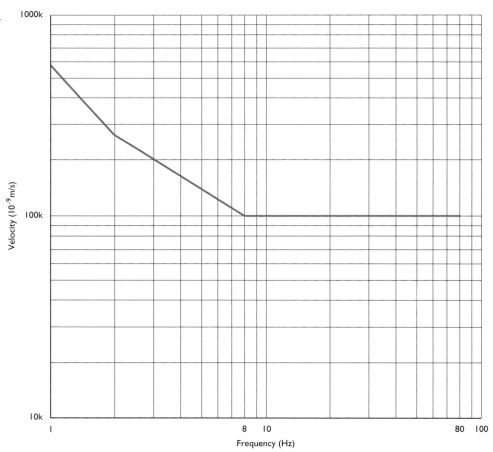

ter per second (100 microns/sec). This velocity is equivalent to a vibration displacement of only 0.0015 millimeter (1.5 microns)—about twice the wavelength of visible light.

For vibration transients lasting one second or less, people are thought to be 10 times *less* sensitive, possibly because the human physiology takes a finite time to sense the minute dynamic forces occurring within one's body.

Equipment. Some specialized apparatus can be affected by floor vibration. Sensitive optical instruments including microscopes and laser interferometers can develop blurred images when subjected to vibration. Certain electronic imaging systems such as electron microscopes and magnetic resonance imagers (MRI) may also experience errors, especially at high magnifications. Some equipment used in integrated circuits manufacturing is affected by vibration amplitudes well below the threshold of human perception.

Models for Building Design

As mentioned earlier in this chapter, building vibration affects the efficiency of both people and equipment. Since the origin and nature of vibration for people and equipment in buildings is somewhat different, each will be discussed separately.

Most of the research in the area of footstep-induced floor vibration began in the 1960s when prefabricated open-web steel joists evolved from their original use in long-span roofs to floor systems. The extremely long spans achievable with these light-weight structural elements triggered serious questions about their vibration serviceability in floors. Over a 15-year period, Dr. Kenneth Lenzen and his coworker Thomas Murray from the University of Kansas developed a comprehensive dynamic floor model, test method, and companion design criterion that addressed the perception of vibration generated by people walking on floors constructed with steel framing. The simplicity of this predictive method is attractive and continues to be used by structural engineers.

The research of Lenzen and Murray was partly based on original investigations into human vibration sensitivity conducted in the 1930s by two German scientists, Reiher and Meister. They discovered that people are equally sensitive to a given *vibration velocity* as opposed to acceleration or displacement. As a consequence, Lenzen and Murray adopted a *velocity* scale to rank degrees of human sensation experienced by a person as another individual walked on the same floor. A floor that responds to footfalls with lower velocity is deemed better.

The Lenzen–Murray method is based on several simplifications that are satisfactory when applied to typical floor sizes constructed with steel framing supporting a concrete-filled corrugated metal deck. Their model may not be appropriate for other categories of floor systems such as light-frame floors with wood joists, floors having small span-to-width ratios, or those with very long spans.

Some equipment manufacturers have published their own vibration criteria. These are often expressed as a *vibration displacement* rather than a *vibration velocity*. Using displacement as a descriptor simply means that the manufacturer is more concerned with disturbances occurring at lower frequencies or, conversely, is less concerned about vibration velocity at high frequencies. An example of such a displacement criterion could be: "…peak displacement between 2 and 5 Hz is not to exceed 2 microns…."

This specification is equivalent to 1000 microinches/sec. at 2 Hz and 2500 microinches/sec. at 4 Hz—a modest requirement suitable for an optical microscope.

Floor Design Criteria. A number of supplemental guidelines for designing low-vibration floor systems have been proposed by the structural engineering community. Murray has expressed the need for adequate energy-absorbing properties (damping) in steel frame floor construction. For years, the Canadian building code has required that wood frame floors meet a minimum stiffness value by limiting the mid-span deflection under a concentrated load. The Swedish building industry has issued a design guideline for residential floor systems that considers four distinct parameters: (1) deflection under a concentrated load; (2) resonant frequency; (3) peak vibration velocity from a single impulse; and (4) average vibration velocity from continuous walking.

Some American consultants have addressed the unique needs of the semiconductor industry by proposing limits for floor velocity induced by human footsteps. Their simplified model is based on the stiffness and resonant frequency for a floor bay. The idea is to design a floor with sufficient stiffness such that a typical walking person generates a maximum vibration velocity deemed appropriate for the given class of equipment. As the equipment becomes more sensitive, the product of stiffness and resonant frequency is increased in order to reduce the vibration velocity. The relationship is:

EQUATION 8.1

$$V = \frac{c}{kf_o}$$

Where V = vibration velocity in inches/second, k = stiffness of floor bay in pounds/inch, f_o = resonant frequency in hertz, and c = a constant depending on units; in this case, $c = 10^4$ pounds/second^2.

These models and criteria attempt to simplify the complex dynamics of a floor system into a series of rudimentary calculations. The process of simplification helps explain why there are so many models and inconsistencies among the models. Each model is reasonable for the class and type of floor system that inspired its development.

Vibration Performance Criteria for Equipment. In the 70 years that followed the introduction of the Reiher-Meister scale, a number of studies has confirmed that human sensation is proportional to vibration velocity. Within the last 20 years there has been a general consensus among specialists that most sensitive scientific equipment responds to *vibration velocity* over the same frequency range as do people. For example, one could state with reasonable certainty that a microscope will exhibit image blurring at some vibration velocity at a frequency somewhere in the range between 8 and 80 Hz. This observed similarity in dynamic behavior between people and equipment may stem from stiffness and mass characteristics common to both the human body and machine components.

A standard format for expressing vibration criteria of sensitive equipment is yet to be universally accepted. In the interim, generalized criteria have been proposed for several classes of devices used in research and semiconductor manufacturing. These criteria are shown as a series of vibration velocity contours that resembles the human threshold of perception.

Figure 8.2 illustrates a family of these generic criteria for various equipment types and functions. The most stringent criterion is a contour entitled, "125 microinches/second" (0.003 millimeter/second). This contour is appropriate for electron microscopes operating at magnifications above 300,000. The criteria are all parallel to, and most are well below, the threshold of human perception (shown in Figure 8.2 as 4,000 microinches/second or 0.1 millimeter/second). The contours with greater amplitudes are intended for less demanding applications.

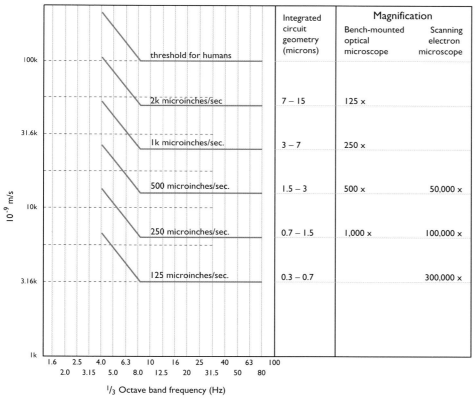

Figure 8.2 Generic criteria for vibration-sensitive equipment. The contours are constant vibration velocity above 8 Hz.

| | Integrated circuit geometry (microns) | Magnification | |
		Bench-mounted optical microscope	Scanning electron microscope
threshold for humans			
2k microinches/sec	7 – 15	125 x	
1k microinches/sec.	3 – 7	250 x	
500 microinches/sec.	1.5 – 3	500 x	50,000 x
250 microinches/sec.	0.7 – 1.5	1,000 x	100,000 x
125 microinches/sec.	0.3 – 0.7		300,000 x

10^{-9} m/s

1/3 Octave band frequency (Hz)

Measurements of Floor Performance

Test Techniques. There are two complementary methods to test the dynamic response of floors. The first technique employs a discrete transient deliberately generated by a person. The second involves a steady-state force from an electro-mechanical device. Each of these techniques has its own advantages and disadvantages. The two methods are described below.

(1) Transient. This test involves a special human footstep called a *heel-drop* shown in Figure 8.3. The heel drop is defined as an impulse generated when an 80 kg (175 lb) person arches his heels up 60 mm (2½ in.) and then freefalls onto the floor. The heel drop concept is based on a model of a floor vibrating as if it were a trampoline. In brief, the peak vibration velocity from

Figure 8.3 The heel drop test. A heel drop performed by a person arching his heels and then free falling onto the floor. The heel drop test generates large forces that are used to evaluate the floor's dynamic behavior. The platform shown in the photo is an electrical-force sensing system.

the heel drop is a kind of figure-of-merit for the floor undergoing the test; the lower the velocity, the better the floor will behave when a person walks upon it. "Better" in this context means that an individual will perceive less disruption when another person walks on the same floor.

(2) Steady State. The second test is performed using an electro-mechanical vibrator that is placed on the floor and generates dynamic forces corresponding to a predetermined electrical signal. Vibration sensors detect the magnitude of the floor's response to the known input force. The floor's response to the stimulus reveals valuable characteristics about the floor's dynamic behavior. These characteristics include several of the resonant frequencies ("modes"), the tendency of the vibration to subside over

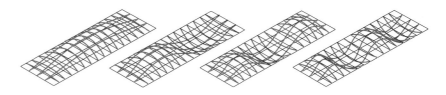

Figure 8.4 Operational deflection shapes.

time ("damping"), and the deformed shape of the floor when vibrating. Some examples of these deformed shapes are shown in Figure 8.4.

A sophisticated mathematical model of the particular floor system can be created from its measured dynamic characteristics. An engineer can then use a computer program to predict the effects of construction modifications such as adding stiffening components or increasing the floor's damping. These modifications can be implemented by incorporating reinforcements or dampers at predetermined locations beneath the floor.

Conclusion

People are quite sensitive to tactile vibration, especially when it affects their whole body. Some laboratory equipment is also vibration-sensitive and its performance may suffer even at vibration amplitudes well below the threshold of human perception. In buildings where people live and work, the likely sources of vibration are rotating mechanical equipment and human footsteps. The common path for vibration transmission is the floor, especially floors supported by structural framing above the earth.

Mechanical equipment is usually fitted with vibration-isolation devices to reduce vibration at its source. Occasionally, sensitive laboratory apparatus is also mounted on vibration isolation tables to improve its performance.

Design professionals use various guidelines to help minimize the undesirable response of a floor to footsteps. These guidelines involve simplifying assumptions; hence, the design methods are not always consistent.

For an existing building with floor vibration problems, it is recommended that field tests be conducted and then evaluated in combination with a mathematical model of the floor system. The model helps the engineer to predict the most effective way to improve the floor's performance by adding stiffeners or dampers.

9

Mechanical & Electrical Systems

Karen E. Decker, P.E.

Space Planning • Noise Criteria Selection • Room Location • Equipment Layout • Noise Control Measures for Main Building Equipment • Fan Systems • Fans • Ductwork • Silencers • Acoustical Plenums • Duct Lagging • Acoustical Louvers • Variable Speed Drives • Active Noise Control • Chillers • Cooling Towers • Boilers • Pumps • Motors • Air Compressors • Noise Control Measures for Air Distribution Equipment • Fan-Powered Boxes • Variable Air Volume Boxes • Fan Coil Units • Diffusers and Grilles • Noise Control Measures for Electrical Equipment • Emergency Generators • Transformers and Uninterrupted Power Systems • Guidelines for Vibration Isolation • Vibration Isolator Characteristics • Stiffness and Static Deflection • Natural Frequency • Damping • Transmissibility • Vibration Isolator Types • Elastomeric Isolators • Springs • Spring Curbs • Air Springs • Vibration Isolation Support Equipment • Housekeeping Pads, Beams, and Stanchions • Inertia Bases • Frames and Rails • Thrust Restraints • Seismic Snubbers • Flexible Connections • Ductwork and Piping Isolation

WHEN designing a building, it is important to control the noise and vibration of its mechanical and electrical equipment. Without adequate consideration, the very equipment that provides thermal comfort and electric power can generate annoying noise and vibration. Proven techniques are available for mitigating noise and vibration from this equipment. The recommended acoustical design sequence for a building project is: (1) select noise criteria for each space in the building; (2) organize spaces to avoid adverse adjacencies of noisy equipment with quiet spaces; and (3) provide appropriate noise and vibration control for equipment.

Building type	Noise criteria, NC or room criteria, RC(N)
Residential	
Private residences	25–30
Apartments	30–35
Hotels/motels	
Guest rooms or suites	30–35
Meeting/banquet rooms	30–35
Corridors and lobbies	35–40
Service/support areas	40–45
Workplaces	
Executive offices	25–30
Conference rooms	25–30
Offices	30–35
Open-plan areas	35–40
Business machines/computers	40–45
Public circulation	35–40
Teleconference rooms	25 max.
Hospitals, clinics	
Private rooms	25–30
Wards	30–35
Operating rooms	25–30
Laboratories	35–40
Corridors	35–40
Public areas	35–40
Courtrooms	25–30
Churches	25–30
Restaurants	40–45
Libraries	35–40
Schools	
Lecture and classrooms	25–30
Open-plan classrooms	35–40
Music teaching studios	25 max.
Music practice rooms	30 max.
Entertainment facilities	
Legitimate theaters	20–25
Movie theaters	30–35
Concert and recital halls	15–20
Recording studios	15–20
TV studios	20–25
Radio studios	15–20
Sound editing rooms	30 max.
Film screening rooms	20–25

Table 9.1 Typical recommended background noise levels in rooms.

Part I of this chapter, *Noise Control,* discusses space planning, criteria, and techniques to control the noise of mechanical and electrical equipment. Part II addresses *Vibration Isolation.*

I. Noise Control

Space Planning

Noise Criteria Selection. As part of the initial planning for a building, noise goals should be selected for each occupied space. Once these goals have been assigned to each space, the most noise-sensitive spaces should be located away from noisy activities and equipment. Occasionally, equipment noise transfer to the outdoors is an issue.

The American Society of Heating, Refrigerating, Air-Conditioning Engineering (ASHRAE) has developed standards for acceptable background noise levels in various spaces. The Noise Criterion (NC) rating provides a single number to describe relative loudness. While NC ratings have been in use since the 1950s, they are gradually being replaced by Room Criterion (RC) ratings. The RC system is slightly more stringent at the low and high frequencies and seems to reflect human perception better.

NC and RC curves are approximately the same for a given value; that is, NC 40 is about equal to RC 40, but the RC values are more descriptive since the numerical rating is followed by a letter to describe the quality of the noise: R for rumbly (excessive bass), H for hissy (excessive treble), and N for neutral. Refer to the latest ASHRAE Guide to learn more about the NC and RC rating schemes. (Table 9.1 includes design guidelines for various spaces, which in part are based on ASHRAE recommendations.)

Room Location. Space planning can be a cost–effective noise control technique. Avoid locating mechanical equipment rooms (MER), emergency generators, and electrical transformer rooms near spaces that require low background noise levels. This recommendation applies to spaces separated both horizontally and vertically. For instance, if a MER is located above a private office, a floating floor may be required in the MER. Had the MER been located above circulation space, the floating floor might not have been needed.

Equipment Layout. MERs should be made large enough to accommodate efficient duct layouts and adequate noise control measures. If the size of the MER allows more than 4.5 m (15 ft.) of duct prior to leaving the room, there are more options for controlling fan noise.

For roof-mounted fans, ample supply and return duct lengths should be provided prior to penetrating the roof. These equipment layout concepts are discussed later in this chapter.

Noise Control Measures for Main Building Equipment

This section addresses noise control considerations for the major pieces of building mechanical equipment, including fan systems, chillers, cooling towers, boilers, pumps, motors, and air compressors.

Fan Systems. Fans propel air usually through a ductwork system. Noise control measures for fans include selecting a quiet fan type, providing ample duct sizes, having adequate duct lengths, and appropriately selecting diffusers and grilles. In addition, duct silencers, plenums, acoustical louvers, duct enclosures, sound-isolating construction (walls, roofs, floor/ceilings), and quiet terminal units are often necessary.

Fans. Although fan selection is often dictated by static pressure requirements or space limitations, the potential noisiness of various fan types should be considered. To a great extent, the noise generated by a fan is controlled by its operating conditions such as air flow rate, fan speed, and static pressure. However, for a given set of operating conditions, the fan types can be ranked in order of increasing noise levels as follows: (1) centrifugal fans with airfoil type blades; (2) centrifugal fans with backward-curved blades; (3) vaneaxial fans;[1] (4) centrifugal fans with forward-curved blades; (5) propeller fans; (6) tubeaxial fans; (7) radial-bladed fans.

The quietest type of fan that will satisfy the operating requirements should be selected whenever possible to reduce the need for mitigation measures. The cost of mitigation may exceed the cost saving for a less expensive, but noisier, fan.

After the fan is selected, it should be oriented so that the transition to ductwork minimizes air turbulence. In general, the first duct elbow should be located at a distance from the fan that is at least 1.5 times the largest duct dimension. Furthermore, this elbow should be radiused a minimum of 15 cm (6 in.) and oriented in the same direction that the fan wheel is rotating as shown in Figure 9.1. Excessive air turbulence noise can be created if the discharge duct is turned in the opposite direction (180 degrees) as shown in Figure 9.2.

Ductwork. There are significant acoustical benefits to providing long duct lengths between fans and the nearest air register particularly if it is internally lined with *duct liner*. In general, a 6 m (20 ft.) long lined duct provides a reduction of 10 dB in fan noise. Furthermore, when a duct has a cross-sectional area less than 0.3 m² (2 sq. ft.), as little as 3 m (10 ft.) of internally lined duct will achieve similar results. However, unlined duct may need to be three to four times as long to be acoustically equivalent.

Fan noise transmitted into a room is generally either *duct-borne* or *breakout* as shown in Figure 9.3. Duct-borne can be described as fan noise that is carried within a duct and then transfers into a room through a register. Breakout can be described as fan noise that passes through the wall of a duct and through the ceiling into a room.

Breakout noise is important to consider even if there is not a duct opening into a sensitive space. It can be beneficial to allow the fan noise to breakout into a non-sensitive space.

A down-discharge fan on the rooftop should be located only near spaces with a noise goal of NC or RC 45 or more. The noise transfer problem is diagrammed in Figure 9.4. A side-discharge fan with long lengths of

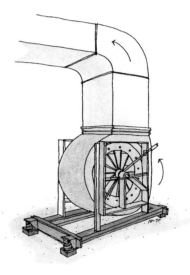

Figure 9.1 Recommended outlet configuration for centrifugal fan.

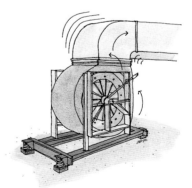

Figure 9.2 Outlet configuration for centrifugal fan that can lead to excessive low-frequency duct rumble.

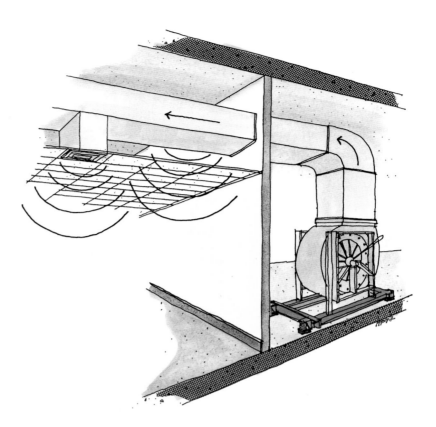

Figure 9.3 Duct-borne and breakout
noise transfer from HVAC system.

rectangular ducts on the roof should be used so that as much noise can break
out as possible, as shown in Figure 9.5.

Round ductwork allows very little breakout noise in contrast to rectangular ductwork which permits the most breakout noise. Flat oval duct
permits almost as much breakout noise as rectangular duct. Internal duct lining and external insulation do not significantly reduce breakout noise.

Besides attenuating fan noise, internal lining also helps control noise
transfer between adjacent spaces served by common ducts, called *crosstalk*.
To reduce the likelihood of crosstalk, it is good practice to locate the main
supply and return ducts above corridors rather than offices. Individually
lined branches can then be extended from the main ducts into each office.

Crosstalk also occurs through return air transfer ducts. To control this
type of crosstalk, an internally lined return transfer duct through the partition is often specified. This duct contains at least one elbow as shown in
Figure 9.6.

Where ceiling-height partitions are used with a sound-rated ceiling
system, the return air transfer is often accomplished with a lined return air
"boot," on the back of the return grille. The required length of an air transfer duct varies with its cross sectional area and the amount of noise control
required. Typically an air transfer duct is 1 to 2 m (4 to 8 ft.) long.

If ducts are undersized, excessive airflow noise may result. In general,
the air speed in ducts should not exceed those listed in Table 9.2, within
12 m (40 ft.) of the room that the duct serves.

To minimize airflow noise, the air speed should be decreased at each
duct split from the fan until a diffuser or return air grille is reached. The

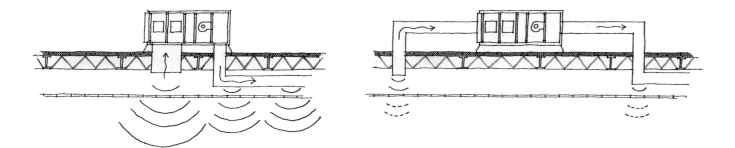

recommended maximum airflow speeds in the final duct approaching the neck of the diffuser or grille are listed in Table 9.3.

Obstructions in the air stream and abrupt transitions can cause air turbulence and result in excessive noise. It is inadvisable to place dampers directly behind the face of a diffuser—locate them a minimum of 3 m (10 ft.) upstream. Large radiused elbows and generously ratioed reduction transitions can minimize air turbulence. Airfoil turning vanes with long

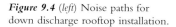

Figure 9.4 (*left*) Noise paths for down discharge rooftop installation.

Figure 9.5 (*right*) Noise paths for side discharge rooftop installation.

Figure 9.6 Internally lined transfer duct above ceiling to control crosstalk.

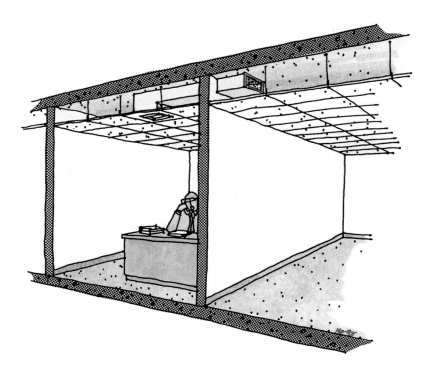

leading edges reduce flow-generated noise at 90 degree mitered elbows. In high velocity ducts, air leakage at the duct joints should be eliminated so that whistling air noise does not occur.

Silencers. Silencers are also called *sound attenuators, mufflers,* or *sound traps.* A silencer installed in ductwork to control fan noise is shown in Figure 9.7. They are usually placed between sections of ducts but can also be located inside an air-handling unit or adjacent to a louver. On the outside, silencers look like ductwork. On the inside, silencers have baffles that run along

Location	NC or RC (N)	Max. speed
Theatre or recording studio	< 25	3.6 m/s (700 fpm)
Conference room or residence	25–30	5.1 m/s (1000 fpm)
Private office	30–35	6.1 m/s (1200 fpm)
Corridor or open plan office	35–45	8.2 m/s (1600 fpm)
Shaft	N/A	10.2 m/s (2000 fpm)
Rooftop	N/A	12.7 m/s (2500 fpm)

Table 9.2 Maximum air speeds in ductwork.

Table 9.3 Recommended air speed in ductwork near diffuser.

Criteria for space	Speed
NC 15 or RC(N) 15	1.3 m/s (250 fpm)
NC 20 or RC(N) 20	1.4 m/s (280 fpm)
NC 25 or RC(N) 25	1.7 m/s (325 fpm)
NC 30 or RC(N) 30	1.9 m/s (380 fpm)
NC 35 or RC(N) 35	2.3 m/s (450 fpm)
NC 40 or RC(N) 40	3.3 m/s (650 fpm)
NC 45 or RC(N) 45	4.0 m/s (800 fpm)

their length. These baffles consist of perforated metal and usually contain fiber fill. Baffles without fill or with encapsulated fill are also available for special clean air environments. In general, the longer the silencer and the thicker the baffles, the greater the noise attenuation.

Silencers that provide the best overall noise mitigation may also induce high static pressure losses, which can reduce the system efficiency. The higher the air velocity through the silencers, the higher the static pressure drop across the silencer. Hence, it is common to see a duct expansion preceding the silencer and a duct contraction after the silencer.

If a silencer is to be located near people, the regenerative noise should be reviewed. Regenerative or self-noise is caused by restricted air flow through the silencer. Regenerative noise is usually not a concern unless the duct velocity is greater than 8 m/s (1,500 fpm), airflow conditions are poor, or the duct serves a quiet room.

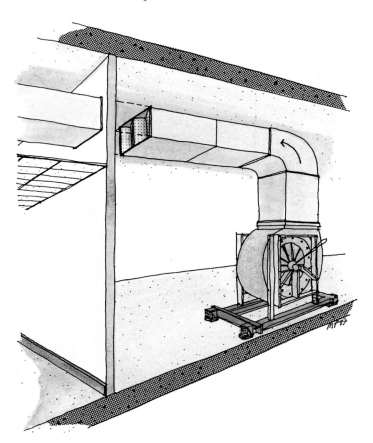

Figure 9.7 Silencer in fan system.

In hospital operating rooms and other sterile environments, duct lining is not permitted. Thus, hospital grade (having encapsulated glass fiber) or packless (having no glass fiber fill) silencers are often utilized.

Acoustical Plenums. An acoustical plenum, shown in Figure 9.8, is often used for noise mitigation in large fan systems with low to medium duct velocities. To be most effective, the interior is surfaced with duct liner board that is from 25 mm to 150 mm (1 to 6 in.) thick. The thicker the liner, the greater the

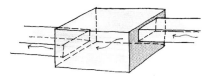

Figure 9.8 Acoustical plenum.

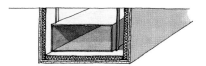

Figure 9.9 Duct lagging using gypsum board enclosure.

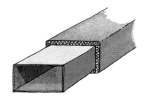

Figure 9.10 Duct lagging using lead wrapped around insulation.

low-frequency attenuation. Attenuation is further increased by offsetting the inlet and the outlet so that there is a break in the line of sight between them, increasing the distance between the inlet and the outlet, and increasing the volume of the plenum.

Duct Lagging. Duct lagging is specified as a part of a design or as a retrofit to solve an existing breakout noise problem. As shown in Figures 9.9 and 9.10, duct lagging may include enclosing the duct in gypsum board or insulation wrapped in sheet lead.

Acoustical Louvers. Acoustical louvers, like duct silencers, have baffles filled with sound-absorbing material as shown in Figure 9.11. In general, the thicker the louver, the greater the noise reduction. Acoustical louvers are 100 to 300 mm (4 to 12 in.) thick and reduce noise by as much as 16 dB. They are typically used in place of standard louvers where noise propagation to the outdoors is an issue.

Acoustical louvers that provide the best overall noise mitigation may also induce high static pressure losses. The higher the face velocity of the air through the louver, the higher the static pressure drop across the louver. The louver area should be specified to meet both the static pressure drop and acoustical requirements.

Variable Speed Drives. Variable drives adjust the fan speed to match the ventilation needs of a room. When the fan slows down, the noise level generally decreases. Mechanical variable speed drives can be noisy while speeds are being changed. Electrical variable speed drives and their cabinets are often noisy. The cabinets should be vibration isolated and never be attached to a partition adjacent to a sensitive space.

It is also important that the fan motor be matched to the variable speed drive to avoid excessive noise and possible damage to the motor. In most systems, electronic filters are required to eliminate those frequencies (speeds) that cause excessive noise.

Active Noise Control. Active noise control is an emerging technology that is still expensive for widespread application. It takes advantage of the physics of superposition of waves by merging unwanted noise with noise that is 180 degrees out of phase (as discussed in Chapter 10).

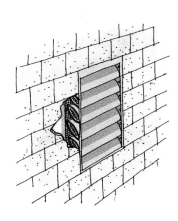

Figure 9.11 Acoustical louver in wall.

Chillers. A chiller is the part of the HVAC system that cools the refrigerant, which in turn cools the air. Most of the noise and vibration is generated by the chiller compressor(s). The tonal noise produced can be intrusive. If chillers are installed directly adjacent to acoustically sensitive spaces, mitigating measures such as floating floors and double-stud partitions will be necessary. Noise mitigation measures for chillers include choosing a quiet chiller type and sound isolating construction (walls, roofs, floor/ceilings). Typically, the noise reducing packages offered by chiller manufacturers are not very effective.

Chillers are listed below in order of increasing noisiness, assuming equivalent performance characteristics: (1) absorption chillers; (2) packaged chillers with centrifugal compressors 280 kW (80 tons) or more[2]; (3) packaged chillers with rotary-screw compressors 90 to 2800 kW (25 to 800 tons); (4) packaged chillers with reciprocating compressors up to 700 kW (200 tons).

Cooling Towers. Cooling towers are used to reject heat from the cooling system. They are used in conjunction with chillers and are typically located outside of the building. Their noise is primarily generated by the fans and is fairly broadband and steady. Since cooling towers are large, the sound emitted from the side where the fans are located may be considerably louder than from the other sides. Therefore, proper orientation of a tower can reduce the noise impact to an adjacent neighbor. It is sometimes necessary to erect a noise barrier between the cooling tower and sensitive areas. Occasionally, silencers or internally lined ductwork at the air inlet and discharge openings are installed.

As always, noise can be reduced by selecting the quietest equipment possible. Types of cooling tower, in order of increasing noisiness assuming equivalent performance characteristics, are: ejector-type, centrifugal-fan, axial-fan, and propeller-fan.

Boilers. The boiler is the heating component of the HVAC system. Much of the noise is generated in the combustion chamber, but with induced- and forced-draft boilers, the fans also generate noise. Common noise mitigation for boilers include choosing a quiet burner type and providing sound isolating construction (walls, roofs, floor/ceilings, barriers).

In order of increasing noisiness, assuming equivalent performance characteristics, boiler types are: natural draft, induced-draft, and forced-draft.

Pumps. Pumps circulate fluids for both the domestic water system and the HVAC system. Pumps generate tonal noise; however, it is unusual for airborne pump noise to be a problem. Common noise reduction measures for pumps include choosing a quiet pump type and sound isolating construction (walls, roofs, floor/ceilings).

Pump vibration transmitted into the building structure is a common problem. Vibration isolation is imperative in most cases.

Motors. Motors are commonly installed in conjunction with driven equipment such as fans and pumps. Motors generate both airborne noise and vibration. Typically, airborne motor noise is not a problem. Structure-borne noise is a more common problem that is controllable with proper vibration isolation.

Air Compressors. Air compressors are often in laboratory buildings or technical shops where pneumatic air is required. They typically have a low-frequency pulsing quality, and their airborne and structure-borne noise can cause problems. Noise mitigation for air compressors include mufflers and sound isolating constructions (walls, roofs, floor/ceilings, enclosures) as well as vibration isolation.

Noise Control Measures for Air Distribution Equipment

Although much smaller than the main building equipment, air distribution components such as fan-powered boxes and supply air diffusers can generate excessive noise. This section highlights air distribution equipment sources and general methods of attenuating their noise.

Fan-Powered Boxes. Fan-powered boxes combine two different types of air. Conditioned air is ducted to a box containing a small fan. Recirculated room air is drawn through an open inlet, mixed with the conditioned air, and then supplied to the space. Figure 9.12 shows a typical fan-powered box.

Fan-powered boxes generate discharge noise, inlet noise, and casing radiated noise. All fan-powered boxes serving occupied spaces should have internally lined duct on the discharge to attenuate the fan noise. The length of lined duct depends on the sensitivity of the space being served, the air volume, and the box size.

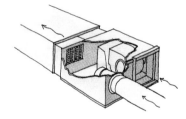

Figure 9.12 Typical fan-powered box.

Most fan-powered boxes have a return inlet open to the ceiling plenum. The open inlet on a fan-powered box is a direct path for fan noise to the room below. In addition, the fan motor is often bolted directly to the casing, causing noise to radiate from the sides of the box. When fan-powered boxes that supply 1700 m³/hr (1000 cfm) are located above a standard acoustical tile ceiling, they typically generate NC 40 in the room below. Therefore, larger boxes should be located over non-sensitive spaces such as corridors. If a large fan-powered box must be located above an office or similar space, steps can be taken to attenuate its noise. The inlet noise can be reduced by attaching an internally lined sheet metal elbow to the box inlet. A secondary casing can also enclose the box as shown in Figure 9.13. This construction takes up room and can make maintenance more difficult. Selecting a quiet box style and locating it appropriately is often the best plan for meeting project noise criteria.

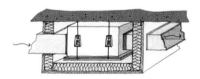

Figure 9.13 Enclosure for fan-powered box.

Variable Air Volume Boxes. Variable air volume (VAV) boxes act as valves that vary the amount of air flowing through a duct. They are essentially an in-line box with a damper inside which is controlled by a thermostat. When less cooling is needed, the damper closes down, reducing the air flow to a room.

Noise is generated by air flowing across the damper and is affected by the pressure drop across the box. If there is high static pressure at the box inlet, the air is forced across the damper at a higher velocity and increased turbulence occurs. Duct-borne noise can usually be attenuated with internal lining in the discharge plenum. Radiated noise can be excessive when a VAV box is located over a space with criteria less than NC 30, when there is no ceiling, or the VAV box is rated at 4500 m³/hr (2500 cfm) or more. It is good design practice to locate VAV boxes over unoccupied spaces like corridors.

Fan Coil Units. Fan-coil units are similar to fan-powered boxes except that the air is conditioned within the unit. Fresh air is supplied through an inlet air duct from a central system or from the outside. A fan draws recirculated room air through an open inlet and mixes it with the fresh air. The combined air is then blown across heating or cooling coils and supplied to the space. A typical floor-mounted fan-coil unit is shown in Figure 9.14.

Figure 9.14 Floor-mounted fan coil unit.

Little can be done to reduce the noise of floor-mounted fan-coil units, except to specify multispeed switches. The airflow and noise level are reduced simultaneously. Discharge noise of a ceiling-mounted unit can be attenuated by using internally lined duct or a silencer. Typically the fans are connected directly to the casing, causing it to radiate noise. For most units, there is a noise reduction option for a metal housing around the fan. Additional quieting can be achieved with an acoustically lined sheet metal elbow or silencer on the inlet and a gypsum board enclosure around the box.

Some fan-coil manufacturers make a quieter unit by vibration isolating the fan, constructing the casing of heavy gauge steel, and internally lining the unit. This quieter unit is larger and weighs more than a standard one, and is often recommended in noise sensitive locations.

Diffusers and Grilles. Supply air diffusers and return air registers or grilles distribute air in a space and are the final components in the HVAC system. They are typically located in the ceiling but can also be in walls, floors, or built into furniture. High-frequency airflow noise can be generated at diffusers and grilles. The noise level is dependent on the speed of the air flowing through the diffuser and its design. Manufacturer catalogs provide NC ratings for diffusers at various air speeds. In general, catalog data is generous in the amount of noise assumed to be absorbed within the room. For small rooms or rooms with multiple diffusers, diffusers should be selected for a catalog rating which is five NC points below the room criterion. For example, if the room has an NC 30 criterion, the diffusers should be selected with a catalog rating of NC 25.

Noise Control Measures for Electrical Equipment
Electrical equipment is associated with providing electrical power to the building. This type of equipment includes emergency generators, uninterrupted power systems, transformers, and switchgear.

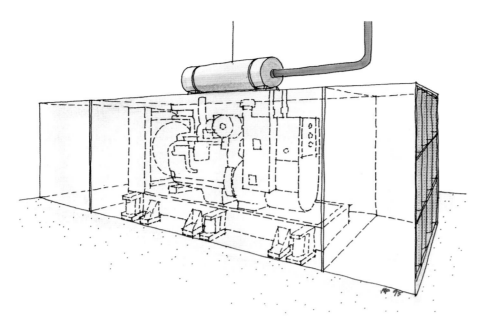

Figure 9.15 A packaged sound attenuation enclosure for emergency engine generator.

Emergency Generators. Emergency generators consist of a diesel engine, a fuel storage system, cooling fans, and exhaust equipment. The generator set changes liquid fuel into electricity when there is a power outage. Emergency generators are probably the noisiest piece of equipment in a building. Fortunately, this equipment usually only operates during periods of testing and emergency. Nevertheless, many cities have ordinances that must be met even under these conditions. One should also consider the short-term impact of the generator's noise and vibration on spaces within the building. Noise sources and transfer paths include cooling air intake, air discharge, exhaust, engine casing, and structure-borne vibration.

Mitigating measures for emergency generators can include sound isolating construction (walls, roof, floor/ceilings, enclosures), exhaust mufflers,

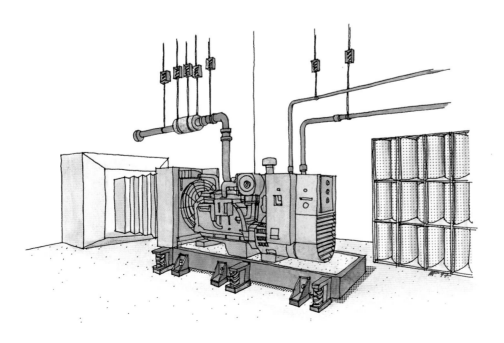

Figure 9.16 Generator inside building.

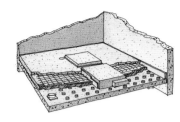

Figure 9.17 Floating concrete floor for a generator.

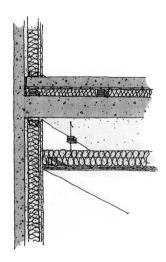

Figure 9.18 Floating floor with isolated ceiling and walls.

Transformer size (kilovolt-amperes, kVA)	Typical noise, level dBA @ 0.3 m (1 ft.)
25–50 kVA	45
51–150 kVA	50
151–300 kVA	55
301–500 kVA	60

Table 9.5 Transformer sound ratings.

and duct silencers. Often emergency generators are located outdoors, either next to a building or on a roof. For these locations, a packaged sound attenuation enclosure can often be purchased (Figure 9.15). This enclosure incorporates silencers on both the air intake and discharge, a muffler on the exhaust, and a panel enclosure. The size and construction of the individual components for the enclosure will vary depending on the amount of sound attenuation required. For very stringent noise criteria, the size of the generator enclosure may be much larger than the generator itself; thus, sufficient space must be allotted.

If the generator is located indoors, the same types of components are required for sound attenuation, including air intake and discharge attenuators and an exhaust muffler. The building itself acts as the enclosure. An indoor installation is shown in Figure 9.16. The sound generated by the emergency generator is enough to vibrate the structure and cause sound to radiate throughout the building.

If it is important that noise criteria be met while the generator is in operation, significant construction may be required, especially if the generator is on an upper floor. In this case, a floating floor as shown in Figure 9.17 would be recommended. The floating floor adds an overall height of approximately 15 cm (6 in.) and additional weight of about 245 kg/m^2 (50 psf). These parameters must be considered early in the design phase of a project so that the building structure can be engineered to support the added weight.

For critical applications, where the generator is located next to a recording studio or teleconferencing room, a fully isolated enclosure may need to be constructed around the generator. This construction could include a floating concrete floor with independent walls resting on the floating floor and an isolated ceiling as shown in Figure 9.18.

Transformers and Uninterruptible Power Supply. The main building transformers reduce the high voltage furnished by the electric utility prior to its distribution within a building. An uninterruptible power supply (UPS) is a D.C. to A.C. converter that operates from large batteries which are continuously recharged by utility power. The UPS generates power free of surges or electrical transients. The UPS also provides temporary backup power in the event of an outage while the emergency generator is being started. UPS is commonly used where reliable computer operation is important.

Transformers and UPSs produce tonal noise. Most transformers are sound rated in accordance with standards prepared by The National Electrical Manufacturers Association (NEMA). These ratings are listed in Table 9.5.

Mitigation for transformers and UPSs include sound isolating construction (walls, roofs, floor/ceilings) and vibration isolation. If transformers are located adjacent to acoustically sensitive spaces, the construction between the spaces must be engineered to attenuate the tonal noise. Also, for large transformer rooms, acoustically absorptive material should be applied to available wall and ceiling surfaces.

II. Vibration Isolation

The rotating or reciprocating motions of internal components in mechanical and electrical equipment produce vibration. If the equipment is not isolated, it can transmit vibration into the building structure. Generally, this equipment produces vibrations of a higher frequency than those which can be felt by humans. The actual physical vibrations produced by this equipment are not generally of concern except in cases where there is sensitive equipment (see Chapter 8). "Structure-borne" noise, caused by vibration, re-radiated by building components, and then perceived as noise, is the issue. Even the smallest fan can produce unwanted structure-borne noise.

Guidelines for Vibration Isolation

Unless there is a concern about physical vibrations, vibration criteria is not established for projects. The criteria for structure-borne noise are the same NC values previously discussed in this chapter.

To reduce the vibration transmitted to the building, equipment is supported by vibration isolators (usually rubber or steel springs). The type of building structure and location of equipment can affect the vibration transmission and effectiveness of the isolators. Equipment on lightweight and long-span floors require a higher degree of isolation since these constructions are less rigid and can be vibrated more easily than massive, short-span constructions. The characteristics and types of isolators are discussed next.

Vibration Isolator Characteristics

When choosing a vibration isolator, its stiffness, static deflection, natural frequency, damping, and transmissibility are considered.

Stiffness and Static Deflection. Every vibration isolator has a stiffness and a static deflection associated with it. The stiffness is a measure of the isolator's resistance to compression. The static deflection is the amount the isolator compresses when the weight of the equipment is applied to it. The static deflection is related to the weight of the equipment and the stiffness of the isolator as follows: $X = MG/K$; where X is the static deflection; M is the weight of the equipment; G is the acceleration of gravity; K is the stiffness of the vibration isolator.

Natural Frequency. Each vibration isolator has a natural frequency. If one pushes down on a weight supported by a spring and then lets go, the rate at which the weight moves up and down around its static position is the natural frequency of the spring.

For critical situations, the natural frequency of the selected isolator should be $\frac{1}{6}$ to $\frac{1}{10}$ of the forcing frequency of the equipment. The forcing frequency is the primary rotational frequency of the vibrating equipment. Also, the static deflection of the isolator should be 6 to 10 times the deflection of the floor when the equipment load is added.

For less critical situations, the natural frequency of the isolator should

Figure 9.20 Elastomeric hanger.

Figure 9.21 Elastomeric mount.

Figure 9.22 Seismic neoprene mount.

Figure 9.23 Neoprene pads.

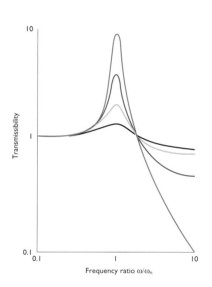

Figure 9.19 Transmissibility for various damping factors.
ω = forcing frequency of the system
ω_n = natural frequency of the isolator.

be $\frac{1}{3}$ to $\frac{1}{6}$ of the equipment's forcing frequency. The static deflection of the isolator should be 3 to 6 times the static deflection of the floor when the equipment load is added.

Damping. Damping is the conversion of mechanical energy to thermal energy. The faster the weight oscillating on the spring comes to rest, the more damping there is in the system. Theoretically, there can be spring systems with no damping. In practice, a vibration isolator always provides some damping. Damping reduces the movement that occurs at resonance, keeping the system more stable.

Transmissibility. Transmissibility is the force transmitted, divided by the exciting force. Figure 9.19 shows the relationship between transmissibility and the ratio of an isolator's natural frequency to the forcing frequency. Resonance occurs when the forcing frequency equals the natural frequency of the isolator. The transmissibility is theoretically infinite for an undamped system.

Vibration Isolator Types
The static deflection for isolators is selected based on the type of equipment, the building structure supporting the equipment, and its proximity to sensitive spaces. Specific isolators and their properties are discussed next.

Elastomeric Isolators. Typically, elastomeric (rubber or neoprene) pads or mounts are used to isolate high frequency vibration. Static deflections typically range from 2.5 to 13 mm (0.1 to 0.5 in.). Elastomeric mounts can be used in compression, shear, or tension. They provide less static deflection but substantially more damping than do spring isolators. Some examples of elastomeric isolators are: (1) Deflected neoprene hangers or mounts used to isolate small fans, fan-powered boxes, small pumps, transformers, and some ductwork and piping (see Figures 9.20 and 9.21); (2) Deflected neoprene mounts with captive steel inserts (Figure 9.22) used when seismic restraint is required; (3) Ribbed or waffled neoprene pads (see Figure 9.23) are often used to support mechanical equipment located on grade such as chillers and boilers with small fans. These pads are also installed beneath spring isolators.

Springs. Springs isolate low-frequency vibration and usually provide up to 100 mm (4 in.) of static deflection. Springs are appropriate for pumps, fans, chillers, compressors, and cooling towers. They provide little damping and tend to transmit high frequencies. For these reasons, they are often combined

Figure 9.24 Spring mount.

Figure 9.25 Seismically rated spring mount.

Figure 9.26 Vertically restrained spring mount.

Figure 9.27 Combination spring and neoprene hanger.

with neoprene pads. Both unhoused and restrained springs are available. With unhoused springs (see Figure 9.24), separate seismic snubbers are required as discussed later.

Although it may seem like a good idea to provide vibration isolation and seismic restraint together in the same device (see Figure 9.25), we have found that the installation is often problematic. It is difficult to size and adjust a group of springs under an unevenly distributed weight so that the springs do not touch their vertical restraining bolts. We almost always recommend unhoused springs with separate seismic snubbers.

Equipment like cooling towers located where wind can create an uplift force are often installed on vertically restrained isolators. This type of spring is shown in Figure 9.26.

A combination spring and neoprene hanger is shown in Figure 9.27. These devices isolate suspended fans, pumps, and sometimes engine exhaust pipes as shown in Figure 9.16.

Spring Curbs. Spring curbs are available for curb-supported rooftop fans as shown in Figure 9.28. These curbs consist of springs between two sets of rails with a watertight and airtight seal all around. Spring curbs are not as effective as individual springs. In acoustically sensitive rooftop locations, we recommend changing from a curb-supported fan to a base-mounted fan on individual springs.

Air Springs. Air springs are used to isolate very low-frequency vibration (down to 2 Hz) and provide 150 mm (6 in.) or more of static deflection (Figure 9.29). Air springs often replace high static deflection springs that do not function adequately. One advantage of air springs is that they do not transmit high frequencies. Two main types of air springs are available: the air bag or bellows type and the rolling lobe. An air compressor provides a constant supply of compressed air to the air springs.

Vibration Isolation Support Equipment

Support equipment such as housekeeping pads are used in conjunction with vibration isolation devices. Support equipment is discussed next.

Housekeeping Pads, Beams, and Stanchions. Housekeeping pads, beams, and stanchions are used to support mechanical equipment. Their main purposes are to provide a stiff, level structure on which the equipment is mounted and to elevate the equipment above the floor or roof. Housekeeping

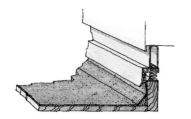

Figure 9.28 Rooftop spring isolation curb.

Figure 9.29 Air spring.

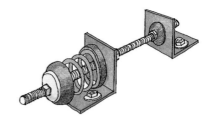

Figure 9.30 Thrust restraint.

pads are slabs of concrete, usually 150 mm (6 in.) thick, on which the equipment and the vibration isolation springs are located. Housekeeping pads provide additional airborne sound isolation between equipment and rooms below due to the increased mass directly beneath the equipment.

Beams are steel or concrete equipment supports. Stanchions are vertical posts that raise mechanical equipment and their support beams a minimum of 460 mm (18 in.) above the roof. This clearance is usually required for maintenance and re-roofing. The distance between the equipment and the roof also increases the airborne sound isolation between the equipment and the rooms below.

Inertia Bases. An inertia base is a block of concrete weighing one to three times as much as the mechanical equipment. Equipment is bolted directly to the inertia base and then the entire load (equipment and inertia base) is supported by vibration isolators. The engine-generator in Figure 9.16 is mounted on an inertia base. Pumps and large fans are often mounted on inertia bases. Inertia bases serve several purposes: (1) They lower the center of gravity, thus reducing the tendency of the load to rock on its springs; (2) They even out the weight distribution of the load making it much easier to size and adjust the supporting vibration isolators; and (3) They limit movement due to lateral forces (thrust) of fans and pumps.

With an inertia base, the total supported weight increases and the movement of the load due to the thrust force is reduced. In addition, a stiffer spring can be used to obtain the desired static deflection. Increasing the stiffness of the spring in the vertical direction also increases stiffness in the lateral direction. The stiffer spring also helps resist the movement due to thrust.

Frames and Rails. Frames or rails serve as equipment bases when an inertia base is not needed and the equipment housing is not rigid enough to support the weight of the equipment (see Figure 9.1).

Thrust Restraints. Thrust restraints are similar to spring hangers except that they are installed horizontally (see Figure 9.30). Their purpose is to limit horizontal movement of fans and pumps. Thrust restraints are often necessary for fans with a static pressure of more than 750 Pa (3 in. water), especially if there is no inertia base. It is important that a pair of thrust restraints be

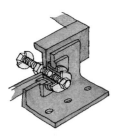

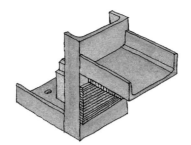

Figure 9.31 Seismic restraint.   *Figure 9.32* Seismic restraint.   *Figure 9.33* Neoprene flexible connector.

positioned symmetrically and are parallel with the axis of airflow. Thrust restraints can be difficult to install properly since there is not always a place to attach them.

Seismic Snubbers. In locations having seismic requirements, equipment must be restrained with devices like snubbers. When equipment is vibration isolated, the seismic snubbers allow the equipment to move. Therefore, seismic snubbers must limit the movement but not interfere with the isolation.

Figure 9.31 and 9.32 show two types of independent seismic snubbers. These devices are both made of steel and are padded with neoprene. On equipment that moves due to thrust, the snubbers should be installed and adjusted when the equipment is operating to be sure that the equipment does not directly contact the snubbers.

Flexible Connections. All duct, pipe, and conduit connections to vibration isolated equipment should be flexible. Flexible connections reduce vibration transmitted into the building, prevent failure of the connection due to fatigue resulting from vibration, and allow for thermal movement.

Flexible duct connections are constructed of rubber or canvas. They are not as effective in reducing vibration when they are rigid and, therefore, should remain slack under all operating conditions. Thrust restraints may be required to prevent fan movement that would otherwise stretch the flexible connection.

There are several types of flexible pipe connections. Rubber hosing is effective in reducing vibration but it cannot withstand extreme temperatures and pressures, and certain types of chemicals. At low pressures, expansion joints are effective in reducing vibration transmission. Expansion joints are made of Teflon or reinforced rubber, and consist of one or two spheres that can expand volumetrically as diagrammed in Figure 9.33. If pressures are high, control rods may be required. Metal hose connectors consist of a corrugated pipe covered with braided metal. They are not particularly effective in reducing vibration transmission and are usually used when temperatures, pressures, and chemicals preclude the use of other types of connectors.

Electrical conduits serving vibration-isolated equipment must be flexible. Corrugated conduit is available for this purpose.

1. Vaneaxial fans are typically quieter than centrifugal fans with forward-curved blades in the low frequency range where noise is the most difficult to mitigate. However, vaneaxial fans are typically noisier in the mid-frequencies. Vaneaxial fans also produce tonal noise that can be annoying.

2. These chillers are usually at their quietest when they are operating at one hundred percent of their rated cooling capacity. They actually become noisier as their operating point approaches zero percent of the rated cooling capacity.

Ductwork and Piping Isolation. Even though flexible duct connections are effective in limiting vibration transmitted from the fan to the ducts, there still is vibration in ducts due to airflow. Ducts are often vibration isolated within MERs and should be suspended from combination spring and neoprene hangers wherever they are adjacent to a noise-sensitive area.

Flexible pipe connectors do nothing to control flow-induced vibrations. Resilient pipe supports can prevent transmission of vibration from the piping system to the building structure. Vibration isolators for piping can be springs or elastomeric mounts. Piping within mechanical rooms and adjacent to noise-sensitive spaces should be vibration isolated.

Conclusion

Mechanical and electrical equipment can generate excessive noise and vibration levels in buildings. The noise and vibration generated by this equipment can cause problems whether the building is a recording studio, single-family home, or warehouse. Noise and vibration can also cause problems for the neighbors. Governmental agencies, private corporations, and building developers have established acoustic criteria for their buildings. In addition, communities have adopted property line noise limits for equipment. Design professionals and building developers should establish criteria for mechanical and electrical equipment and develop means by which to meet the established criteria. The ASHRAE Guide can provide additional information about noise and vibration control for mechanical and electrical equipment.

10

Active Reduction of Noise

Jason R. Duty

Definition and Description • *Building Applications* • *Ducting Systems* • *Pipe Systems* • *Enclosures*

IN everyday life, we all encounter noise that we wish we could eliminate: talking in the next room, traffic noise from outside, the hum of a central ventilation system, and the rumbling of structural vibrations. Elimination of this noise can conceptually be achieved either passively or actively. Various forms of passive noise control, such as retrofitting structures, are described throughout this book. This chapter discusses systems which actively reduce noise. In theory, active noise control systems work by measuring a propagating sound wave and electronically producing a wave of equal magnitude and of opposite phase. This second wave adds to the first, canceling it and eliminating the unwanted noise in the system. So instead of muffling the sound by adding sound-absorbing or sound-isolating materials as done in passive systems, active systems use only sound waves to reduce the noise. This chapter provides a brief overview of active noise-control systems and their applications.

Definition and Description

Active noise–control systems are constantly adapting to changing noise elements. Figure 10.1 shows the basic construction of an active system, consisting of a source sensor, an adaptive filter, a filter output, and a control sensor. Initially, the source sensor measures the input signal, which is fed into the adaptive filter; then the filter produces a filtered signal, which is added to the input signal to create the output signal. If the filtered signal is of equal magnitude and opposite phase as the input signal, the output signal will be zero. Finally, the control sensor measures the output signal and adjusts the adaptive filter so that the power of the output is maintained at zero.

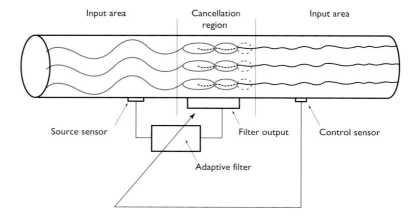

Figure 10.1 Basic construction of an active noise control system.

A fetal EKG or heart monitor is an example of an active noise-control system. With a traditional EKG, a sensor is placed near the heart to detect the patient's heartbeat. This same procedure, however, will not work when trying to detect a fetal heartbeat. The problem arises because the mother's heartbeat can be detected throughout her whole body. As a result, a sensor placed near the fetus would detect a combination of the fetal and mother's heartbeat. Thus, it is necessary to have an EKG capable of removing the mother's heartbeat from the detected signal. And since the mother's heartbeat is a constantly changing signal, the EKG must adapt its filter accordingly. Figure 10.2 illustrates the problem.

The adaptive filter in the fetal EKG is set up so that the source input is placed directly over the mother's heart. The filter will only cancel the signal sent to its source input. As a result, when the filter output is added to the combined signal, only the mother's heartbeat is filtered out, and the fetal heartbeat signal is sent to the display unit. As you can see, an active noise-control system can be designed to remove specific sounds from the overall noise environment. A similar active noise-control system is utilized with teleconferencing (see Chapter 12).

Building Applications

Active noise-control systems can be effective in simple applications, such as HVAC ducts. Multidimensional problems, such as removing ambient noise inside an entire room, however, are much more complicated. A network of adaptive filters is required, each with their own grid of output speakers or vibration oscillators, plus input and control sensors. The filters, in turn, work against each other, trying to not only eliminate the unwanted ambient noise in the room, but also the unwanted cancellation signals from the other filters. An effective way of handling multidimensional noise

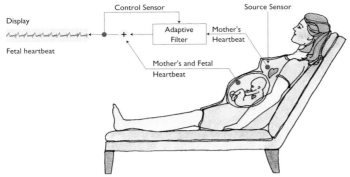

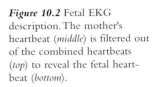

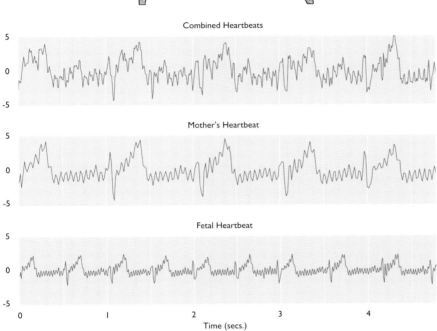

Figure 10.2 Fetal EKG description. The mother's heartbeat (*middle*) is filtered out of the combined heartbeats (*top*) to reveal the fetal heart-beat (*bottom*).

reduction with active systems has not yet been found. A better alternative is to eliminate the noise at the source.

Compared to passive noise-control systems such as in-duct silencers or duct lining, active noise control systems are relatively expensive. In most applications, an active system can cost three to ten times the amount of a comparable passive system. Because of the above mentioned issues, active noise-control systems are rarely used. For example, in the San Francisco Bay Area, in 1997, there are no known buildings where they have been implemented.

Ducting Systems

Active controls for low-frequency reduction in ducting systems are commercially available. These systems do not work well in the presence of excessively turbulent airflow. Active silencers work best when duct velocities are less then 7.6 m/s (1500 fpm) and there is smooth, evenly distributed airflow.

Active controls do have important applications in the reduction of sound from chimney stacks of turbine-driven power plants. Traditionally, these stacks consist of a network of large passive silencers, which create back pressure on the plant's exhaust systems. The back pressure causes the systems to use large amounts of fuel. A system which incorporates an active silencer to reduce low-frequency noise, allows for the reduction in size of the medium- and high-frequency passive silencers which are still required. With this network of silencers, fuel consumption is reduced and the sound radiation pattern can be modified to mitigate fan noise.

Pipe Systems

In a typical pipe system, pressurized fluids can stiffen flexible connections, allowing noise transfer. Active noise-control systems work well in these situations. Water pumps, for example, can generate tones which resonate through the length of the pipe. In power plant water pipes, a 20-40 dB increase in sound pressure has been measured due to this resonance. M. J. Brennan and others designed a non-intrusive fluid wave actuator to apply pressure to the exterior of a large diameter thin-walled pipe.[1] The external pressure alters the modal structure of the pipe that results in up to 40 dB of noise reduction. Figure 10.3 is a chart of the dynamic fluid pressure in decibels verses frequency.

Figure 10.3 Pressure measured using the hydrophone with and without control.

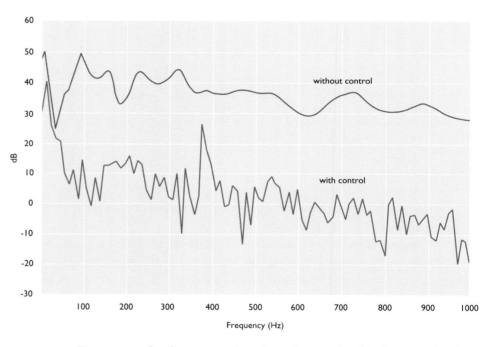

Brennan and others state that there is a noticeable increase in the vibration of the pipe walls at some locations. As a result, reduction of sound pressure at a particular location does not guarantee control at other locations. Thus, the use of more than one actuator is recommended, which leads to an increase in cost. As of now, there are no known installations of active vibration control for pipe systems. However, this technology does have a potential to find wide application in the future.

Enclosures

The interest in using active noise-control systems to reduce the ambient sound levels in enclosures (such as cars, buildings, and aircraft) has increased over the past few years. As previously mentioned, it is difficult to actively remove noise from a large space. When the listener is in a single known location such as an automobile passenger compartment, the potential for the technology increases. This possible development in automobiles could carry over to other enclosures such as aircraft.

Conclusion

The development of effective active noise-control systems is in its infancy. As technology develops and research continues—specifically, faster computers, better algorithms, and working multidimensional systems—active systems will be used more often for specific noise-control applications.

Notes

1. M. J. Brennan, S. J. Elliott and R. J. Pinnington, "Active Control of Fluid Waves in a Pipe." *ACTIVE 95*, pp. 383, 394.

11

Sound Amplification Systems

Thomas J. Corbett
Robert B. Skye
Thomas A. Schindler, P.E.

Basic Elements of Sound Amplification Systems • Design Objectives for Sound Amplification Systems • Application Examples • Equipment Racks • Microphones • Loudspeakers • Digital Systems

PRIOR to electronic sound amplification, performers and lecturers were almost completely dependent upon the success of a particular architectural design to reinforce their music or voices to a level that could be heard by an audience. Electronic systems are now used to reinforce speech, live music, and playback of prerecorded audio. In this chapter, the design objectives necessary for a successful sound amplification system are reviewed, along with considerations for specific applications. The goal of a successful sound amplification system design is to provide high quality audio for as many people as possible in a given location. Before continuing with a discussion of what "high quality" means in this context, the basic functional parts of sound amplification systems are reviewed.

Basic Elements of Sound Amplification Systems

Figure 11.1 shows the four basic elements or blocks of almost any live sound amplification system. The overall process is the conversion of acoustic energy into electrical "signals," and then back again into acoustic energy. Devices that perform this energy conversion are known as *transducers*. The function of each block is detailed as follows:

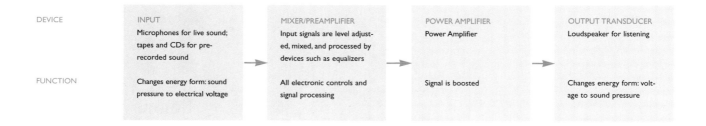

DEVICE	INPUT	MIXER/PREAMPLIFIER	POWER AMPLIFIER	OUTPUT TRANSDUCER
	Microphones for live sound; tapes and CDs for pre-recorded sound	Input signals are level adjusted, mixed, and processed by devices such as equalizers	Power Amplifier	Loudspeaker for listening
FUNCTION	Changes energy form: sound pressure to electrical voltage	All electronic controls and signal processing	Signal is boosted	Changes energy form: voltage to sound pressure

Figure 11.1 The four basic parts of a sound amplification system. The arrows show the signal flow from acoustic energy to an electrical signal and then back into sound waves.

The first section of the sound amplification system involves one or more signal sources or inputs. A signal source can be a transducer such as a microphone, or a playback device such as a compact disc or audio cassette player. Transducers generate the signal source by converting an acoustical signal to an electrical signal. Microphones change air movements into a corresponding electrical pattern. Playback equipment converts a sound that is stored on optical (compact disc) or magnetic (audio cassette) media to an electrical signal.

The second section of the sound amplification system is the mixer/preamplifier. Three basic functions occur at this control section: (1) various input sounds are level-adjusted for loudness and signal compatibility; (2) these same sounds are mixed or combined and assigned to amplification and output sections; and (3) the sound signals are processed for various effects. Signal processors are devices that perform specific modifications on signals before or after they are mixed together. Signal processing includes sound equalization for control of timbre, addition of artificial reverberation, or other special effects.

The third and fourth sections function together to take the level-adjusted, combined, and processed signals from the mixer/preamplifier section and convert the result back into acoustic energy. The power amplifier takes the low-voltage signal levels from the mixer output and produces a signal with voltage and current level sufficient to power a loudspeaker.

Loudspeakers are transducer devices that convert voltage back to acoustic energy. You may have seen a loudspeaker cone move back and forth or vibrate at low frequencies. This is a visible example of how a loudspeaker pushes air molecules to create sound waves similar to those that were received at the input transducer. Additional information about loudspeakers is provided later in this chapter.

Design Objectives for Sound Amplification Systems

Sound amplification systems and room acoustics design should both be considered at the onset of a new project. A sound amplification system added to an existing space, however, must consider the existing acoustics. The criteria for success depends on the context, but usually the requirement is that music and speech be optimized for sound quality and intelligibility, and be evenly distributed throughout the audience space.

It is possible to develop a successful design by examining the following as criteria for quality: (1) uniform sound coverage; (2) intelligibility and clarity; (3) high fidelity and frequency bandwidth; (4) dynamic range and peak sound levels; (5) operator usability; and (6) control of feedback.

Uniform sound coverage refers to the relative sound level at each listening position. A well-designed sound amplification system is capable of generating a sound field that varies no more than ± 3 dB throughout the room. Different criteria apply for the level in specialized circumstances. For instance, an evacuation command given over a public address system in an airport lobby or at a sporting event would need to have a relatively high level, since it is imperative that the announcement be heard over higher background noise. Public address announcements should be a minimum of 15 dB above the crowd noise.

Intelligibility refers to the ability of a listener to understand speech, independent of the loudness level. In a given ambient, the amplified level and intelligibility are related. This is because a low amplified level will be "masked" by ambient noise. Masking refers to the ability of one sound to overwhelm another. Intelligibility can also be diminished by reverberation. When reflections are loud enough or late enough in time, they have the effect of masking the direct sound. Industry practice is to examine the ratio of direct-to-reverberant sound at each listener position, and then to design the room acoustics and the sound amplification system to maximize the amount of direct sound in order to optimize intelligibility.

High fidelity means "very faithful," in the sense that the sound at the loudspeaker output is a "faithful" reproduction of the sound at the input. High fidelity has become synonymous with "high quality." Note that in an era of digital audio and surround-sound systems, the average person's awareness of what is possible must be taken into account as the baseline for fidelity. In the past, a sound amplification system that was loud and intelligible was deemed adequate. Now, however, one with acceptable fidelity must also sound "natural." This is particularly important for music, but holds for speech as well. Fidelity encompasses frequency bandwidth, power handling (maximum loudness before clipping or other distortion), and electronic noise in the absence of signal.

One way to achieve high fidelity is to design the sound amplification system so that the frequency response is matched between the input and output blocks. The sound amplification should be as linear as possible, and not significantly change the spectrum of the sound source. High fidelity is also measured in terms of what isn't present in a sound. The sound amplification systems should not distort, "hum," or rattle. The lack of these disturbances is another measure of the fidelity of a sound amplification system.

Usability refers to the human-factors aspect of the design of a system. The system should be easy for the end user to successfully operate. Simplicity is a virtue in the design of the "human interface." In many cases, controls and devices should be easily understood by both professionals and non-professionals. Like any type of design, the human interface for a sound system can even be intuitive and elegant as a result of combined simplicity and functionality.

Feedback is the "howling" sound sometimes heard from a loudspeaker. It occurs when a microphone picks up sound from a loudspeaker in the same room and amplifies it. "Gain-before-feedback" is the technical measure of the sound level available for productive use within a room before feedback. Reverberation lowers the gain-before-feedback.

Application Examples

It is possible to classify types of sound amplification systems according to their application and function. Here are six examples of common sound amplification applications: (1) speech reinforcement; (2) live music reinforcement; (3) pre-recorded media playback; (4) recording and broadcast studios; (5) teleconferencing systems; and (6) voice-lift.

Speech reinforcement is a part of almost all sound system designs, but is of particular importance in lecture halls, courtrooms, conference rooms, and contexts where a person giving a presentation is the center of attention. For adequate intelligibility and with minimal listener fatigue, a good speech reinforcement system design requires that the amplification system provide a sound pressure level at least 15 dB above the ambient noise level.

Speech reinforcement systems require microphones and loudspeakers to be integrated within a design that is optimized for a particular environment. A classroom, for instance, will have very different design requirements than a sports arena.

The following is a guideline for determining when a speech reinforcement system may be needed. Assuming that a person is using a normal voice level in a quiet room, a listener up to 12 m (40 ft.) away should be able to hear without amplification. At distances between 12 to 18 m (40 to 60 ft.), the need for reinforcement is a function of the loudness of the person speaking, room shape, background noise, reverberation time, and the arrangement of seats—whether on a slope (as with the Greek theater discussed in Chapter 1) or on a level surface. Beyond 18 m (60 ft.), a sound amplification system is typically required.

The main distinction between a speech reinforcement system and a live music amplification system is that the former is optimized around speech frequencies. A music system must be capable of reproducing vocals and musical instruments over a wider frequency response and dynamic range. Dynamic range refers to the power capability between the maximum and minimum sound pressure level that a sound source or system is capable of producing. Sounds that exceed the dynamic range are said to be *clipped,* while sounds below the dynamic range are said to be in the *noise floor* of the system.

A distinction should be made between "live electronic" and "live acoustic" music. Live electronic music refers to popular music of various styles, such as rock, jazz, and pop, where the overall sound is associated by the audience as being amplified. In this case, reinforcement of vocals (and in large halls, of all the instruments) is expected by the audience, because the amplification hardware is seen and heard. Amplification of live acoustic music refers to contexts where the audience does not expect amplification—for instance, classical music, opera, plays, and some theater productions. Live acoustic music amplification requires that the microphones, loudspeakers, and other hardware be as unobtrusive as possible. Many people are amazed to discover that their enjoyment of a classical music concert or opera was actually augmented by hidden loudspeakers, designed to subtly increase level and to provide uniformity and clarity.

Another type of sound amplification system is one that is used for pre-recorded media playback from hardware such as CD players, video players, or computers. Generally, these systems are similar to live music reinforcement systems in terms of bandwidth, but microphones are not a part of the system. In addition to audio/visual presentations, low-level playback systems are often installed for background music in restaurants, shopping malls, zoos, and hotel lobbies. Such systems are even installed in homes, to distribute music throughout the environment and provide ambiance. At some zoos, sound is played that is both comforting to the animals and suggestive of wild environments to humans. A concealed distributed loudspeaker system is usually appropriate in these contexts. Background systems can also be used for paging.

Sound systems for live broadcasting and recording typically require a mixer operator to combine the audio. The mixed audio output is sent to a transmitter for broadcast and/or to a recorder for archiving. Studios for recording and broadcasting often involve at least two rooms: one for the performer or talker, and another for the recording engineer. These two rooms must be acoustically isolated from each other, and therefore require additional sound amplification systems. One such system is called a *monitoring system*. This allows the recording engineer to listen to or monitor the sound being recorded using loudspeakers or headphones and then to also selectively play the tape back to the performer.

Another type of sound amplification system is the *talk-back system*. The talk-back system provides the recording engineer or control room operator with the ability to give commands directly to the performers. The talk-back system is reserved for "live" communications between the two rooms and is used to direct the production's technical staff. Intercoms are a type of talk-back system used in theaters and production studios by the stage manager to communicate to various locations such as the dressing rooms, manager's office, and box office.

Teleconferencing systems involve participants at two or more remote locations engaged in a conversation typically linked by telephone wire. The goal is to allow the participants to communicate easily and naturally as if they were all in the same room. One challenge for teleconferencing sound systems is to maintain usability by non-experts. The end users typically want a "transparent technology" that involves minimal effort on their part. It is a challenge to provide a design that simultaneously offers sufficient coverage, intelligibility, and simplicity for each conference site. (A further discussion of teleconferencing systems is presented in Chapter 12.)

One type of speech reinforcement system applicable for conferences or other situations with larger rooms and multiple participants is known as a *voice-lift* or *mix-minus* system. Multiple speakers and microphones are distributed around the room to provide equal coverage to groups of participants. The system is engineered so that microphones do not feed sound to nearby loudspeakers, thus eliminating a potential feedback loop between microphones and loudspeakers. With the loudspeakers arranged into several zones, anyone can clearly hear the presenter in even the most unusual architectural setting. This system is appropriate for courtrooms, council chambers, and distance-learning classrooms. The combined output of all microphones can be broadcast or recorded.

Equipment Racks

The placement of hardware elements such as loudspeakers and microphones must be designed into the acoustical and architectural layout of a space. Hardware elements, especially large equipment racks and loudspeakers, should be placed as unobtrusively as possible.

For aesthetic and security purposes, it may be necessary to locate an equipment rack in a closet. On the other hand, usability may dictate that the equipment be easily accessible to control playback level. Figure 11.2 shows a typical equipment rack, containing mixers, amplifiers, and other user-operated devices. An additional design consideration is ventilation for the equipment; if fans are used, they must not contribute to the background noise in the meeting space. Equipment rack placement and serviceability can be a challenge that frequently involves the architect, acoustical engineer, sound system designer, and mechanical engineer.

Microphones

All microphones have directivity that is measured and displayed as a "polar pattern." Chapter 2 has additional information on directivity. It is important to select the microphone whose characteristics meet the application requirements, which are dependent upon directional sensitivity and audio coloration.

An omnidirectional microphone is equally sensitive 360 degrees around its element. This microphone is not used in reverberant spaces for speech, because it would be too sensitive to noise and feedback. However, the omnidirectional microphone creates a very natural sounding signal that is very forgiving of the speaker's head movements and mouth direction.

Cardioid and hyper-cardioid microphones are directionally sensitive and provide improved gain-before-feedback in a reverberant room. Cardioids provide excellent low-frequency response and can enhance the speech characteristics of a presenter close to the microphone. Figure 11.3 compares the polar patterns for omnidirectional, cardioid, and hyper-cardioid microphones. Refer to the manufacturer's literature for specific sensitivity patterns.

For speech reinforcement applications, all that may be required is a microphone mounted on a floor stand. This might be appropriate for some situations, but often a presenter requires a lectern to hold notes. A rigid or flexible "gooseneck" microphone can be mounted on a podium. In many

Figure 11.2 An equipment rack containing amplifiers, mixers, and other sound devices installed in an equipment room.

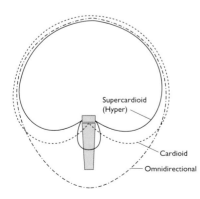

Figure 11.3 Polar patterns for three microphones.

cases, especially where a person wants to move away from the podium, a wireless lavaliere microphone is used. This consists of a small microphone and transmitter that is attached to the clothing. Care must be taken with any wireless technology to minimize interference from radio and television frequencies (RF interference). In particular, the wireless microphone frequency must be selected to prevent "stray" signals (such as an emergency vehicle broadcast) from being heard over the PA system simultaneously with the speech. When properly installed, however, these systems offer great mobility for lecturers.

Wireless hand-held microphones allow freedom of movement, including passing the microphone among people during a question-and-answer session. On the other hand, wireless microphones are more expensive, require batteries, and are generally less reliable. Wired microphones do not usually suffer from RF interference, but their cables can be awkward: they tether a user to a location determined by their length. Loose cable on the floor can also be a tripping hazard.

Whether or not a microphone is wireless, any hand-held microphone can cause problems for inexperienced users. If someone talks into the side of the microphone (outside the directional pattern) and then into the top (within the directional pattern) the level can change radically. In addition, hand-held microphones must be kept at a relatively constant distance from a person's mouth so that undesirable level changes do not occur. Figure 11.4 shows a hand-held microphone, along with its recommended usage.

Figure 11.4 A hand-held microphone, with a (fictional) directional pattern of the microphone indicated. In reality, these directional patterns are more complex, and the response changes as a function of frequency.

One factor that is particularly important when designing sound amplification systems is to consider the level of the spoken voice in the particular application. A trained speaker will project his voice at a louder level, while a reluctant witness in a courtroom may whisper. In both cases, the sound amplification system should be designed to deliver an acceptable level to listeners.

An important consideration for wired microphones is the location of signal input receptacles. Receptacles mounted on plates can be inconspicuously placed in the lower part of a wall or in the floor. In some installations, receptacles are covered when not in use. In an A/V equipped room, receptacle plates are located to provide an adequate number of access points for various types of user applications within the room. Receptacle plates are part of the early phases of a room design because of the need to locate conduit, route signals, and coordinate with the location of power receptacles.

Loudspeakers

Loudspeakers are typically arranged to provide the best coverage for the room occupants. This coverage occurs when the loudspeaker is within the visual sight-line of the listener. Among the different types of loudspeaker systems are: (1) central loudspeaker cluster; (2) distributed loudspeakers; and (3) special purpose systems such as for film sound and theater.

A central loudspeaker cluster groups all of the required loudspeakers at one location in the room, as shown in Figure 11.5. This type of system

provides the best match between the visual location of a performance and the audio spatial image. It also provides the most uniform timbre and loudness for the audience in different seating locations. For these reasons, the central cluster is usually the first choice for a loudspeaker layout design, particularly for speech reinforcement.

A central loudspeaker cluster must be elevated enough to achieve an adequate "throw ratio." The throw ratio is the ratio of the distance from the loudspeaker to the seating locations in the last row, relative to its distance from the first row. The throw ratio should be less than 2.0. In a large room with a low ceiling, where the throw ratio is more than 2.0, the central cluster would provide poor uniformity. The sound would be too loud in the front and too quiet in the back.

In rooms with low ceilings, the distributed system works well because it uses many loudspeakers distributed throughout the room. Each loudspeaker puts out a lower sound level than that of a central loudspeaker cluster. Uniform coverage is achieved by appropriate loudspeaker placement and power output. One way of mounting the loudspeakers is flush into the ceiling, as shown in Figure 11.6. A diagram of the coverage achieved by this type of system is shown in Figure 11.7.

Distributed loudspeakers can also be located in desks or on the backs of chairs. Loudspeakers can also be installed onto the backs of pews in a place of worship, as shown in Figure 11.8. Distributed loudspeaker systems are often used when the acoustical environment is very reverberant or where there are no mounting locations for a cluster. The short distance between each loudspeaker and the listeners allows operation at a relatively

Figure 11.6 A distributed loud-speaker system (compare to Figure 11.5). The speaker coverage is directed downward toward the audience, and overlaps slightly.

low sound level. Distributed systems offer very high speech intelligibility and work well in boardrooms, courtrooms, and council chambers.

Specialized loudspeaker systems can include multichannel film sound playback in "surround" mode, playback of stereophonic material, and countless other types of applications. Some of these systems are reviewed in Chapter 12.

Digital Systems

As computers have found applications in every aspect of our lives, they have also become important in sound systems. In the past, computers were limited to control and monitoring functions. Today, mixing and signal processing functions of sound amplification systems can be performed on a computer platform. The systems allow all signal-processing functions such as equalization, mixing, and level adjustments to be done in the digital domain. Virtual devices can be created and wired together on a computer screen and controlled using a computer mouse. The benefits of these *digital signal processing* (DSP) based systems include: (1) easy reconfigurability—device functions are merely redrawn on screen or deleted as required; (2) reduced hardware and space requirements; (3) better signal-to-noise ratio; and (4) high fidelity.

These "all-in-one" solutions have not completely replaced the need for outboard audio hardware devices. Many of the functions required by sound system designers are not yet available in a single DSP device. Also, as with any new technology, there is resistance to change by people concerned about reliability and sound quality. Eventually, DSP-based single platform systems are expected to replace many of the separate hardware devices in use today.

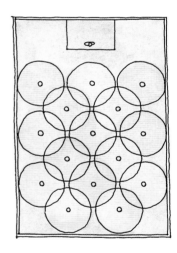

Figure 11.7 Coverage of distributed loudspeaker system shown in Figure 11.6.

Figure 11.8 A pew-back system, utilizing distributed loudspeakers in the backs of the seats.

Conclusion

During the programming phase of a project when functional requirements are defined, it is important for all of the participants to realize that there is no single "right" sound system. A careful usage analysis can best guide the choices for applying design criteria including uniformity, intelligibility, high fidelity, and usability. Because there is an increasing demand for high quality audio in multimedia presentations for education, business, and other applications, an early integration of sound design considerations into a project will help achieve successful results.

12

Audio Visual (A/V)

Thomas J. Corbett
Kenneth W. Graven, P.E.
David R. Schwind, FAES

The Relationship Between the Presenter, the Audience, and the Image (Sight Lines) • *Functional Requirements of the Space Being Designed* • *The Presenter's Operation of Equipment* • *Types of Presentation Equipment* • *Front and Rear Projection Screens* • *Comparative Imaging* • *Audience Viewing Area* • *Lighting* • *Heating, Ventilating, and Air-Conditioning Systems Noise* • *Receptacles* • *Loudspeakers* • *Planning for the Future* • *Appendix: Integrating Audio with Moving Pictures*

PRESENTATIONS take a variety of forms—actors, musicians, attorneys, marketers, and educators all have a message to deliver. The more an audience is involved in a presentation both visually and auditorily, the more effective is the presentation. This chapter describes design issues involving audio/visual technology for conference rooms, council chambers, courtrooms, boardrooms, video conference rooms, distance-learning classrooms, and entertainment video facilities (home theater). The Appendix contains information for those readers who are interested in learning more about how audio is integrated with moving pictures.

I. Design Issues

When designing a space for audio/visual presentations, the factors to consider are: (1) the relationship between the presenter, the audience, and the image (sight lines); (2) the functional requirements of the space being designed; (3) the presenter's level of involvement in operating the presentation equipment; (4) presentation formats: slides, video, computer graph-

Figure 12.1 End stage configuration.

Figure 12.2 Thrust stage configuration.

ics, projectors, and control systems; and (5) front and rear projection screens. These factors are discussed in the following sections.

Sight Lines

Everyone in the audience should be provided with a clear view of the information. Different theater styles suggest ways that presenters and the audience can be arranged. Some common theater configurations are:

(1) An end stage, where the presenter faces the audience, typically standing to one side behind a podium, frequently on an elevated stage. The projection screen is usually centered as shown in Figure 12.1.

(2) A thrust stage, where the presenter is partially surrounded by the audience who see not only the presenter but other audience members. Figure 12.2 shows how this arrangement provides for greater intimacy. Some of the audience on the sides have a limited view of the central screen. Multiple screens can be used for better visibility.

(3) Corner stage, where the presenter has an audience on two sides as diagrammed in Figure 12.3. One wide projection screen can be adequate, but two screens may be better.

(4) Full arena, where the audience surrounds the presenter as shown in Figure 12.4. Multiple screens are necessary for everyone to see the presentation.

Functional Requirements of the Space Being Designed

Audio/video presentations occur in a variety of venues; the following is a discussion of five room types:

Conference and board rooms. Conference and board rooms are typically designed to promote discussion with conferees facing one another across a table. A screen is generally located on one end of the room as shown in Figure 12.5.

Figure 12.3 Corner stage configuration.   *Figure 12.4* Arena configuration.

Teleconferencing is a common function in these rooms. (Case studies of conference rooms and board rooms are presented in Chapter 17.)

Large audiences: auditoriums and courtrooms. Large audience rooms such as an auditorium or a courtroom typically have fixed seating. In auditoriums, a projection screen is at the front of the room and the audience is arranged for good viewing of the projected image. In courtrooms, the projection screen is located where it can be best viewed by the jury. A sound system provides playback audio to the audience.

The auditorium used for video or film display is arranged with the audience seated near the centerline of the room. If a podium is used on a stage at the front of the room, then the stage must be large enough for the podium to be located to the left or right of the screen. If the stage is used without the screen, then the podium may be located at a center stage position against a backdrop of either drapes or a wall.

Speech reinforcement is typically required in courtrooms. Increasingly, various types of audio/visual media are being used for presentation of evidence. Large screen displays are required for viewing by the judge, jury, and spectators. Video monitors are sometimes brought into the courtroom; however, installed screens help preserve the formality of the space. The lower courts, in particular, use video along with the installed audio systems. Higher courts rely less on video and projection and depend more on speech reinforcement. (A courtroom case study is presented in Chapter 17.)

Teleconference room. Teleconferencing systems can be just audio, or both audio and video. The most basic form of audio conferencing could involve speakerphones. Audio conferencing presently transmits over standard telephone lines. Video conferencing uses analog or digital telephone lines to provide an electronic audio/video signal between groups at each conference site. Cameras and video monitors are required at each location. A high

Figure 12.5 Conference or board-room configuration.

speed video telephone service such as ISDN is recommended to connect Codecs (Compression/Decompression device) for high speed video transfer over the telephone lines.

When a room is designed for video conferencing, it typically has one of two floor plans for effective front camera imaging. One plan has a V-shaped table opened toward the camera to allow more conferees to be viewable and to also allow attendees to see a central monitor. The lens for this video conferencing unit must have a wide enough field of view to capture all of the room occupants; however, when the video signal is transmitted, those people at the far end of the table will appear smaller.

A second video-conferencing floor plan features a smaller curved table where all participants are equidistant from the camera. This configuration avoids the problem of participants located in the center appearing smaller in the broadcast image. This room configuration is appropriate for a small group but is not effective for a large group unless additional rows of seating are placed behind the table. For these additional people to participate effectively, they must be elevated on a platform to be seen by the camera and to see the monitor.

For both types of floor plans, microphones should be located along the length of the table. Optimally there should be one microphone per person. As a minimum, there should be one microphone for every two people at the table.

Ceiling-mounted microphones for teleconferencing are less desirable because they pick up room reverberation and noise that reduce intelligibility at the distant receiver. Wireless microphones are seldom used because signals are not confidential and are susceptible to interference.

Distance-learning classrooms. Distance learning is a hybrid between a presentation environment and a video-conferencing environment. An instructor at the front of the room presents material to, and interacts with, both local and distant participants. Both local and distant participants can ask questions.

Video monitors are typically located in the front and toward the rear of the instruction room. The video monitors at the front of the room display instructional material. The video monitor(s) towards the rear of the room allow the instructor to see the distant students while maintaining eye contact with local participants.

In its simplest form, the instructor must be capable of controlling the presentation equipment along with monitoring the transmitted image. Alternatively, some distance-learning environments provide a technician to control the cameras and media presentation.

The distance-learning setting is similar to the video-conferencing environment in that the instructor must stay on camera. This fixed position usually means that the instructor is seated at a desk in the front of the room. Some instructors, new to teaching in a distant-learning environment, object to being seated at all times. If this immobility is a problem, then a technician can operate the camera while the instructor moves about the room.

Automated camera systems are not recommended for on-camera tracking. Although, they follow the movements of the presenter, they do not offer control of features like zooming and framing.

Home theater. Home theater attempts to bring the cinema experience into residences. This technology is spreading beyond the residence into numerous business and entertainment facilities. There are significant challenges in bringing the full range of motion picture presentation into the home. These challenges include picture size, resolution, sound quality, and the sound format. The appendix to this chapter presents information about loudspeaker placement and viewing so that audio is effectively integrated with the picture.

The terms *5.1 channel sound* and *surround sound* describe audio playback formats. Most music recordings are presented using a stereo speaker system, while home theater is comprised of six channels. Two of the channels are the stereo right and left loudspeakers; an additional center channel is placed above, below, or behind the visual image. Two surround sound channels are located to the sides and rear of the listener(s) and provide ambiance and sound effects. This constitutes five of the six channels referred to in a 5.1 channel sound system (the "5" in 5.1); the .1 refers to the limited bandwidth provided by an additional subwoofer channel located in the front of the room. The 5.1 format is derived from the 70 mm six-track magnetic sound format used for motion pictures. Most commercially viable digital sound formats for motion pictures have adopted this six-channel format. The digital versatile disk (DVD), the next form of the laser disk, has 5.1 channel sound.

Surround sound is another type of theater audio format. Dolby Laboratories incorporated the surround sound format using four channels of sound, three behind the screen (right, center, and left) and a single channel for the surround field. However, since the optical sound track on film only had stereo capability, it was necessary to encode the center channel and surround channel with a matrix similar to those used in the early

1970s for quadraphonic sound. Stereo VHS tapes are frequently encoded using this four-channel sound format. Although the array of loudspeakers is similar, the advantage of the 5.1 channel sound over surround sound is that the channels in 5.1 are independent and discrete.

The Presenter's Operation of Equipment

Audio/visual presentation equipment can either be operated by a presenter or by an attendant. Discussions and considerations of both scenarios follow.

Operation of equipment by a presenter. With the emergence of integrated remote control systems, a wide variety of equipment can be easily controlled by a presenter from a system control panel. Only equipment that requires media insertion such as VCRs or document cameras need to be close at hand. A VCR, for example, might be located at the lectern. Increasingly, a presenter's work station is an island of electronic playback devices as shown in Figure 12.6. Some lecterns have a side-mounted graphics table for presenting flat artwork and objects. This graphics table is designed to operate with a camera mounted on it or above it in the ceiling.

Motorized lecterns can be adjusted to accommodate a standing or sitting individual. However, a motorized lectern does not have any space for playback devices since the pedestal of the lectern contains the elevation mechanism.

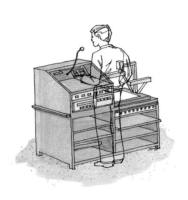

Figure 12.6 A/V lectern.

Operation of the equipment by an attendant. In this scenario, a projection room contains most of the playback devices. An operator loads playback media and operates the equipment according to the needs of the presenter. This relationship allows the presenter to better focus on the material being presented. A remote control system is often installed so that if there is no operator, the presenter can pre-load the materials and then operate from the presentation location.

The layout of equipment racks is important for efficient operator function. Figure 11.2 shows a full-height equipment rack, which could be recessed into a wall near the user. A typical full-height equipment rack has about 2 m (77 in.) of usable space. Operable devices such as VCRs are mounted between standing and seated eye heights or about $3/4$ to $1 2/3$ m (30 to 65 in.). The upper and lower rack areas are used for passive electronic devices, such as power amplifiers, which do not require operator adjustment.

Types of Presentation Equipment

The seven most common electronic media devices involved in an audio/visual presentation are: (1) videotape playback; (2) document camera stand; (3) slide-to-video converter; (4) video projector; (5) personal computer; (6) video monitor; and (7) control system. A description of these components follows:

Videotape playback. Until the 1980s, 16 mm film projectors and 35 mm slide projectors were commonly used for visual presentation. Film projectors have mostly been replaced by videotape cassette players due to ease of handling, fewer mechanical problems, and quieter operation. Videotape formats commonly used today include VHS/SVHS, 8 mm, and Hi-8. Each format provides performance trade-offs including price, quality, availability, and compatibility.

Document camera stand. The stand shown on the right side of the lectern in Figure 12.6 is capable of displaying flat artwork, transparencies, and three-dimensional objects. The size of flat artwork is typically limited to the size of a standard piece of paper. With marker and paper, the camera stand replaces a blackboard, using a projected video image. In many respects, the camera stand is better than a blackboard because the instructor can continue to face the students. A blackboard must also be erased to make room for new work, whereas with the camera stand, new paper can be used to renew or enlarge the work surface.

Slide-to-video converter. Slides are still used for presentations due to their high resolution image and brightness. For certain applications where it is not possible to project slides optically, such as video conferencing, a slide-to-video converter is used. The carousel style gravity-drop tray is placed on top of the slide-to-video converter. The slide-to-video converter is usually located immediately adjacent to the presenter rather than at the rear of the room. Because the image is scanned by a television camera, the resolution is reduced to that of video.

Video projectors. Figure 12.7 shows three possible video projector locations for a presentation space. Location #1 is for use with a front projection screen. Normally the projector would be located 1.3 to 1.5 projection screen widths away from the screen. The projector can be on the floor or suspended from the ceiling. This type of installation is relatively straightforward but has a disadvantage of placing the projector in the room along with its heat and noise.

Location #2 is for rear projection. While the projector should be the same distance from the screen as in front projection, it is possible to alter the apparent throw distance using mirrors to reflect the projection beam. The mirrors save floor area in the rear projection room. The disadvantage of mirrors is that they can add aberrations, may collect dust if not cleaned regularly, and can degrade the image. A mirror-folded optical path is also more difficult to focus.

Location #3 is in a separate projection booth. This configuration uses a more expensive light valve projector, which produces a brighter image on the screen. Light valve projectors are typically larger, generate more noise, cost more, and weigh more than the CRT type.

In all three locations, it is necessary to locate the video projector along the centerline of the projected image for correct horizontal placement. Vertical keystoning (distortion of the image due to the projector lens not being on axis with the center of the screen) can be somewhat cor-

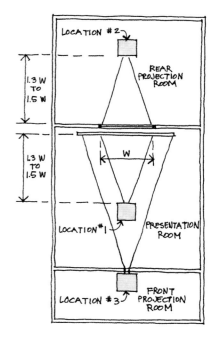

Figure 12.7 Three typical video projector locations.

rected electronically by adjusting the video projectors. Table 12.1 summarizes the comparative features of the four most common video projectors.

Table 12.1 Comparative summary of projector features.

LCD = Liquid Crystal Display
CRT = Cathode Ray Tube
DLP = Digital Light Processing
NTSC = National Television Systems Committee (15.75 kHz)

(*Denotes the number of on-screen pixels for both horizontal and vertical axes. **Denotes the horizontal scan line rate capabilities. A higher number means greater resolution.)

	Projectors			
	LCD	CRT	Light valve	DLP
Typical display resolution	NTSC video to * 1024 x 768 ** (48 kHz)	NTSC video to 2000 x 1600 (130 kHz)	NTSC video to 1280 x 1024 (63 kHz)	NTSC video to 1024 x 768 (63 kHz)
Brightness (ANSI lumens)	350–5,000	200–275	2,500–12,000	1,200–5,000
Light source	Metal halide lamp	CRT	Xenon	Metal halide lamp Xenon lamp
Comments	Pixelated images Available at low cost	Flexible aspect ratios Low light output No pixelization High color depth	Very bright High maintenance	Newer technology High cost Low pixelization High color depth

Personal computers. Computers are increasingly used for audio/visual presentations. Computer programs allow a presenter to take prepared slide formats, add titles and standard images, and create an integrated presentation. A presenter can easily transport the presentation using a lap-top computer and an LCD (Liquid Crystal Display) overhead-projection panel. The two devices can fit into an briefcase. The overhead projector can also be carried since fold-down mast overhead projectors are relatively small. The software-generated presentation is quickly replacing other media due to its flexibility and quality. Many presenters prefer to plug their computers into an existing video projection system.

Video monitors. A video monitor is the most common visual display device. Televisions and computer monitors use a CRT, where the picture is drawn by an electron beam onto a phosphor-coated glass tube.

For an audio/visual application, one must often choose between a video monitor or the video projector. A direct-view monitor offers a sharper, brighter image and is a self-contained unit not requiring a separate projection screen. The primary disadvantage of direct-view monitors is that they have a limited viewable image size and can be heavy and cumbersome.

With the emergence of new lightweight monitors such as plasma displays, the size disadvantages of direct-view monitors using CRT technology will be less relevant. Both plasma displays and LCD panels are finding their way into audio/visual applications because their thin profile allows the panels to be placed inside lecterns, daises, or directly on a wall, where a rear projection room or ceiling-hung video projector is not possible. As these technologies evolve, they are expected to replace the CRT technology in home, computer, and audio/visual applications.

Control systems. Because of the complexity of many audio/visual systems, remote control systems are often employed to integrate the control of audio/visual devices with other room functions, such as lighting, motor-ized projection screens, and drapery, into a unified system. With a remote control system, all audio/visual devices and many room functions can be operated from a single control panel.

Control panels can be wireless, portable with wire, or permanently installed into a wall or millwork. Remote control panels can vary from mechanically-activated push button types to software-based touch screens. It is the function of the remote control system to make the audio/visual system as easy as possible for the users to operate. It is therefore important for the system designer to consider the needs and abilities of the users. The user needs will determine the design and layout of the control panel so that controls are easy to use and understand. The panel layouts should not be intimidating to a novice or inhibiting to a more advanced user. Because of the complexity of many systems and the varying needs and abilities of each user, however, it is often not possible to determine what the best panel design is until after the system has been in use for a time. Therefore, soft-ware-based touch panels are frequently used in the design of a complex system. Not only does the touch panel allow easy expandability, but con-trol buttons can also be renamed, relocated, and duplicated in a short time. In this way, the remote control system can evolve as the systems' and users' abilities grow.

Front and Rear Projection Screens

Front projection screen. With front projection, an image is projected on the front of the screen and reflected back to the audience. The projector can be located in a separate projection room or in the same room as the pre-senter and the audience. The main advantage to using front projection screens is that they are relatively inexpensive to purchase and install, and require little space. Their disadvantage is that ambient light is scattered off the screen in the same way that a projected image is scattered. This is why front projection screens exhibit poor contrast unless the room is dark. Perforated permanent screens are utilized in large rooms with loudspeak-ers installed behind the screen to integrate the audio with the image.

Rear projection screen. A rear projection screen is a framed, translucent "win-dow" installed in the front wall of a room. The window surface facing the audience is coated with a light-diffusing material. An image is projected onto the back side of the screen from a room adjacent to the presentation room. Because the image is projected from behind, it is reversed electron-ically. If slides are involved, the slides must be reversed left to right in the tray, when compared to orientation for front projection.

While more costly, a rear projection screen has advantages over a front projection screen:

(1) It provides a more effective display in high ambient light environ-ments. The rear projection screen scatters the projected image to the view-ers' eyes while reflecting less ambient light from the room environment.

(2) The presenter at the front of the room does not need to stand to one side of the projected image. In a front projection environment, when the presenter walks in front of the screen, a shadow is cast onto the screen. In a rear projection environment, the presenter has full use of the front of the room, because the image is still present on the screen and not projected onto the presenter.

(3) A more integrated presentation environment is established architecturally, and projector noise is eliminated in the presentation room.

When a rear projection screen is installed, a front projection screen may also be installed for use with an overhead projector or other front-projected images. Ambient light incident on the front of a rear projection screen will not generally be reflected back to the audience. For this reason, a laser pointer cannot be used in a rear-projection environment and other pointing methods are required.

II. Additional Design Issues

Comparative Imaging

The foregoing discussion has primarily addressed a single-image display. Many times, however, two screens are used to compare images such as "before" and "after" diagrams. Another example would be during computer training, where one image demonstrates finger position on the keyboard and the other shows the computer display.

Although the screen is no taller, it must be made wider for two images. If two screens are used, space can be provided between them for a desk or podium. With two screens, however, it is difficult to have a good view of both at the same time.

Audience Viewing Area

Good viewing area refers to the space in a room that provides an undistorted view of the projection screen. Five factors are taken into account to maximize the prime viewing area of a room:(1) screen gain; (2) screen size and position; (3) image aspect ratio; (4) viewing distance; (5) stage (or platform).

Screen gain. All front projection screens have a gain factor. The gain factor is a numerical representation of the dispersion of reflected light. The function of a screen is to scatter all light incident upon any point of the screen to the seating area.

A screen gain of 1 means that light incident on the screen would be uniformly reflected from it. A front projection screen with a gain of 1.3 is typical for a wide viewing area. A gain of 1.3 will provide good viewing for audience members seated up to 50 degrees off axis. If the gain increases to 2, then the good viewing area narrows to 30 degrees off axis. A longer, narrower room can effectively use a higher gain screen as more of the light is reflected along the axis of the optical path. Very low light output video projections have used a gain as high as 8.

With newer plastic rear projection screens, a complex arrangement of lenses is engraved into the surfaces of the projection screen to collect incident light. This not only helps the screen appear much brighter, but it also improves image uniformity because the screen will appear as bright at the edges as it does towards the center.

Some plastic screens also have a lenticular surface on the front, which helps to scatter the projected light along the horizontal axis. Horizontal scattering spreads light efficiently so that light energy otherwise wasted in vertical dispersion is refracted left and right, making the screen appear brighter for viewers seated off axis. While the brightness is enhanced by these screens, they have size limitations and are more costly than glass screens.

Screen position. The top of the screen should be located at least 152 mm (6 in.) from the top of the wall to minimize reflections off the ceiling. In a home theater environment with a single row of seats, the bottom edge of the screen can be as low as 460 mm (18 in.) off the floor. In an educational or commercial setting, however, the bottom edge of the screen should not be lower than 1.2 m (4 ft.) from the floor to provide a clear sight line to the entire screen from any seat in the room.

Screen aspect ratio. The ratio of a screen's width to height is called the *aspect ratio.* The correct aspect ratio for a screen is governed by the type of media being used. For the majority of screens, a 1.33:1 (width to height) aspect ratio is used. This is the current aspect ratio of television as well as most computer displays. The common screen aspect ratios are listed in Table 12.2.

For a 35 mm slide, the aspect ratio is 1.49:1. Therefore, a 3.7 m (12 ft.) wide screen is 2.4 m (8 ft.) high. If all slides are horizontal, a fixed focus lens could fill the entire screen with the projected image. If vertical slides are also to be displayed, then equipping the slide projector with a zoom lens would allow the image to be vertically shrunk to fit the screen.

The aspect ratio for television is called the NTSC standard. By the year 2000, the High Definition Television (HDTV) aspect ratio is expected to be broadcast within the United States. NTSC broadcasts will be discontinued after a period of simultaneous broadcasting of both standards. The wider HDTV aspect ratio (1.78:1) does not fully capture the academy standard cinema film (1.85:1), but it does project more of the image than the standard NTSC format.

Viewing distance. The minimum recommended screen height is calculated by dividing the distance from the screen to the farthest seat in a room by eight for video or by six or less for data-grade projection. Thus, if a room is 21.3 m (70 ft.) long, the screen height should be a minimum of 2.7 m (9 ft.) for video presentations and as much as 3.7 m (12 ft.) for viewing data. After the screen height is determined, the screen width is calculated using the information in Table 12.2. Figure 12.8 shows the room layout considerations for optimum viewing.

Format	Aspect ratio
NTSC TV	1.33:1
35 mm slide	1.49:1
Digital high definition TV (HDTV)	1.78:1
Academy standard film Anamorphic lens	1.85:1
Panavision® and Cinemascope®	2.35:1

Table 12.2 Aspect ratios of common visual formats.

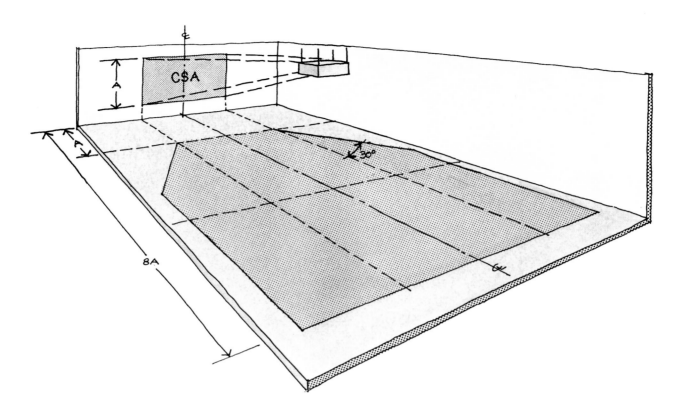

Figure 12.8 Seating layout based on height of screen.

Stage (platform). A stage elevates the presenter to facilitate sight lines. The stage should be high enough to allow at least the presenter's head to be seen from the back row. Tall platforms may reduce intimacy with the audience. If the presenter cannot be elevated, the audience can be sloped to provide a clear sight line. If neither the presenter nor the audience members are elevated, then those people seated beyond the first row may have an obstructed view of the presenter.

Lighting

Lighting for A/V presentation rooms includes general lighting for the audience seating area and presenter lighting for the podium and front-of-room area.

General lighting. Many A/V rooms need to accommodate meetings, discussions, classes, and other activities in addition to A/V presentations. The lighting system must provide proper conditions for both "lights on" and "lights off" use of the room.

For "lights on" activities, carefully designed fluorescent lighting is often the best choice because of its high efficiency and long lamp life. The lighting should be designed to produce approximately 30 footcandles measured horizontally at a height of 760 mm (30 in.) above the floor. Because these rooms often are used for extended periods of time, care should be taken to avoid the tiring effects of ballast hum, lamp flicker, and glare through the use of electronic ballasts, improved-color T8 lamps, and low-glare fixtures. Additional lighting for the walls of the room relieves the uncomfortable "cave" effect of sitting in a room with a dark surrounding.

"Lights off" activities usually require some form of note-taking illumination. This lighting does not have to be bright, but it should provide fairly even coverage of the entire seating area, with no spill light on the projection screen. Lighting for note-taking can be provided by dimmable incandescent downlights, by making the general fluorescent lighting dimmable, or by circuiting the fluorescent fixtures so only some of the lamps remain on during A/V presentations. Windows should have blackout shades so the room can be made completely dark.

Presenter lighting. The presenter must be visible to people in the room and also to viewers of video. Spotlights with tungsten-halogen lamps and louvers, internal shutters, or "barn doors" can be aimed so they light the presenter at a podium while avoiding spill light on the screen.

Borrowing from the lighting techniques used in broadcast TV studios, the presenter should be lit with "key" light, "fill" light, and "back" light. Key and fill lights are positioned 45 degrees to the left and right of the presenter, and 40 degrees above horizontal as measured from the presenter's face. The key light is brighter than the fill light. The back light is positioned above and slightly behind to light the top of the presenter's shoulders and head. All three lights are needed to make the presenter appear natural on video and to separate the presenter's image from the background.

These lighting techniques can be expanded for the entire front-of-room area if the presenter will walk away from the podium or if the area will be used for a panel discussion during an A/V presentation.

Other lighting issues.

(1) Lighting controls. Room lighting should be controlled by a dimming system with preset scenes for "bright" and "dim" settings of the general lighting and independent controls for the front-of-room and podium lights. The preset dimming system allows easy selection of predetermined "looks," and it also allows the lighting to be controlled by a touchscreen or other A/V equipment via an A/V lighting interface.

(2) Emergency lighting. Adequate emergency lighting must be provided to allow safe exiting. Circuiting some of the general lights as "night lights" that remain on all the time should be avoided because it often will be desired to use the room with all lights off. Separate, normally turned off, emergency lights with battery packs or an automatic emergency transfer switch to bypass the dimming system are two other options.

(3) Lighting for document cameras. Document cameras require supplemental lighting. This lighting should be separately controlled from other lighting, and it can be circuited so it turns on automatically whenever the document camera is in use.

(4) Lighting in projection rooms. Projection rooms require bright fluorescent work lights for maintenance and dimmable track lighting for use during presentations. Switches for work lights should be located so they cannot accidentally be turned on by someone walking into the pro-

jection room during a presentation. An illuminated sign reading "Projection in Progress—Do Not Enter" should be installed outside the service door of a rear projection room. This sign can be controlled by the A/V system so it operates whenever the rear projector is in use.

Heating, Ventilating, and Air-Conditioning Systems Noise

In a presentation environment, the HVAC system noise should not exceed NC 30. It can be very tiring for a presenter to talk loudly enough to be heard over a noisy HVAC system, a projector fan, or other ventilating devices for the A/V equipment. It is also tiring for the audience to separate speech from the background noise.

Receptacles

The presentation area is surrounded by receptacles for devices used by the presenter: microphones, audio, video, and computer connections. These receptacles are mounted on walls or recessed into the floor to minimize visual impact and to provide convenient access. Wiring is routed so tripping hazards are avoided. Individual receptacles must be labeled for users to identify.

Loudspeakers

Two separate sound systems: playback and speech reinforcement, each with its own loudspeakers, are not unusual in a presentation room. A distributed ceiling loudspeaker system provides uniformity of speech coverage in a conferencing environment. Playback loudspeakers are usually located in the front of the room adjacent to the projection screen to provide directional realism that relates to the projected image. Front loudspeakers may also provide directional realism for the presenter.

To be effective, front wall-mounted loudspeakers should be pointed toward the rear of the room but not at the rear wall. Loudspeakers must have their high-frequency elements aimed directly at the listeners, with the axis of the coverage extending to the farthest listener.

Planning for the Future

Changing broadcast television from the current analog NTSC standard to the expanded digital HDTV standard not only means that frequency allocations will change within the broadcast spectrum but television receivers will change to receive digital rather than analog signals. Non-broadcast video will also change. Current video signal formats will be replaced by digital. The clarity of a computer screen will be brought to television as the two technologies converge.

Any current audio/visual installation, therefore, should consider these changes in the video format. Rear-projection installations may benefit by installing an HDTV aspect ratio screen. Because the HDTV screen is wider, it may be difficult to install later. To replace a smaller screen with a larger one could also mean upsetting the architectural design at the front of a room.

Front-projection screens are more easily changed because they typically recess-mount into a lay-in ceiling. A projection screen that can accommodate HDTV is almost large enough to accommodate two reduced size NTSC images in a side-by-side projection format. There are potential benefits to installing a larger aspect ratio screen even though HDTV is just beginning.

Conclusion

During the last forty years, there have been constant changes in projection media and equipment. The rate of change is accelerating. In the 1950s, 16 mm films and slide projections were the primary presentation tools. While these continue to be useful for international distribution of images, video has emerged as a primary presentation technology. Soon there will be a shift to digital video. The devices may change, but the architectural considerations for viewing and listening will likely remain. The presenter's relationship to the audience and the audience's relationship to the screen are issues that normally do not change except with a change in use.

Digital TV (HDTV) and computing technology are likely to merge. The internet is likely to play a central role in both communications and entertainment. It is already possible to conduct video conferences over the internet. (View our web site at cmsalter.com.)

Appendix. Integrating Audio with Moving Pictures

Sound with picture should be integrated by the listener as a complete experience. In other words, the audio should be auditorily localized to the picture. Therefore, it is advisable to locate the screen channels similar to the best motion picture theater (Figure 12.9).

The trend in cinemas has been toward larger screens. The largest image width typically results in a 50 degree sub-tended angle from the viewer to the edges of the screen so that the picture is not large enough to be in the viewer's peripheral vision. For screen sizing purposes, the viewer is generally placed in the center of the audience. If right and left screen channel loudspeakers are placed near the edges of the screen, a 45 degree sub-tended angle results.

It is common practice to locate stereo speakers at two corners of an equilateral triangle with the listener positioned at the third point as viewed from above in plan and create a 60 degree sub-tended angle from listener to loudspeakers. This loudspeaker and seating layout is shown in Figure 12.10.

When listening with picture, it is advisable to reduce this angle to 45 degrees for right and left screen channels. This locates the speakers closer to the edges of the picture for better picture and sound integration. Home theaters occasionally utilize the projector and perforated screen typical of a cinema. More commonly, a direct view monitor, large screen TV, or rear projection screen is used. All of these have the disadvantage of not being able to locate audio in the visual image, unlike the perforated cine-

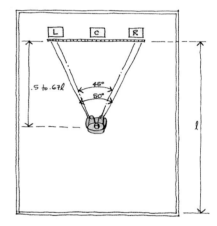

Figure 12.9 Optimum viewing and listening location.

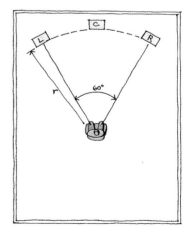

Figure 12.10 Optimum listening location for stereo sound (without viewing).

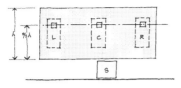

Figure 12.11 Location of loudspeakers behind perforated
screen.

Figure 12.12 a, b, and c
Possible loudspeaker layout
with solid screen.

ma screen. Locating the right and left channels at 45 degrees and close to the screen as shown in Figure 12.9 helps improve the illusion.

Generally when locating screen channel loudspeakers behind a per-forated screen, it is advisable to locate the high-frequency driver or tweet-er at an elevation that corresponds to two-thirds of the screen height. The mid-frequency loudspeakers are therefore located near the center of the screen, helping to localize dialogue. Dialogue is usually reproduced through the center screen channel (see Figure 12.11).

When using a solid screen, such as a direct view monitor or rear pro-jection screen, it is necessary to locate the loudspeakers associated with the screen as close as possible to the picture. Figures 12.12 a, b, and c show some possible configurations. The layout shown in Figure 12.12a is preferred because the left and right channel loudspeakers are closest to ear height.

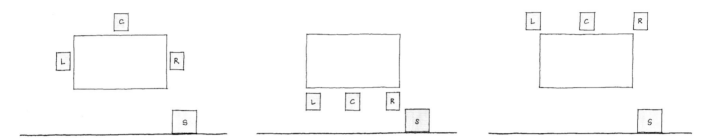

Figure 12.13 depicts a typical surround sound loudspeaker layout for a cinema or other large room. Generally speaking, the surround channel loudspeakers should be located at a height equivalent to two-thirds of a room's ceiling height, although this can depend on the room width and lis-tening location. In most cases, the speakers are evenly spaced on the side walls and may follow the floor slope to maintain a constant distance above the floor. The forward-most surround speakers should never extend in front of the first row of seating. Surround loudspeakers should not be placed in the corners of the room. The number of surround speakers is determined by the power-handling requirements, that is, the peak sound pressure levels necessary throughout the space, and the directivity or cov-erage provided by the loudspeakers. In most good cinemas, the sound level from the main channel loudspeakers varies only 3 to 4 dB from front to rear seating rows. It is desirable to maintain a constant main-channel to sur-round-channel sound pressure level ratio in the theater. To accomplish this, it is necessary to taper the surround array so that it also provides reduced sound pressure levels toward the rear of the theater. Tapering the sound level can be accomplished electronically using separate amplifiers or by spacing the array closer together near the front of the theater and increas-ing the spacing at the rear.

There are two types of surround sound available for the home. The first uses only two surround speakers located to the sides of the listener (Figure 12.14), or slightly biased toward the rear of the room using a loud-speaker which radiates sound directly to the listener.

Since one of the objectives of surround sound is to envelope the listener, the second surround sound loudspeaker type uses either dipole or tripole radiating loudspeakers. These loudspeakers radiate toward the front, rear, and sides of the listener (Figure 12.15). While these surround speakers are elegant in their simplicity, a fairly reverberant listening room is required.

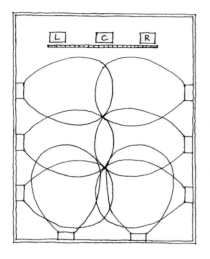

Figure 12.13 Typical locations of surround sound loudspeakers showing their coverage pattern.

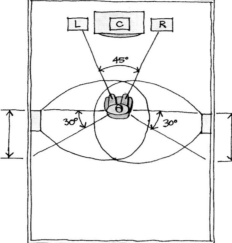

Figure 12.14 Plan view of home theater showing typical surround loudspeakers and their coverage pattern.

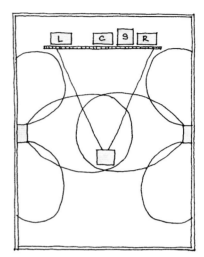

Figure 12.15 Plan view of home theater showing radiating tripole surround sound loudspeakers and their coverage pattern.

Surround sound and home theater are becoming increasingly prevalent. Perhaps a better term would be Mini-Theater, since this technology is commonly being installed in presentation rooms of all types and sizes. The advantage over simple stereo listening is that dialogue is localized to the image center and is no longer a phantom, but a "hard" image. In addition, the surround loudspeakers immerse the listeners in the sound field. These attributes enhance listener involvement with the presentation.

13

Acoustical Simulations in the Multimedia Age

Kenneth W. Graven, P.E.

Multimedia-Based Calibrated Audio Demonstrations • Room Acoustic Simulations • Cost Benefits of Acoustical Simulations • Audio Simulation Case Studies

FOR many years, architects have built physical models of their designs. Advances in technology now allow architects to design and model their projects using computers. Acoustical engineers also use this technology to demonstrate the predicted acoustical outcome of a room design. No longer restricted to numeric descriptors, acoustical engineers can now create aural simulations of what will occur in a completed room.

There are two catergories of acoustical simulations: calibrated audio demonstrations and room acoustics simulations.

Multimedia-Based Calibrated Audio Demonstrations

The simplest type of acoustical simulation is the calibrated audio demonstration. Calibrated audio demonstrations are accurate recreations of the sound levels created by sonic events, such as jet flyovers or traffic noise. Often, the sonic events that need to be recreated are reduced in level by an architectural element such as a wall, door, window, or barrier; each simulation is uniquely tailored to account for the acoustical effects of these elements. Calibrated audio demonstrations can also demonstrate levels of speech privacy, the effects of sound masking, and the quality of ventilation noise.

Acoustical consultants present these audio simulations in courtrooms, at public hearings, and other venues. In its simplest form, these demonstrations can be created with a tape deck, equalizers, an amplifier, and a loudspeaker playback system. While audio demonstrations are not new, recent developments in digital and computer technology have allowed them to become more streamlined, accurate, and to have random accessibility. Computer-driven multimedia allows random access to graphics, animations, video, text, and audio information that has been stored on a hard disk. Random access allows the sounds to be played in any order. For example, if the simulation consists of four scenarios, each one can be played in different succession. If the simulations were taped (instead of on computer), the consultant would have to rewind or fast forward each time the client wished to make an audible comparison. Random access reduces delays and allows the opportunity to compare samples while each is fresh in the listener's mind.

Thus, real-time, interactive simulation is created using multimedia. The user controls the characteristics of what is being seen and heard. For example, if a random access simulation contained different wall constructions for a room, the client would be able to compare the effects of these various treatments at the push of a button. Furthermore, multimedia's use of a variety of visual elements such as graphics, animations, and video can convey additional insight into the acoustical issues at hand.

Room Acoustic Simulations
Another important category of acoustical simulation involves room acoustics. These simulations usually demonstrate the effects of reverberation and echoes on the quality of speech. These demonstrations might also illustrate how a concert hall's shape and design finishes affect the quality of the acoustics, or how an overly reverberant room diminishes speech intelligibility. Room acoustics simulations, although not always entirely accurate or realistic, can convey useful information about a room's design prior to construction, thus helping to avoid costly mistakes.

Unlike calibrated audio demonstrations, room acoustic simulations would not be possible without the advances in digital audio technology. Digital signal processing (DSP) technology has become so common that even the average layperson can recreate various room ambiances on home stereo systems, ranging from the ambiance of a large cathedral to that of a small nightclub. However, these simulations are generic in nature and are of little use to the design professional. More advanced systems are available to acoustical consultants that allow aural recreation of a room's reverberation based on the room's size and shape. First, a computer plots the many possible sound paths and reflections within a room and then creates a mathematical model of this sound energy. This model is then transferred to a DSP-based reverberation processor where an anechoic recording can be made to sound as if it were recorded in a room. Headphones or a specially designed room can be used to hear the simulation. Both options are explained in the following sections.

Using Headphones to Recreate an Acoustic Environment. Often, acoustically simulated auralizations are recreated using binaural headphones. Headphones can simulate a controlled environment that is difficult to achieve with loudspeakers. If the signal delivered to the headphone accounts for the shape of the outer ear (which helps a listener localize a sound) this method can deliver stunningly real effects. Headphones, however, sometimes deliver unnatural effects. The sound source will unnaturally move as the head of the listener turns, detracting from the listener's experience. The intended spatial effect can vary slightly from listener to listener because every person has a differently-shaped outer ear.

The Presentation Studio. An alternative to using headphones for acoustical simulation purposes is to use a room equipped with multiple loudspeakers that envelope the listener in sound. Such a room would have to be adequately quiet, with virtually no reverberation and unwanted noise. Charles M. Salter Associates built such a room, which is called "The Presentation Studio" (see Figure 13.1). The studio is designed to create a variety of realistic acoustical demonstrations for clients. Within the studio, accurate calibrated audio demonstrations are presented to a maximum of fifteen people. In addition to the audio and acoustic qualities of the room, the studio also provides large screen computer projection and state-of-the art multimedia capabilities, including cinema-quality surround sound.

Figure 13.1 The presentation studio in Charles M. Salter Associates' office.

Cost Benefits of Acoustical Simulations

If any technology is to be fully realized, it must have a practical application and it must be cost effective. An acoustical simulation can give decision makers the opportunity to compare different construction options. This comparison process can then expedite the decision-making by providing tangible evidence on which to base decisions. Finally, acoustical simulations are cost effective because the results of various construction alternatives can be explored without incurring the actual construction costs; this "safe exploring" reduces the likelihood of embarrassing design mistakes, client and tenant complaints, and costly lawsuits.

Audio Simulation Case Studies

The remainder of this chapter consists of three case studies which illustrate the value of acoustical simulations. These include a calibrated audio demonstration used in a lawsuit against the City of San Francisco's cable car system, a simulation of speech privacy resulting from various wall constructions, and a demonstration of the effects of reverberation on speech intelligibility.

Cable Car Case Study. In 1992, a suit was filed against the City of San Francisco by a couple who owned a home next to a cable car line. The suit claimed that retrofits to the line had caused a noticeable increase in the noise levels coming from the cable cars. Whether or not there was actually an increase in noise was in dispute. Charles M. Salter Associates (CMSA), serving as an expert witness for the City, maintained that regardless of any actual change in noise level, the couple's main problem derived from the fact that their house was poorly isolated from noise.

The double-hung windows in the plaintiffs' home were in poor repair—the wood window frames were rotting and there were cracks in the glass. It was CMSA's belief that if a jury could experience the sound level of the cable car line through these windows and then through windows of improved constructions, they would understand that the plaintiffs were taking an unreasonable position.

A calibrated audio demonstration was prepared to represent four levels of window construction. The first demonstration was a digital audio tape (DAT) recording measured in the bedroom of the house experiencing the highest noise level (see Figure 13.2). This recording was played back to the jury at the exact calibrated level at which it was recorded. Three conditions were simulated using audio filters and level attenuators to demonstrate what the proposed improvements could achieve. The improved constructions consisted of the following: (1) gasketing the existing windows; (2) replacing the window glass with 10 mm ($^3/_8$ in.) thick laminated glazing; (3) replacing the existing windows with double windows separated by 100 mm (4 in.) air space.

By gasketing the existing windows, CMSA predicted a 5 dB reduction. For the laminated glazing demonstration, a 14 dB reduction was simulated, and for the double windows, a 21 dB reduction was simulated. The

Figure 13.2 Inside plaintiffs' bed-room during acoustical measurements.

demonstration using the plaintiffs' existing windows, gasketed, showed the jurors that even minimal efforts on the plaintiffs' part would have resulted in a noticeable noise reduction. The sound level demonstrations of the noise transfer through the laminated and double windows were barely audible. The demonstrations were effective in illustrating the city's case.

The Apple Computer Research Campus. In this demonstration, the effects of various wall and ceiling constructions on speech privacy in enclosed offices were recreated. Apple desired that adequate speech privacy be obtained in each office in their new facility. Due to the facility's vast size, even the slightest change in construction detailing could cause a significant cost increase; thus, any increased level of speech privacy could potentially cost thousands of dollars.

During the building's design phase, the client was offered three choices for wall types. These included: (1) uninsulated ceiling height wall with standard acoustical tile ceiling (STC 33); (2) insulated ceiling height wall with a high transmission loss acoustical tile ceiling (STC 36); (3) insulated slab-to-slab wall and an additional layer of gypsum board on each side (STC 45).

Because the clients could not visualize the relationship between each recommended treatment and the associated benefits and costs, they had difficulty making a selection. To help the decision-making process, an acoustical simulation was recommended. The simulation sought to demonstrate the level of speech privacy that each wall/ceiling type would provide by using a multimedia computer, large screen video display, and the audio capabilities of the Presentation Studio.

Using programmable equalizers, CMSA made audio recordings of conversational speech and filtered them to approximate the noise reduction of each construction type. The filtering spectrum was based on actual field measurements of the transmission loss of similar assemblies. A multimedia program was then produced with on-screen buttons that would not only randomly access audio demonstrations from the hard disk, but would also display a visual architectural graphic of the wall/ceiling assembly being demonstrated.

In order to make the simulation as accurate as possible, pink noise (random noise containing equal levels of all audible frequencies) was sent to overhead loudspeakers to create the effects of ventilation noise. With this setup, CMSA was able to demonstrate to the architect, cost estimator, owner representative, and end user the audible result of each construction option. The demonstration encouraged a quick decision. In fact, the decision to use the STC 36 assembly was reached within an hour after the simulation. Thus, a decision that had been delayed for over a month was made quickly and easily after the simulation, demonstrating the power of tangible evidence.

The Apple Computer Research & Development Atrium. For its new R&D campus, Apple envisioned a sophisticated facility with urban amenities like an espresso bar in the large entrance atrium. Apple's idea was that this facility could be a place where workers could "hang out" as if they were at an outdoor cafe. It was also anticipated that this space could be used for meetings and lectures. There was some concern that the facility, with its large volume and numerous sound-reflecting surfaces, would not provide good speech intelligibility and might also be noisy due to reverberation buildup. The decision to add sound-absorbing finishes, however, was not easily made because of aesthetic and functional constraints.

To give the project designers and end users a sense of what the reverberation would be like in this facility, an acoustical simulation was prepared (see Figure 13.3). To understand the effects of the reverberation on speech intelligibility, the simulation included samples of recorded speech at distances of 3 and 9 m (10 and 30 ft.) from the talker. The listener received more of the direct signal from the talker at 3 m (10 ft.). At 9 m (30 ft.), the listener was standing deeper in the reverberant field of the room, and thus experienced more degradation in speech intelligibility. This demonstration was presented for two reverberation times: (1) the "as designed" reverberation of nearly 5 seconds, and (2) a proposed reverberation time of 2.2 seconds, which would be achieved by adding acoustical treatments.

The demonstrations were produced using two stereo digital reverberation processors. Samples of pre-recorded speech were randomly accessed from

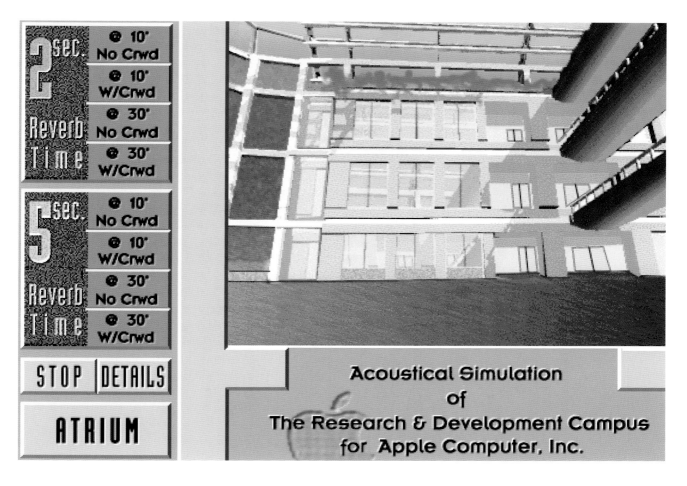

a computer hard disk using a multimedia interface. It became quite clear to all attendees of the demonstration that additional sound absorption would be necessary. Following the demonstration, CMSA was asked to recommend sound absorption options for the atrium. Once again, a decision that had lagged for weeks was made within an hour after the simulation was presented.

Figure 13.3 Acoustical Simulation of the Apple Computer Research & Development Atrium.

Conclusion

Computer technology will play an increasing role in the prediction of the acoustical characteristics for designed facilities. The ability to manipulate and exchange materials within an architect's computer-aided design, and then to listen to the results is now a reality. Design professionals will need to recognize how these tools can be cost effective and beneficial to a project. For the acoustical engineer, this technology provides greater opportunities to communicate with clients and to achieve the main goal of acoustical consulting—to inform and educate the design community on the important role acoustics play in all successful projects.

14

Audio Forensics

John C. Freytag, P.E.

Audio Enhancement and Dialog Recovery • Tape Validation and Authentication • Voiceprint Analyses • Sound Source Identification • Audibility Analyses • Case Study I—What Was Said? • Case Study II—Could She Have Heard It?

AUDIO forensics is the process of establishing the acoustic history of an event or sequence of events. Acoustic history must follow an accepted set of technical and ethical standards as required for legal evidence. Most audio forensics work deals with tape-recorded material for consideration as evidence.

Audio forensics first became widely known in 1974, with the now famous "Nixon tapes" of White House telephone conversations. The primary issue with the tapes was not who was speaking, but the "chain of custody," and degree and type of manipulation that may have been involved. This audio forensic investigation determined which tape recorders were used in recording and editing, and whether the tapes were originals or copies.

Today the role of audio forensics has expanded to include: audio enhancement and dialog recovery; tape validation and authentication; voice (speaker) identification/voiceprint analysis; sound source identification and location; audibility analyses.

These processes are routinely used in examining electronic surveillance recordings made by law enforcement. Such surveillance techniques include telephone wiretaps, body recorders or transmitters, other concealed microphone techniques, and, often, unconcealed tape-recorded interviews. With the increased focus on the enforcement of drug smuggling and trafficking laws, Racketeer Influenced and Corrupt Organizations (RICO) statutes, and "three strikes and you're out" type laws, the demand for audio forensic services has increased dramatically over the past few years. Concurrent with this attention to audio forensics has been an even more dramatic growth in the technology available for audio forensics, particularly with the arrival of digital recording, processing, and editing techniques. Currently, no standards have been established for audio evidence, although draft standards are now being formulated.

Audio Enhancement and Dialog Recovery

The audio enhancement and dialog recovery processes are typically referred to as *denoising*. The objective is to separate one sound source from another, generally to recover dialog. The obscured dialog may have been caused by deliberate sound masking through selection of a noisy recording location or by a variety of recording problems. Noisy recording areas include loud reverberant public spaces such as bars and restaurants, street corners, and factories.

Most recording problems arise from poor recording technique. Such technique often results in a low signal-to-noise ratio from inadequate input (recorder) gain (volume or level setting). The recording process often causes signal-to-noise degradation since any other noise in the airborne sound field contaminates the information transfer. Quality recording requires matched input and output levels for good signal-to-noise ratio. All tape-to-tape transfers must be done on a dubbing tape deck or with patch cables from output to input, with monitored recording levels, to balance input and output levels. Other recording problems include low batteries on DC-powered recorders, causing recorders to slow down, dirty tape heads, misaligned or damaged heads, use of improper recording tape, incorrect recorder settings, and improper use of AC-power units.

Information that was recorded with noise (or unwanted information) can be recovered, at least in part. Information which was not picked up by the recording cannot be recovered. Thus, recordings with a hopelessly low signal-to-noise ratio may be enhanced somewhat, but never fully restored. This is also true with distorted or overloaded recordings.

Denoising techniques are divided into the following categories: (1) compressors, limiters, and noise gates—to separate signals by amplitude level, or to discriminate signal from noise; (2) filters—to separate signals by frequency, such as to isolate the speech frequencies from the background noise frequencies or to eliminate a tone or hum; and (3) adaptive filtering—to separate stationary (periodic) signal components, such as those from machinery, from non-stationary (random) components, such as those from speech.

A variety of analog and digital equipment currently exists for audio enhancement. The primary market for such equipment is not for audio forensics, but rather the large audio and video post-production processing markets, which still encounter significant problems with quality recordings, despite their substantial investment in training and equipment. The recording and film industries have developed the technology for audio forensics.

A fourth, and potentially very effective, denoising technique exists with stereo recording. Here a recording is made with the signal–plus–noise on one channel and only the noise on the other channel. These two signals, arriving at slightly different times, are recorded simultaneously on parallel channels. Processing the two signals by trial–and–error time delay (determined from cross correlation) effectively subtracts the known noise signal from the signal–plus–noise channel. The result is a substantially enhanced signal. The potential for more successful body–wire surveillance exists with quality stereo recording, particularly when incorporated with high dynamic range digital audio tape (DAT) recording.

Tape Validation and Authentication

The primary issues regarding audio evidence are: (1) Is the tape an original? (2) What is its origin and chain of custody? and (3) Has it been altered since its original recording? A key issue with evidentiary tapes is a nontechnical one: the motive and opportunity for tape tampering. Because the field of audio forensics is in its infancy, no standards currently exist for audio forensic evidence. However, the Audio Engineering Society, Working Group 12 (WG-12) has developed draft standards, currently under review, for audio forensic evidence. Essentially, common sense dictates the following guidelines for all audio evidence: (1) The evidence be obtained by reliable, unbiased means and persons; (2) The history of evidentiary recording be carefully documented, noting times, locations, equipment, witnesses, etc.; (3) The evidence be stored securely, preventing the opportunity for tampering; and (4) The parties responsible for the recording be available and prepared to testify about the recording.

The issue of authentication generally deals with: (1) whether the tape is an original or a copy; (2) whether the tape has been edited since its original recording; (3) whether the reported time, location, and equipment are correct.

Authenticity investigations involve both physical inspection of analog tape and wave analyses. The question of original recording may first be evaluated by examining the analog tape for evidence of overdubbing or "drop-in" recording (recording over previous information). Over-dubbing is detected by wave analyses and proven by using magnetic developing fluid and a microscope with a camera to produce a photographic image of the magnetic recording. This image reveals the physical imprint of the erase head preceding the record head and the start/stop signature of the recorder. Evidence of recorder erase head and closely spaced start/stop signatures indicate overdubbing.

Overdubbing is initially detected by wave analyses, an electronic examination of the magnetic time history of the signal as it is recorded. Analog recorders modulate a low-frequency signal proportional to the speed of the capstan driving the tape over the erase and record heads. Since no capstan is perfectly round, the eccentricity of the capstan produces a low-frequency signal referred to as "wow." Overdubbing on the same or another recorder produces another low-frequency signal, which may be detectable by frequency analysis. The two low-frequency signals produce new "beat frequency" equal to the difference of the two wow frequencies. Careful overdubbing, using the same high quality analog recorder, may be performed by a knowledgeable person without detection by either physical or wave analyses.

Editing, like the tape copy analysis, is discovered by physical examination and wave analysis. Analog editing is characterized by start/stop signatures, which may be viewed by photographic development of the magnetic tape. Each recorder has a characteristic start/stop signature that is controlled by the spacing of the erase and record heads and the inertial properties of the tape transport, which dictate start and stop speeds. Additionally, wave analyses show the start/stop signature is also governed by how long the recorder has been off prior to its last start/stop signature, since tape recorder record heads have a characteristic settlement time after use.

Editing may be performed by either a physical start/stop operation of an analog tape recorder, physically splicing the tape (and subsequently copying to a previously unrecorded tape), or digital editing. Physical splicing of tape avoids detection by photographic development and by start/stop signature wave analysis. However, it is generally detectable by a change in recording level, a change in background noise character, or by a phase change in a tonal component continuing throughout the tape. The source of such tonal components may be from the AC-power source, from the recorder bias oscillator, or from a background noise source or recording artifact.

Other authentication issues are often detectable by critical listening to the tape and by a variety of analyses. Background noise is generally discernible and may be used to help identify a recording location. Besides audible background noise sources, it is possible to determine if a recording was made in a reverberant room or outdoors. Sometimes too much acoustical information causes a tape to be discredited, as with the case of a reported telephone conversation carrying information above 4 kHz, the upper frequency limit of analog telephone equipment.

Digital recording and editing equipment has rendered many physical tape development and wave analyses techniques obsolete. Just as digital sound recording, mixing, and editing techniques have greatly facilitated audio forensic identification, powerful tools for circumventing detection have also become readily available. At this time the author must concede the possibility for undetectable copying and editing by an experienced person using state-of-the-art digital equipment.

This possibility should not eliminate the use of this modern day investigative technique any more than one would eliminate the use of cameras, since one cannot deny the possibility of copying and editing photographs. In fact, it is accepted that the contents of a recording are not necessarily the truth but only a witness to what was said.

Tape Validation and Authentication

Implicit in the term "voiceprint" is the existence of a voice measurement that uniquely characterizes the speaker, as with a fingerprint. In fact, the reality is much more complex because voice signatures are influenced by psychological and physiological factors as well as by recording equipment and environment. In some states, the status of voiceprints for speaker identification is more like that of polygraph (lie detector test) than of fingerprints. That is, the Court gives circumstantial credence to someone willingly taking and passing a voiceprint elimination examination, but does not grant full technical credibility to expert evidence of matching voiceprints. Nonetheless, voiceprints are widely used today in federal cases and are admissible in most state courts.

A voiceprint is a spectrogram of a speaker's voice. A spectrogram (as shown in Figure 14.1) relates the linear frequency on the y-axis (vertically) with time on the x-axis (horizontally). Signal amplitude is in the z-axis, directly off the page, and is indicated by the relative darkness or lightness of an area (its intensity on the optical gray scale). The first spectrograms used a constantly sweeping filter on a tape loop of the audio material, producing a paper stripchart recorder. These early spectrograms, now more than forty years old, are still in common use, although digital spectrograms may be taking over. Color spectrograms are now available from a variety of instruments and are used in speech research, but are not yet standardized nor widely accepted for courtroom use.

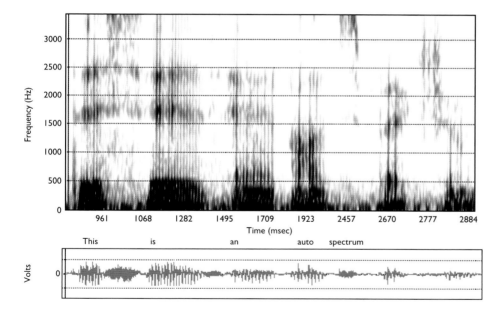

Figure 14.1 Sample of a spectrogram.

The phonetic alphabet is comprised of 94 individual sounds. Every spoken word in every language can be broken down into its phonetic components, although no single language uses the entire phonetic alphabet. In principal, each individual has a unique way of pronouncing each element of the phonetic alphabet. Any recorded phrase, by this reasoning, could be broken into its phonetic components and the speaker identified by matching the phonetic elements with a library of recorded phonetic elements from the candidate speaker. Each individual has a certain fundamental (lowest) frequency, speech rate, harmonic and bandwidth characteristics associated with his or her speech.

This type of pattern-matching is the basis for voiceprint analysis, but the application is more complex than this ideal model. In fact, an individual's speech characteristics may vary substantially from one day to the next, from stress, head cold symptoms, and other factors. More importantly, each individual intones the phonetic alphabet somewhat in the speech context; that is, we change the way in which we pronounce things according to the preceding and succeeding sounds. Expert training and experience are required to discern which elements of voice spectrograms vary for the same speaker and which remain largely constant. Thus far, automated signal-processing algorithms have been unsuccessful in identifying a speaker. To date, voiceprint identification remains the province of the Justice Department and a handful of voiceprint experts across the United States. With the tremendous strides being made in speech recognition, artificial voices, and digital signal processing, voiceprint identification technology may advance rapidly in the future.

Two primary types of voiceprint analyses are typically presented as evidence: (1) random context speech, whereby a voice sample is examined to determine a match with another speech sample; (2) elimination whereby the speaker cooperates by repeating the passage that he/she is alleged to have recorded for a voiceprint comparison.

Random context speech is more difficult to analyze because similar speech patterns must be searched and analyzed between samples, whereas elimination uses the same context material for spectrogram matching. With elimination, several voluntary speech samples, called exemplars, are taken at different times and compared with each other as well as with the alleged context.

Sound Source Identification

Techniques exist to identify a particular sound source and its location. A sound source may be identified through a family of digital signal processing techniques known as signature analyses. These techniques allow for quantitative digital comparison of audio signals when recorded at different times, distances, etc.

Signature analysis was first employed in industrial applications to evaluate the condition of rotating equipment. An example is automobile engine testing on a Detroit assembly line where a new engine is run up to a fixed speed, and the sound spectrum digitally recorded and compared to a master (properly assembled) engine. If the signature analysis of the two spectra match, the engine is accepted; if not, the spectrum of the new engine is compared to a library of malfunctioning engine spectra to diagnose the problem.

Such signature analysis techniques may be employed in audio forensics to determine sound sources. These sources may be very subtle and drowned out by other noise on the tape. Signature analysis is useful in analyzing flight recorder data where information such as engine speed and load conditions, airspeed, and warning-system status may be accurately determined from auditory recordings.

Sound source location may be determined if multichannel auditory information is available or a range of possible locations can be identified from single channel recording. The most dramatic example of auditory sound source location is the gunshot location systems in operation in several U.S. cities. These systems determine a gunshot location, within a hundred meters (a few hundred feet), a few seconds after the gunshot event. By knowing the location of microphones within the city and the time (in milliseconds) that the shot is recorded at each location. It is straightforward to calculate the origin of the shot from the geometry of the microphone locations and the speed of sound in air.

Gunshots, or other impulsive noise, also reflect off room surfaces such as walls, floors, and ceilings. A single-channel gunshot recording may be analyzed to determine the time between the shot recording and successive reflections from interior surfaces. The time between reflections directly relates to the distance between reflected surfaces. By carefully analyzing reflection times, and knowing the recorder location and room geometry, it is possible to determine possible gunshot locations. As with any forensic analysis technique, the concept is straightforward but source localization is not always possible. The actual analysis is often complicated by a poor recording, complex surfaces, poor acoustical reflection properties, and interposing objects or structures.

Audibility Analyses

Audibility analyses are conducted to determine if specific auditory information may have been heard by a person with average hearing ability at a particular time. Such analyses determine whether a warning device, such as a fire alarm, was audible or if a conversation might have been overheard under a particular circumstance.

Audibility analyses always assess the magnitude of the sound source in question with respect to the background noise (unwanted disinformational sound). Audibility always involves a signal-to-noise ratio assessment; information may be readily heard in a quiet environment that would be indiscernible in a louder environment. The issue is whether the information in question was masked by the background noise. This masking effect is not restricted to loud background noise environments, but applies equally well in very low background environments since our auditory processes automatically tune down to a range of hearing commensurate with the background noise environment.

The effects of sound masking cannot be directly analyzed by comparing overall levels from the sound source and background noise. Several factors complicate the analysis: (1) Sounds whose primary frequencies are close to the frequency of the background noise are more easily masked; (2) Sounds with strong tonal components are more difficult to mask than broadband noise; (3) The informational content of masking noise is significant, as are temporal elements (repetition) and other factors; (4) Auditory processes have an acclimation time requiring several minutes for a listener to adapt from a louder or quieter environment.

The best way to conduct an audibility analysis is to simulate the entire sound environment (signal and background noise) in a controlled environment and perform listening tests using an unbiased jury. This requires a sound-studio-type environment to isolate the listening area from all outside noise sources. While courtroom noise environments are superior to most public presentation areas, they may be unacceptable for conducting a calibrated sound simulation. Background noise from ventilation systems and exterior activities often renders courtrooms unacceptable for listening tests. It may be possible for the expert witness to conduct an audibility simulation outside the courtroom and report the test results in court.

Case Study I—What Was Said?

A police interview tape was analyzed for a case involving a fatal shooting by the police during a drug arrest. The interview tape was garbled at a critical point, and an analysis needed to be performed to determine what was said. The two possible interpretations of the critical dialog were:

"He's seeing a gun...," or
"He's stashing a gun...."

The recorded section of the tape was first enhanced, using digital adaptive predictive deconvolution, to remove background noise. While the tape was clearer and the second interpretation seemed to sound most likely, two interpretations of the critical dialog remained possible.

A voiceprint was then made of the two alternative dialog interpretations, using the ideal phonetic pronunciation. The voiceprints were then compared with a voiceprint of the enhanced dialog section, and the second interpretation was demonstrated to match more closely than the first, confirming the subjective critical listening of the enhancement.

Case Study II—Could She Have Heard It?

During a murder case, a surprise witness, nearly 1.6 km (1 mi.) away, came forward claiming to have distinctly heard two gunshots at 1:15 AM on the night of the murder. She had not come forth earlier because the newspapers had reported the time of death at around 10:00 PM, when she was elsewhere. It was only when the defense had established a considerable range in the possible time of death that she contacted authorities. The defendant was accused of committing the murder at approximately 10:00 PM and traveling to another location where he was verifiably seen at 1:15 AM. Thus, the credibility of the surprise witness and her ability to hear gunshots at that time became critical issues.

The approach to investigate the gunshot audibility was to simulate the conditions as closely as possible. A weapon of the same caliber was used at the murder location on a still night at 1:15 AM, with calibrated noise level results measured using a DAT recorder at the witness location. The gun was fired at prescribed times with the source gunshot recorded (via radio transmission) on one channel of the DAT and the receiver gunshot recorded on the other channel. This stereo recording enabled precise measurement of propagation time and provided verification of the gunshot noise source. The initial results were inaudible but detectable on the DAT recording.

The gunshot sound level on the DAT tape was carefully analyzed along with the background noise. The sound level of the gunshot was computed for the night of the murder by considering the difference in temperature, humidity, and wind conditions. More importantly, the background noise conditions for the night of the murder were known to be below those on the simulation night. The resultant computed adjustments showed that the gunshots indeed could have been heard on the night of the murder.

Conclusion

Audio recording, editing, and voice simulation technologies are rapidly advancing. This brings greater opportunities for collecting quality audio evidence, for technically tampering with evidence, and for detecting falsified evidence. Increasingly, audio evidence is an adjunct to video-recorded evidence. The judicial systems face an ongoing challenge in admitting valid audio evidence while rejecting questionable material.

15

Design
& Construction
Issues

Thomas A. Schindler, P.E.

Initial Concept • Programming • Schematics • Design Development • Construction Documents • Value Engineering and Cost Reduction • Bidding • Construction Submittals • Construction • Project Completion/Post-construction

MANY design professionals associate acoustical engineering exclusively with the design phase of a project. However, acoustical engineering can play an important role in all phases, beginning with the project's conception and continuing through its construction. Considering acoustics throughout the entire process can make the project more cost-effective. If critical acoustical issues are not addressed early enough, then significant, often costly, design or construction may need to be redone in order to avoid major functional deficiencies. The acoustical engineer's comprehensive involvement in a project's success is illustrated in this chapter. The consulting services that each phase might require are discussed.

Initial Concept Phase

During the initial concept phase, the basic objectives of a project are determined. Pre-design acoustical cost estimates are often required to determine a project's feasibility. An analysis of financial resources and preliminary cost estimates can be a powerful tool for determining the design direction.

Acoustical considerations can substantially impact a site location and space adjacencies. In assessing a possible site for a project, it is often necessary to determine both the likely impact of the project's construction and operation on the surrounding community, and the potential impact of the

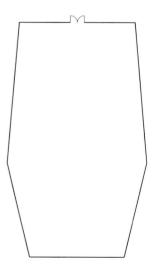

Initial concept phase sketch.

surrounding environment on the project's feasibility. An example of the former consideration is a proposed project located in a residential area, which includes a truck loading dock and an emergency generator. The presence of these facilities may generate unwanted noise, complaints, even lawsuits. Typically, estimates can be made of the noise levels produced by mechanical and electrical equipment and other noise sources. By comparing the estimated noise level of the proposed equipment to the existing noise level at the site, the project's acoustical impact on the surrounding community can be predicted. This type of analysis is included in an environmental impact report, which is often required by local, state, or federal agencies for all projects that have the potential to significantly affect the environment. (For more information on environmental impact reports, refer to Chapter 5.)

Existing noise levels can affect the feasibility of a project. For example, transportation noise (highways, railroads, aircraft) could adversely affect projects that require low noise levels such as a recording studio. Greater intrusive noise levels from the surrounding environment could require more extensive sound isolation constructions at a higher cost.

Programming Phase

During the programming phase, project acoustical criteria are established by determining what the acoustical quality of each space needs to be. For example, how loud or quiet should a space be, should it be absorptive, should it transmit speech effectively? Such criteria are often identified through meetings with the prospective users of the facility. Determining the project acoustical criteria is a critical starting point for the acoustical and audio/visual (A/V) design of a facility. Typical acoustical performance parameters include ventilation system background noise levels, sound isolation, and room acoustics. The functional requirements of an audio/visual system include both the operative characteristics of the system and the qualitative aspects of the system, including the fidelity of the sound and the quality of the visual images. Along with budget considerations, these requirements determine the scope and complexity of the acoustical and audio/visual systems.

Schematic Phase

During the schematic phase, the arrangement of rooms within a facility is determined by the architect and owner. Spaces with low background noise criteria, such as conference rooms and bedrooms, and spaces with high background-noise levels, such as mechanical equipment rooms, should not be located next to each other. By separating quiet spaces from noisy spaces, costly constructions can be avoided. Room shape and volume should also be addressed for acoustical compatibility with a room's function.

Design Development Phase

During the design development phase, appropriate acoustical assemblies and details are determined based on the developing architectural and engi-

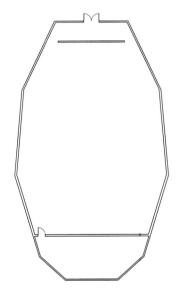

Schematic design phase.

neering design documents. These include types of wall and floor/ceiling constructions, sound-rated doors and windows, and acoustical room finishes. The need for specialized constructions, such as floating floors, is also assessed during this phase. Based on the preliminary design of the mechanical and electrical equipment, noise reduction strategies are also developed to achieve the established project noise criteria.

During the design development phase, functional diagrams are drawn and specific equipment requirements are established for audio/visual systems. Functional diagrams graphically illustrate system components and their relationship to one another. The location of the audio/visual equipment is also identified on the architectural plans.

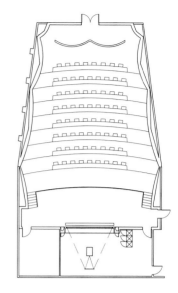

Design development phase.

Construction Documents Phase

During the construction documents phase, every element of the project and its assembly is delineated on drawings and in a specification manual that is organized into seventeen divisions. As is the case with many architectural and engineering systems, adequate detailing is required for the successful implementation of acoustical devices and constructions. Typical details include cross sections of wall constructions, wall/ceiling intersection details, and vibration isolation systems. Specification information for acoustical systems and devices is a very important element that must be included in the construction documents. Acoustical input is commonly provided for the sections of the project specifications listed in Table 15.1.

Typical audio/visual design documents completed during this phase include the system functional diagrams, control system diagrams, receptacle plate details, equipment rack layouts, and the audio/visual/video specification sections. These specifications are included in Division 11: Equipment, of the project manual, so that they can be bid on by a qualified audio/visual contractor who will act as direct subcontractor to the general contractor. In most cases, it is inappropriate for an electrical subcontractor to bid on the A/V portion of a project that involves anything more complicated than a simple audio system. Due to the specialized requirements of installation, fabrication, and adjustment of audio/visual/video systems, most electrical subcontractors are not capable of installing these systems. At best, the electrical subcontractor will overbid the job; at worst, the electrical subcontractor will underbid the audio/visual work and then attempt to coerce an audio contractor to work at a low margin, thus making a compromised installation more likely.

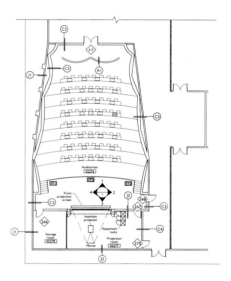

Construction documents phase.

Value Engineering and Cost Reduction Phase

Value engineering and cost reduction are employed to reconcile the facility design with the construction budget. The objective of value engineering is to reduce the cost of a device or system through evaluation, without reducing its overall value to the owner. Often the pressure to dramatically lower cost results in reduced function of a device or system; this process would then more accurately be referred to as cost reduction.

Table 15.1 Sections of construction document specifications which require acoustical input by section number.

Section	Title
03250	Concrete Accessories: concrete floating floor systems
06100	Rough Carpentry: wood floor installation techniques to reduce floor squeaks
07200	Insulation: material used in sound isolating construction (e.g., fiberglass batt insulation)
07900	Sealants: acoustical sealant and installation information
08100	Metal Doors and Frames: acoustically rated metal doors
08200	Wood and Plastic Doors: acoustically rated wood doors
08300	Special Doors: acoustically rated doors of unusual size and/or operation
08500	Metal Windows: acoustically rated metal windows or minimum air leakage performance for non-acoustically rated windows
08600	Wood and Plastic Windows: acoustically rated wood and plastic windows or minimum air leakage requirements for non-acoustically rated windows
08650	Special Windows: acoustically rated windows of unusual configuration
08700	Hardware: hardware and specialty devices for acoustically rated doors
08800	Glazing: specification of glazing thicknesses and lamination requirements
09100	Metal Support Systems: resilient channels for support of wall and ceiling sheathing, and resilient framing attachments and accessories
09250	Gypsum Board: specification of gypsum board types and thicknesses and application information for multilayer gypsum board installation
09500	Acoustical Treatment: specification of acoustical tile ceiling systems
09550	Wood Flooring: wood flooring with resilient underlayment

During this phase, the selected acoustical and audio/visual systems should be reviewed and alternate, less costly system configurations considered. Reduced function and benefit must be clearly explained to the owner at this time so that expectations are appropriately reduced.

Bidding Phase

During the bidding phase, contractors submit bids based on information in the contract documents. The qualifications of specialty subcontractors are evaluated and the bids are reviewed by the architect, the engineers, and the owner to help ensure that the contract documents have been correctly understood by the bidders. The bids and contractor qualifications should be reviewed by the acoustical engineer and audio/visual engineer.

Construction Submittals Phase

During the construction submittals phase, the contractor submits various detailed drawings and material information to the acoustical consultant for review. These submittals are reviewed to verify that the specification requirements have been satisfied. Sometimes, materials substituted by the contractor for acoustical and audio/visual systems may not meet the specification requirements. There can be subtle differences between acoustical materials that may not be evident to someone who lacks experience with these materials.

Construction Phase

Since many acoustical construction systems are not commonly used, it is important that observations be made to verify that the contractor is installing the equipment and construction assemblies in the specified manner. Inspections are often required during framing installation, gypsum board application, sealing of sound-rated constructions, and the installation of noise and vibration reduction devices, such as vibration isolators, resilient channels, and floating floors.

Review of the installation of audio/visual systems is also beneficial during the construction phase including the A/V conduit systems. These are typically installed much earlier than the actual audio/visual devices themselves.

Project Completion/Post-Construction Phase

To verify that the acoustical equipment and assemblies have been installed in accordance with the construction documents, observation or testing of these systems can be conducted at project completion. This phase is essential to detect any weaknesses in the installation—such as a poor door seal adjustment—before the user occupies the building.

Certain operational tests of audio/visual systems should be conducted by the design consultant upon completion of the audio system. The specifications often require that the A/V contractor conduct preliminary testing to show that the systems are nominally within the operational bounds. The design consultant then verifies these results on behalf of the owner.

Conclusion

Acoustical assemblies and audio/visual systems are integrated into architectural, and mechanical and electrical systems of a building. It is important, therefore, that the acoustical design is represented in every phase of the project. For the best integration of form and function, acoustical and audio/visual systems need to be considered as systems similar to mechanical and electrical systems, rather than considered as furniture or add-on fixtures.

09650	Resilient Flooring: specification of padded vinyl flooring
09680	Carpet: specification of adequate carpet/pad weight and thickness
09950	Wall Coverings: specification of prefabricated wall panels and stretched fabric acoustical wall systems
10200	Louvers and Vents: acoustical louvers (sometimes specified in Division 15: Mechanical)
10850	Operable Partitions: material and performance requirements for acoustically rated operable partitions
11130	Audio/Visual Equipment: specification of audio/visual systems
13080	Sound, Vibration, and Seismic Control: seldom used for acoustical devices in current practice. Usually limited to specialty seismic control devices
14200	Elevators: specification of noise reduction devices for elevator equipment
15060	Piping Systems: specification of resilient supports and installation techniques
15160	Vibration Isolation and Seismic Restraints: detailed materials specification for vibration isolation systems and installation requirements
15180	Mechanical Insulation: internal acoustical duct lining
15400	Plumbing: specification of resilient attachments and installation requirements
15820	Fans: specification of maximum fan noise levels
15840	Ductwork: specification of duct construction and installation requirements
15870	Air Inlets and Outlets: requirements for diffuser noise
15890	Duct Sound Attenuators: performance & installation requirements
16050	Basic Electrical Materials and Methods: requirements for conduit and equipment installation; vibration isolation devices for transformers

16

Costs
& Benefits

Philip N. Sanders

Wood Studs • Metal Studs and Resilient Channels • Staggered Stud and Double Stud Construction • Concrete Masonry • Operable Partitions • Accordion Walls • Panel Walls • Floor/Ceilings • Decoupling Ceiling from Floor • Floor Finishes • Windows • Doors • Outdoor Barriers and Enclosures • Mineral Fiber Board • Glass Fiber Panels • Matte-Faced Insulation • Spray-Applied Materials • Other Sound-Absorbing Panel Systems • Specialized Acoustical Products • Ductborne Mechanical Equipment Noise • Vibration Isolators • Air Flow Noise • Sound Systems

PLANNING is the most cost-effective method for meeting the acoustical needs of a project. For example, a single stud wall with batt insulation in the cavity might achieve acceptable acoustic separation between a mechanical room and a restroom. But a concrete block wall or double stud construction might be required to separate the same mechanical room from a corporate board room. Similarly, if a music practice room is shaped to diffuse sound rather than reflect it back and forth between parallel walls, fewer add-on acoustical finishes will be required.

This chapter has four parts which discuss: (1) *Sound-Insulating Assemblies,* including walls, operable partitions, floor/ceiling assemblies, windows, doors, and outdoor barriers; (2) *Sound-Absorbing Materials;* (3) *Mechanical Noise and Vibration Control;* and (4) *Sound Systems.* Each part includes approximate costs and a discussion of costs and acoustical benefits. Costs for materials and labor vary with location, economic cycles, the bidding climate, quality of construction documents, and building type. The costs pre-

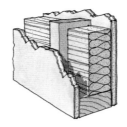

Figure 16.1 Single stud wall.

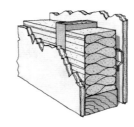

Figure 16.2 Single stud wall with resilient channels.

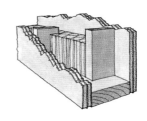

Figure 16.3 Staggered stud wall.

sented here are for materials and labor, unless otherwise stated. The intent of this chapter is to help readers make broad cost/benefit decisions relating to acoustical design. Published cost data books are good sources of detailed cost information, but, ultimately, a qualified contractor's bid will provide the most accurate cost estimate.

I. Sound-Insulating Assemblies

The costs presented in this section assume a small to medium-sized project consisting of 900 to 2,800 m² (10,000 to 30,000 sq. ft.) of office or commercial space, union labor, painted finish, and use of materials readily available in the vicinity of the project. The sound insulation ratings are based on laboratory tests. See Chapter 7 for detailed coverage of the technical aspects of sound-insulating assemblies.

Wood Studs

A wood stud partition with no insulation and no acoustical sealant at the perimeters has a Sound Transmission Class (STC) rating of about STC 29, and costs $60 to $75.50/m² ($5.50 to $7/sq. ft.) of wall surface.

An additional $5.50/m² ($0.50/sq. ft.) spent on caulking the perimeters and sealing all penetrations increases the STC by about 6 points to a maximum of STC 35. The additional cost of $5.50/m² ($0.50/sq. ft.) to include 76 mm (3 in.) thick glass fiber batt insulation in the stud cavity increases the STC by about 2 points (see Figure 16.1). Sealing and insulating a wood stud partition increases the STC by about 8 points—a significant improvement—and adds about 16 percent to the construction cost.

Metal Studs and Resilient Channels

Walls built with light gauge metal studs or resilient channels perform better than simple wood stud walls. A wall framed with 0.6 mm thick (25 gauge) metal studs costs less than a comparable wood stud wall by $5.50/m² ($0.50/sq. ft.), and the acoustical performance is significantly better. Drawbacks to light gauge studs include height limitations and limited capacity to support cabinetry. The acoustical benefit of metal studs diminishes with increased stiffness. Thicker metal studs (1 to 1.3 mm thick

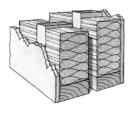

Figure 16.4 Double stud wall.

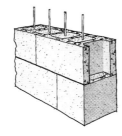

Figure 16.5 Grouted cement block wall.

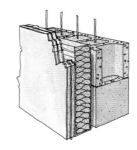

Figure 16.6 Combination masonry & stud wall construction.

or 20 to 18 gauge) provide little acoustical advantage over wood and cost about the same.

Figure 16.2 shows a wood stud wall with the gypsum board on one side attached using resilient channels. The additional cost of $8/m^2 ($0.75/sq. ft.) for resilient channels can improve the performance of a wood stud wall by more than 10 STC points: a dramatic improvement. The acoustical performance of walls with resilient channels decreases if they are not installed properly. The quality control and supervision required to achieve a correctly built construction can add to the cost of these partitions.

Staggered Stud and Double Stud Construction

Staggered stud and double stud walls cost more than single stud walls with resilient channels and take up more space. However, they can provide greater sound insulation and require less supervision to achieve the desired results.

Figure 16.3 shows a staggered stud wall whose acoustical performance is comparable to the resilient channel wall in Figure 16.2, but the cost is $8/m^2 ($0.75/sq. ft.) more. The base plate of this wall is 140 mm (5½ in.) wide, with 38 x 89 mm (2 x 4) studs staggered 200 mm (8 in.) on center or 300 mm (12 in.) on center.

Double stud walls, shown in Figure 16.4, are the most acoustically effective drywall assemblies because the two independent sets of base and head plates separate the two sides of the walls. The cost for double stud walls is $97 to $107/m^2 ($9 to $10/sq. ft.). Adding layers of gypsum board improves sound insulation, but with diminishing returns beyond two layers per side.

Adding interior layers of gypsum board between the two sets of studs degrades the acoustical performance of a double stud wall (see Chapter 7). Such interior layers are not recommended for sound insulation, but are sometimes required by fire codes.

Concrete Masonry

Concrete block (CMU) walls can provide better low-frequency sound insulation than drywall assemblies due to their mass, but must be painted or sealed and must have their cells filled with grout (Figure 16.5).

Superior low-frequency performance is not always indicated by STC ratings, which include only the frequency bands from 125 Hz to 4 kHz. Figure 16.6 shows a combination CMU/drywall partition that provides very good low-frequency sound insulation and is suitable as a movie theater demising wall. The STC rating for this composite wall is only slightly higher than the rating for a double stud drywall partition, but low-frequency performance is much better. The composite wall costs $65 to $107/m² ($6 to $10/sq. ft.) more than a double stud wall with insulation and one layer of gypsum board on each side. The additional cost and weight of CMU or composite CMU/drywall partitions may be justified when low-frequency sound attenuation is required, such as between movie theaters with digital sound. Table 16.1 summarizes costs and sound insulation ratings of the wall types discussed above.

Table 16.1 Data summary for walls.

Walls	Cost range	STC	Other considerations
Wood studs, no insulation	$60–$75.50/m² ($5.50–$7/sq. ft.)	29	
Wood studs with caulk and batt insulation (Figure 16.1)	$71–$86.50/m² ($6.50–$8/sq. ft.)	37	
Light gauge metal studs with caulk and batt insulation	$65–$80/m² ($6–$7.50/sq. ft.)	49	Light gauge studs have height limitations and cannot support heavy cabinetry
Wood studs with resilient channels, caulk and batt insulation (Figure 16.2)	$78–$94/m² ($7.25–$8.75/sq. ft.)	49	The installation of resilient channels requires extra care
Staggered wood studs with caulk and batt insulation (Figure 16.3)	$86–$102/m² ($8–$9.50/sq. ft.)	49	51 mm (2 in.) thicker than a single wood stud wall
Double wood studs with caulk and batt insulation (Figure 16.4)	$97–$108/m² ($9–$10/sq. ft.)	59	Approximately twice as thick as a single stud wall
CMU with grout, paint and drywall on studs with batt insulation (Figure 16.6)	$160–$215/m² ($15–$20/sq. ft.)	63	Heavy weight and high cost. Appropriate for cinema multiplexes or other locations where low-frequency sound insulation is important

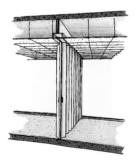

Figure 16.7 Accordion type operable partition.

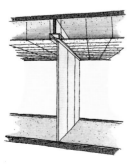

Figure 16.8 Panel type operable partition.

Operable Partitions

Operable partitions subdivide a large room into two or more smaller ones. Accordion (Figure 16.7) and folding panel partitions (Figure 16.8) are the most common types. Vertically folding flat panel walls that fold up into a storage position at the underside of the roof deck are also available, though more expensive. (Table 16.2 summarizes information about operable partitions.)

The cost of an operable wall is not limited to the cost of the installed product, but also includes the cost of a plenum barrier extending from the top of the wall to the underside of the structure above. Without a plenum barrier, sound isolation is limited by the acoustical performance of the ceiling. Figures 16.7 and 16.8 show plenum barriers installed above both accordion type and flat panel type operable partitions. The installed cost of a plenum barrier above an operable wall is $54 to $107 m² ($5 to $10/sq. ft.) of barrier surface area.

Operable partitions	Cost range	STC	Other considerations
Accordion partition (Figure 16.7)	$430–$750/m² ($40–$70/sq. ft.)	35–44	
Manual panel partition (Figure 16.8)	$750–$970/m² ($70–$90/sq. ft.)	40–50	May be cumbersome to erect and store, but manually compressed seals tend to work well
Electric panel partition	$970–$1,200/m² ($90–$110/sq. ft.)	40–50	Easy to operate, but perimeter gaskets may not seal as well as those on manual partitions
Custom walls (glass panels, doors)	$1600+/ m² ($150+/sq. ft.)	35–45	Doors and windows may limit acoustical performance
Plenum barriers	$54–$108/m² ($5–$10/sq. ft.)	49	Must be installed in the ceiling space above operable partitions in order to achieve the full sound insulating benefit of the operable partition

Table 16.2 Summary data for operable partitions.

Accordion Walls. Most accordion walls are not tested for sound isolation, and are intended for visual rather than acoustic privacy. However, some manufacturers have made modifications to their standard products and can achieve laboratory sound isolation ratings between STC 35 and STC 44 for a cost of $430 to $750/m² ($40 to $70/sq. ft.). The installed acoustical performance of both accordion and panel type operable partitions fall short of laboratory ratings because of variations in field conditions. Manufacturers typically acknowledge this and report anticipated field performance, as well as laboratory ratings, in their literature. The metric for installed acoustical performance is the *Noise Isolation Class,* or NIC (see Chapter 7). The NIC rating for sound-isolating accordion partitions is in the range of NIC 30 to 37.

Panel Walls. Panel operable walls, commonly used to divide large hotel ballrooms and other large rooms, provide better sound insulation than accordion partitions because they are heavier and their perimeter seals are more effective. However, even the best models have only moderate sound isolation ratings (STC 50 laboratory, NIC 42 installed). This is adequate for most business gatherings such as luncheon meetings with a guest speaker, but not for a dance party adjacent to a business conference. Higher ratings can be achieved by installing two operable partitions parallel to one another with an air space separating them.

Operable partitions can be electrically or manually operated. Electrically operated doors move into position and back into storage automatically with the flip of a switch, while manually operated doors may take twenty minutes or more to move into place. Manually operated walls are more reliable for sound insulation than electrically operated doors, because they have special hardware for compressing the perimeter seals.

The installed cost for a high quality, manually-operated, flat panel partition is $750 to $970/m² ($70 to $90/sq. ft.). Electrically operated walls cost \$970 to \$1,200/m^2 (\$90 to \$110/sq. ft.). Custom partitions incorporating glass panels and doors can cost \$1,600/m^2 (\$150/sq. ft.) or more.

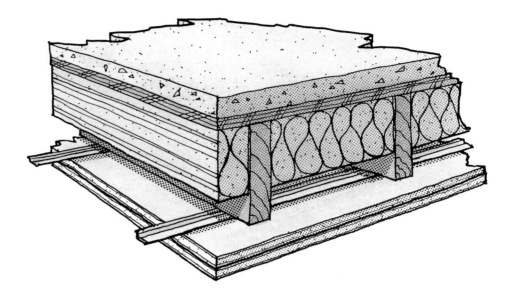

Figure 16.9 Wood framed floor/ceiling construction.

Construction	Cost
19 mm to 38 mm (3/4 in. to 1 1/2 in.) gypsum or cement based topping	\$16 to \$21.50/m^2 (\$1.50 to \$2/sq. ft.)
152 mm (6 in.) batt insulation	\$8/m^2 (\$0.75/sq. ft.)
Extra layer ceiling board	\$11/m^2 (\$1/sq. ft.)
Resilient channels	\$8/m^2 (\$0.75/sq. ft.)

Table 16.3 Costs of components in floor/ceiling construction above.

Floor/Ceilings

Floor/ceiling assemblies perform two acoustical functions (see Chapter 7). Like walls, they provide acoustical separation between adjacent spaces (airborne sound insulation), but they also reduce the sound of footfalls and other impact sounds from an upper floor (impact insulation).

Impact insulation and airborne sound insulation can be upgraded by decoupling ceilings from the structure and by altering floor finishes. Since floor/ceilings are integral to the building structure, we present incremental costs for upgrading assemblies during initial construction rather than costs for entire assemblies.

Decoupling Ceiling from Floor. A base assembly consisting of a plywood subfloor, joists, and 13 mm (1/2 in.) gypsum board screwed to the joists can be upgraded from STC 37 to over STC 58 by adding a lightweight topping slab, batt insulation, resilient channels, and a second layer of gypsum board as shown in Figure 16.9. Table 16.3 outlines the cost of each upgrade.

Figure 16.10 shows a resiliently suspended gypsum board ceiling using wire hangers and neoprene isolators. The cost of adding the resilient isolators to a suspended gypsum board ceiling is \$21 to \$31/m^2 (\$2 to \$3/sq. ft.), including labor.

Floor Finishes. Using carpet and pad on the floor, instead of hardwood or ceramic tile, is the simplest and least expensive way to improve impact insulation; however, building owners and occupants often want floors with hard finishes and good impact insulation. A number of resilient underlayment systems are available to provide a cushion beneath hard floors. Which one to choose depends on the floor finish (e.g., ceramic tile or hardwood), the underlying floor construction (e.g., concrete slab or plywood floor), the acoustical criteria for the project, and the individual preferences of the designers and builders. Adding a resilient underlayment to a floor finish system costs $32 to $75/m² ($3 to $7/sq. ft.), including installation, and adds between 6 to 51 mm (¼ and 2 in.) to the floor height.

Some special cases—for example, a highly sensitive recording studio, or a mechanical equipment room located above office space—require a secondary concrete slab that is physically separated from the structural slab. Figure 16.10 shows one of several floating floor assemblies. The incremental cost for adding a floating slab assembly is $215 to $325/m² ($20 to $30/sq. ft.). Table 16.4 summarizes the costs of various elements in floor/ceiling assemblies.

Floors and ceilings	Cost range	Comments
19 mm (¾ in.) to 38 mm (1½ in.) gypsum or cement	$16–$21.50/m² ($1.50–$2/sq. ft.)	A floor/ceiling assembly consisting of the base assembly described in the text (plywood subfloor, wood joists, one layer of gypsum board attached to the joists) has an STC 37 rating. When upgraded by the combination of first 4 items at left, the STC rating will exceed 58.
152 mm (6 in.) batt insulation	$8/m² ($.75/sq. ft.)	
Second layer of gypsum board	$11/m² ($1/sq. ft.)	
Resilient channels	$8/m² ($.75/sq. ft.)	
Resilient isolators in suspended ceiling	$21–$31/m² ($2–$3/sq. ft.)	
Resilient floor underlayment beneath hard floor finish	$32–$75/m² ($3–$7/sq. ft.)	Can improve IIC rating in the upgraded assembly above from IIC 38 to IIC 55.
Floating slab	$215–$325/m² ($20–$30/sq. ft.)	Used where a high degree of air-borne and structure-borne sound insulation is required.

Table 16.4 Data summary for floors and ceilings. (Note: costs in table are incrementally added to the cost of the base assembly.)

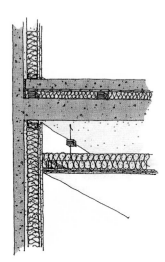

Figure 16.10 Floating floor construction with isolated ceiling and walls. Note: ceiling is suspended on wire hangers and neoprene isolators.

Windows

In most wall/window combinations, the walls are much more effective sound insulators than the windows. Therefore, upgrading the acoustical performance of an exterior facade usually requires upgrading the windows.

The first incremental improvement beyond single pane monolithic glazing is laminated glass. The interior film in laminated glass not only holds the glass together when it breaks, but also damps the resonance of monolithic glass (discussed in Chapter 7) and reduces high-frequency sound transmission. Laminated glass costs about $30/m² ($3/sq. ft.) more than monolithic glass, and increases the cost of typical residential windows from

Table 16.5 Data summary for
sound-rated window glazing.

Windows	Cost (materials only)	STC	Comments
Typical window with plate glass	$140/m² ($13/sq. ft.)	25	
Typical window with 6 mm (¼ in.) laminated glass	$170/m² ($16/sq. ft.)	32	Polymer inner layer damps resonances.
Acoustical window 6 mm (¼ in.) laminated glass, 50 mm (2 in.) air space, 5 mm (³/₁₆ in.) plate glass	$270–$375/ m² ($25–$35/sq. ft.)	42	
Acoustical window (e.g., studio control room window with large airspace)	$540/m² + ($50/sq. ft.) +	50+	

about $140 ($13/sq. ft.) to $170/m² ($16/sq. ft.) for materials. The STC rating increases from about 25 to 32.

The next incremental improvement is a double pane window with a large air space. Contrary to intuition, standard thermal pane windows, with 6 mm to 13 mm (¼ in. to ½ in.) air spaces between thin panes of glass do not necessarily provide better acoustical performance than single pane windows, and can even be less effective than monolithic or laminated glass of the same total thickness. To significantly improve sound insulation, the air space between panes should be at least 50 mm (2 in.).

A sound-rated window with one pane of 6 mm (¼ in.) laminated glass, a 50 mm (2 in.) air space, and 5 mm (³/₁₆ in.) glass has a rating of STC 42, and costs $270 to $375.60/m² ($25 to $35/sq. ft.) for materials: nearly twice the cost of a good quality residential thermal pane window.

Windows with larger air spaces and heavier glass can further improve sound isolation to STC 50 or higher, and cost $540 m² ($50/sq. ft.) or more for materials. Costs and other data about windows are summarized in Table 16.5.

Doors

It is possible to get good sound insulation from a simple solid core wood door by using gaskets to seal the perimeter. This kind of system is appropriate for apartment entry doors, conference rooms, or offices. However, proprietary sound-rated doors for critical spaces such as recording studios can cost as much as $2,500 each. Between these two extremes are doors that are appropriate for mechanical rooms, film screening rooms, and other applications where good sound insulation is important, but recording studio sound insulation is not required. Unless otherwise noted, all costs in this section exclude lock sets, frames, handles, and installation.

Figure 16.11 shows a solid core wood door with perimeter smoke seals and a door bottom shoe or sweep gasket. Such a system, including the door panel and gaskets, but excluding frame and hardware, costs about $125 for materials and has a rating of STC 30.

The next incremental improvement is a solid core wood door that comes packaged with gaskets and has a laboratory-tested rating of STC 38. This kind of door costs about $400. A 44 mm (1¾ in.) thick insulated metal door fitted with gaskets performs similarly for about the same cost.

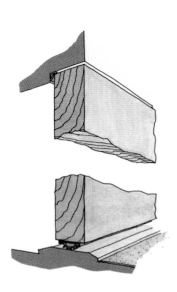

Figure 16.11 Gasketed door.

Commercial grade adjustable acoustical gaskets and door bottom drop seals are more effective than fixed smoke seals and door bottom sweep seals, and cost about $150 per door. An insulated metal door fitted with adjustable gaskets and drop seals costs between $600 and $800.

Proprietary sound-insulating metal doors can provide ratings of STC 45 to 53 or higher. The cost is $1,500 to $2,500 per door depending on the application, and the degree of sound insulation required. Proprietary sound-insulating doors typically include frame, gaskets and hinges, but not lock sets, handles, or thresholds. The cost and other data about doors are summarized in Table 16.6.

Outdoor Barriers and Enclosures

Outdoor barriers and enclosures control noise from mechanical equipment and highway traffic. They can be constructed of nearly any solid material, such as concrete, wood, or metal. The material used is less important than location and height. (See Chapter 5 for a discussion of barrier design.)

Concrete masonry walls, which are the most common sound barriers for highways, cost about $108/m^2 ($10/sq. ft.). Equally effective wood barriers can be constructed for less than half the cost, but are not as durable.

Sound barriers and enclosures can be made more effective by using a sound-absorbing finish on the noisy side, but it is difficult to find an appropriate material that resists exposure to the weather. Some spray-applied materials like acoustical plaster are appropriate as discussed in Chapter 6. Pre-manufactured barrier panels—usually made in a sandwich configuration consisting of perforated sheet metal on one side, sound-absorbing glass fiber in the middle, and solid sheet metal on the other—are made for outdoor applications, and cost $215 to $270/m^2 ($20 to $25/sq. ft.) for materials. This cost data is summarized in Table 16.7.

II. Sound-Absorbing Materials

Sound-absorbing finish materials are classified by Noise Reduction Coefficient or NRC rating. This rating represents the percentage of sound energy absorbed and is discussed in Chapter 6.

The variety of sound-absorbing materials on the market can be made more manageable by grouping them into a small number of functional materials that have a variety of different decorative facings as described in the following sections.

Mineral Fiber Board

Mineral fiber is the core material in typical acoustical ceiling tiles, which are available in different sizes, colors, and patterns for adhering directly onto ceiling surfaces or for installing in ceiling grids. The range of costs is $21 to $160/m^2 ($2 to $15/sq. ft.) installed, depending on options such as special edges, colors, or decorative patterns imprinted in the boards. The least expensive is perforated 30 cm x 30 cm (12 in. x 12 in.) glue-on tile. More expensive types offer flexibility in the shape of the suspension grid,

Doors	Cost range	STC
Solid core wood, smoke seals, and door bottom shoe	$125 each	30
Solid core door packaged with acoustical gaskets	$400 each	38
Acoustical gaskets added to a door	$150/door	
Insulated metal door packaged with gaskets	$600 to $800 each	38
Proprietary acoustical door	$1,500 to $2,500 each	45+

Table 16.6 Data summary for doors.

Outdoor barriers & enclosures	Cost range
Concrete masonry (installed)	$108/m^2 ($10/sq. ft.)
Prefabricated plenum panels (materials only)	$215–$270/m^2 ($20–$25/sq. ft.)

Table 16.7 Data summary for outdoor barriers and enclosures.

higher sound insulation (STC) ratings, and a variety of colors. Sound absorption ratings range from NRC 0.5 to 0.7.

Glass Fiber Panels

Glass fiber panels 48 to 112 kg/m³ (3 to 7 lb/cu. ft.) are available with lightly painted finishes, fabric covers, or perforated vinyl covers. All are available in a variety of sizes, or custom shapes, with mounting systems for walls or for ceilings as shown in Figure 16.12.

Stretched fabric systems are field-assembled in variations of fabric-covered glass fiber panels. They consist of frames attached to walls or ceilings with glass fiber panels inside the frames and fabric stretched over the frames.

Costs vary depending on the type of facing and the thickness of the panels. Lightly painted panels, made to fit in a lay-in ceiling grid, cost $32 to $65/m² ($3 to $6/sq. ft.) installed. Fabric covered panels and stretched fabric systems cost $130 to $215/m² ($12 to $20/sq. ft.) installed. Fabric covered panels in custom shapes and sizes may be $375/m² ($35/sq. ft.) or more.

Noise Reduction Coefficients range between 0.80 and 1.00 for panels that are 25 to 50 mm (1 to 2 in.) thick. Performance depends on thickness of the panels and the material used for the decorative facing.

Matte-Faced Insulation

Matte-faced glass fiber with a density of 24 to 48 kg/m³ (1.5 to 3 lb/cu. ft.) makes an excellent sound-absorbing core for custom decorative wall and ceiling systems. For example, it can be placed behind fabric, wood slats, or perforated metal. It is often used unfaced or behind black muslin in film studios and movie theaters. Its matte black color makes it difficult to see behind the decorative facing. Without a decorative facing, the cost is $11 to $21/m² ($1 to $2 sq. ft.) for materials. Fabric, wood slat, or perforated metal facings can increase the cost to between $70 and $215/m² ($7 to $20/sq. ft.) or more for materials. Acoustical performance depends on the type of facing used. NRC ratings for unfaced 25 mm to 50 mm (1 to 2 in.) thick insulation range between 0.8 and 1.00.

Spray-Applied Materials

Spray-applied materials include cellulose and cementitious materials commonly known as acoustical plasters. The acoustical performance of these materials depends on their thickness, porosity, mix of the material, and quality of installation. The installed cost is $32 to $92/m² ($3 to $8.50/sq. ft.). Acoustical performance is in the range of NRC 0.50 to 1.00.

Cementitious spray-applied materials come in a variety of densities with sound-absorption ratings that decrease as the density increases. The higher density materials, however, are very durable. The appropriate balance of sound absorption and durability must be determined on a project-by-project basis.

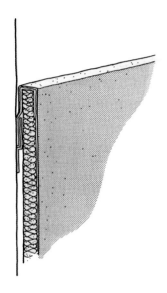

Figure 16.12 Fabric-wrapped glass fiber wall panels.

Other Sound-Absorbing Panel Systems

Other sound-absorbing materials include wood fiber panels and open cell foams. As with mineral fiber and glass fiber panels they are available in a variety of colors, shapes, and facings. Each material has its own sound-absorbing characteristics, and each is appropriate for certain circumstances. Costs range from $32 to $215/m² ($3 to $20/sq. ft.).

Specialized Acoustical Products

There is a variety of products specifically designed to modify room acoustics by diffusing or absorbing sound in particular ways. These products are used most often in auditoriums, recording studios, music practice rooms, control

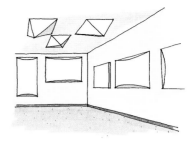

Figure 16.13 Sound-diffusing panels.

Figure 16.14 Prime-root diffuser.

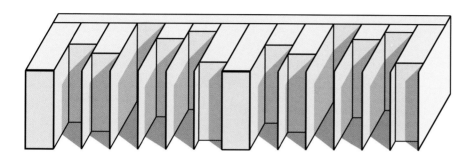

rooms, and home media rooms. Some examples of specialized acoustical products include pyramidal or bowed diffusing panels (Figure 16.13); quadratic residue or prime-root diffusers (Figure 16.14); field-constructed custom diffusers (such as bowed plywood); modular absorbers that serve the same purpose as surface applied sound-absorbing materials, but can be placed in a room and moved around to achieve the desired effects; masonry units with resonant cavities (Figure 16.15); products for adjustable acoustics including theater drapes and rotating triangular panels (reflecting, diffusing, absorbing).

Since these products are specialized, their costs tend to be high. The least expensive commercially manufactured quadratic residue diffusers, for example, cost about $215/m² ($20/sq. ft.) for materials. The most expensive, made of exotic tropical hardwoods, can cost more than $5,400/m² ($500/sq. ft.). Table 16.8 is a data summary for sound-absorbing materials.

III. Mechanical Noise and Vibration

Ductborne Mechanical Equipment Noise

Locating mechanical equipment far from noise-sensitive spaces is the best way to reduce equipment noise transmitted along the duct. Unfortunately, space limitations and the cost of long duct runs sometimes makes this impractical. The strategies used to reduce ductborne noise in these cases are discussed in Chapter 9. The costs of three of those strategies, internal duct lining, duct silencers, and acoustical plenums, are as follows:

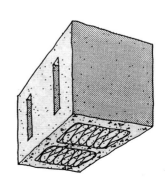

Figure 16.15 Acoustical masonry unit.

Table 16.8 Data summary for
sound-absorbing materials.

Acoustical finish materials	Cost range	NRC	Comments
Mineral fiber ceiling tiles	$21–$160/m² ($2–$15/sq. ft.)	0.50–0.70	Installed cost
Glass fiber ceiling panels	$32–$65/m² ($3–$6/sq. ft.)	0.70–1.00	Installed cost In lay-in grids, they are more absorptive than other ceiling products. Often used in open office areas.
Fabric-covered panels and stretched fabrics wall system	$130–$215/m² ($12–$20/sq. ft.)	0.80–1.00	Installed cost
Fabric-covered panels— custom shapes and sizes	$375+/m² ($35+/sq. ft.)	0.80–1.00	Installed cost
Duct liner board for use behind decorative facings	$11–$21/m² ($1–$2/sq. ft.)	0.80–1.00	Material cost only
Spray-applied materials	$32–$92/m² ($3–$8.50/sq. ft.)	0.50–1.00	Installed cost
Other materials (wood fiber boards, open cell foams, etc.)	$32–$215/m² ($3–$20/sq. ft.)	0.40–1.00	Material cost only
Modular diffusers, modular absorbers, and other specialty acoustical products	$215–$5,400/m² ($20–$500/sq. ft.)	N/A	Material cost only

Internal duct lining is the least expensive means of attenuating duct-borne noise. Duct liner board costs $11 to $21/m² ($1 to $2/sq. ft.). However, since duct liner takes up space inside the duct, its use also adds to the cost of larger ducts.

Duct silencers (Figure 16.16) typically come in 915 mm, 1,525 mm, 2.1 m, and 3 m (3, 5, 7, or 10 ft.) lengths and provide significant sound attenuation over a short distance. Silencer types include:

(1) Standard rectangular silencers that have a thick, sound-absorbing glass fiber fill behind perforated metal. They vary in price from $300 to $700 depending on size and sound attenuation. Round silencers range from $600 for a 0.91 m (3 ft.) long, 30 cm (12 in.) diameter model to $3,000 for a 1.82 m (6 ft.), 91 cm (36 in.) diameter model.

(2) Hospital-grade silencers that have the glass fiber fill encased in a thin mylar bag to prevent any fibers from escaping into the air stream. These cost 15 percent to 20 percent more than standard silencers of comparable length and cross sectional area. The mylar bag slightly reduces high-frequency performance.

(3) Packless silencers are often used for laboratory fume hoods, or where the air in the duct contains chemicals, fumes, or moisture. Packless silencers do not have any glass fiber fill, but instead rely on acoustic resonance in a series of chambers inside the silencer. These range in price from $350 to $2,000.

Acoustically lined sheet metal plenums (Figure 16.17) are useful for attenuating low-frequency sound (see Chapter 9), but they take up a lot of space. The installed cost of sound-attenuating plenums is $38 to $92/m² ($3.50 to $8.50/sq. ft.) of surface area. This amounts to about $250 for a relatively small 305 mm x 458 mm x 610 mm (2 ft. x 3 ft. x 4 ft.) lined plenum. The sound attenuation obtained from such a plenum is comparable

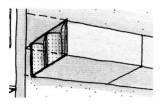

Figure 16.16 Duct silencer.

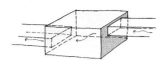

Figure 16.17 Acoustical plenum.

Figure 16.18 Neoprene pads.

Figure 16.19 Neoprene mount.

to that of a small silencer at high frequencies, but could be significantly better at low frequencies, depending on the shape and size of the plenum and the locations of the inlet and outlet. Prefabricated modular plenum panels cost $215 to $270/m² ($20 to $25/sq. ft.) for materials.

Electronic active noise reduction (see Chapter 10) is a fourth strategy for reducing duct-borne noise, but it is rarely used since it is relatively a new technology and the cost is still too high to make it practical for widespread use.

Vibration Isolators

The vibration isolators most commonly used for mechanical equipment can be divided into three categories:

(1) Neoprene pads add approximately 5 percent to the cost of the equipment to be isolated (Figure 16.18);

(2) Neoprene mounts and hangers add approximately 10 percent to the cost of the equipment to be isolated (Figures 16.19 and 16.20);

(3) Spring isolators add 15 to 20 percent to the cost of the equipment to be isolated, including accessories such as flexible duct, pipe, electrical conduit connectors, and thrust restraints (Figures 16.21 through 16.23).

Seismic spring isolators (Figure 16.24), which fall into the third category, are often specified because they combine vibration isolation and seismic restraint in one device, and are therefore less expensive than open coil

Products for reducing duct-borne noise	Cost range	Comments
Duct liner	$11–$21/m² ($1–$2/sq. ft.)	Material cost only
Standard silencers	$300–$3,000 each	Product cost only
Hospital-grade silencers	15%–20% more than standard silencers	Used where absorption of contaminants would be problematic
Packless silencers	$350–$2,000 each	Used where the airstream may contain corrosive chemicals (e.g., fume hood exhausts)
Field-fabricated plenums	$38–$92/ m² ($3.50–$8.50/sq. ft.)	Installed cost
Modular plenum panels	$215–$270/m² ($20–$25/sq. ft.)	Material cost only

Table 16.9 Data summary for silencers and plenums.

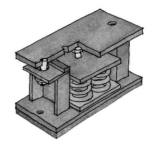

Figure 16.20 Neoprene hanger.

Figure 16.21 Open spring isolator.

Figure 16.22 Vertically-restrained spring isolator.

Vibration isolation	Cost range
Neoprene pads	5% of cost of equipment to be isolated
Neoprene mounts	10% of cost of equipment to be isolated
Spring isolators	15% to 20% of cost of equipment to be isolated
Seismic restraint	5% of cost of equipment to be isolated
Inertia bases	$1,000 to $5,000+ each

Table 16.10 Data summary for vibration isolation devices.

springs with separate seismic restraints. Open springs with outboard seismic restraint devices can add an additional 5 percent in materials and labor to the cost of the equipment to be isolated, bringing the cost of vibration isolation to between 20 and 25 percent of the cost of the equipment. However, housed seismic isolators are more difficult to align and adjust during installation, and more difficult to examine and correct if a vibration isolation problem is identified after building occupancy. The additional cost for open springs and outboard seismic restraints is justified for sensitive installations and where a vibration problem could be costly.

Concrete inertia bases (Figure 16.25) consist of steel frames filled with concrete. The cost is typically $1,000 to $5,000, depending on size and thickness, although very large inertia bases for heavy equipment, such as diesel generators, can cost up to five times this amount. The costs for vibration isolation devices are summarized in Table 16.10.

Air Flow Noise

Air flow noise is generated by obstacles in the air stream, such as dampers, right angle bends in the ducts, diffusers, grilles, and registers. The higher the velocity of the air when it encounters the obstacles, the higher the sound level generated. Locating dampers away from diffusers, maintaining low air velocity, creating smooth air flow conditions and selecting appropriate diffusers, grilles, and registers are keys to reducing airflow noise.

Proper damper location, low air velocity, and smooth air flow conditions add cost to a project by requiring larger and longer ducts. Quiet diffusers, designed for low resistance to airflow, can cost five to six times more than standard diffusers. The costs for terminal devices are summarized in Table 16.11.

IV. Sound Systems and Electronic Sound Masking

Sound system costs vary over a wide range depending on the size of the system, its functionality, and its flexibility. Chapter 11 covers some of the options available and the process of sound system design. Chapter 19

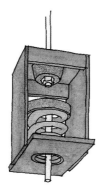

Figure 16.23 Spring hanger.

Figure 16.24 Seismically rated housed spring isolator.

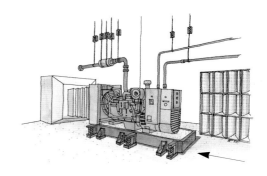

Figure 16.25 Concrete inertia base.

includes a discussion of electronic sound-masking systems. In this chapter, we present approximate costs for typical sound systems for two different project types and the approximate costs for sound-masking systems.

A typical sound reinforcement system for a 300-seat auditorium or church, including a central loudspeaker cluster, five microphone inputs, and cassette and CD playback capabilities would cost $50,000 to $75,000 installed.

A sound system for a 15,000 seat sports arena, with a central speaker cluster, distributed fill-in speakers for upper level seats, concourse speakers, and a high degree of flexibility to accommodate a variety of events, would cost $250,000 to $500,000 installed.

The costs for sound systems are summarized in Table 16.12. Sound-masking systems require low cost electronics and materials that can be procured and installed for approximately $21/m² ($2/sq. ft.) for a 930 to 1860 m² (10,000 to 20,000 sq. ft.) office. The cost is higher for small offices, but could be as low as $13/m² ($1.20/sq. ft.) for large installations.

Conclusion

With the exception of sound systems, acoustical design elements are integrated with other building systems—such as partitions, HVAC, and finishes—and do not occupy a separate line in the construction budget. The costs of these elements include materials and labor, indirect costs such as loss of floor space due to increased partition thickness, and the cost of increasing duct sizes to accommodate acoustical lining, etc. Each acoustical design element provides a benefit that should be balanced against its direct and indirect costs.

Diffusers, grilles, & registers	Cost range
Sidewall diffusers	$25 each
Perforated ceiling diffusers	$45 each
Drum louver diffusers (quiet)	$140 each

Table 16.11 Data summary for terminal devices.

Sound systems	Cost range
PA system for 300-seat church or auditorium, including five microphone inputs, cassette and CD playback	$50,000– $75,000
Large arena system with central cluster, distributed fill-in speakers, concourse speakers, and flexibility to accommodate various events	$250,000– $500,000
Electronic sound masking	$13 to $21/m² + ($1.20 to $2/sq. ft. +)

Table 16.12 Data summary for sound amplification systems.

17

Case Studies

Charles M. Salter, P.E.

VIRTUALLY all building projects have acoustical design issues. When constructing or renovating a building, the most common concerns are: room acoustics—controlling excessive sound buildup in occupied spaces; sound isolation—minimizing noise transfer from noisy spaces to quiet spaces; and, mechanical and electrical equipment noise and vibration control—preventing excessive mechanical noise within occupied spaces, as well as controlling noise emissions to the outdoors and neighboring locations. It also concerns plumbing noise transfer into occupied areas and selecting light fixtures that are sufficiently quiet for the space they serve.

This chapter provides a sampling of the thousands of projects worked on by Charles M. Salter Associates, Inc.. These projects are organized into 32 separate categories: Amphitheater; Amusement Park; Aquarium; Boardroom; Broadcast Studio; Classroom; Concert Hall; Conference Room; Convention Center; Council Chamber; Courtroom; Dubbing Stage; Environmental Noise Analysis; Hospital; Hotel; Multi-Family Housing; Single-Family Housing; Industrial Facility; Laboratory; Library; Museum; Office; Presentation Studio; Recording Studio; Restaurant; Retail Complex; Scoring Stage; Screening Room; Sound Reinforcement System; Dramatic Theater; Motion Picture Theater; Worship Space (Place of Worship.)

Each case study is illustrated with a photograph and followed by a brief discussion about the specific acoustical design issues faced during the course of the project.

1 AMPHITHEATER

Hollywood Bowl

Location Los Angeles, California •
Original Architect Lloyd Wright
(Frank Lloyd Wright's son) • *Original Acoustical Consultant* Vern
Knudsen

THE historic Hollywood Bowl opened in 1922. As one of the larger American outdoor theaters, it requires a powerful sound reinforcement system. The seating capacity of the amphitheater has been increased as part of several expansions and is now an impressive 18,000. The last row of seats is over 150 m (500 ft.) from the stage. During its early years, Bell Laboratories installed a three-channel audio system as an experiment. Over the next 50 years, various renovations were made to the Bowl. Notably, spherical diffusers were installed over the stage (under the direction of acoustician, Abe Meltzer and architect, Frank Gehry) to improve the on-stage acoustics and prevent excessive sound focusing on the stage itself.

In 1988, the monaural audio system was modified to provide an improved three-channel sound system, which upgraded the quality of sound by improving fidelity, power handling, and spatial localization. Stereo cross-feed and delay were introduced into the center channel loudspeaker cluster in order to maintain accurate acoustical imaging and perspective for seats very close to the stage.

In order to improve power handling and reduce phase interference from closely spaced loudspeaker horns, electronic comb-filtering was applied. This helped reduce the perception of echoes. Nevertheless, on-site tests revealed a strong echo was created by sound reflecting off a portion of a masonry wall, which had been constructed to reduce traffic noise and separate concessions from the seating area. Sound-absorbing materials were recommended that were suitable for outdoor installation.

The audio system has since been redesigned to accommodate 32–output channels driving 140,000 watts of power, with global computer automation. The current system, designed by Joseph Magee and built by Sound Image Inc, features active tuning compensation for changes in atmospheric conditions. Attending a concert at The Hollywood Bowl remains a unique and memorable experience.

Camp Snoopy

Location Minneapolis, Minnesota • *Architect* Hope Design Group, San Diego, California

KNOTT'S Berry Farm's Camp Snoopy is a fully enclosed family entertainment complex that is part of the Mall of America in Minneapolis, Minnesota. The 28,000 m² (7-acre) theme park has 23 rides and attractions and is the largest indoor facility of its kind in the United States. The pre-construction acoustical design objective was to provide a comfortable theme park environment that would not be excessively loud. The glass roof, concrete floors, and the noise generated by rides, people, and the large waterfalls that were included in the park's design, created acoustical challenges.

During the design phase, acoustical measurements were conducted in a similarly enclosed space in Alberta, Canada, as well as at the open-air Camp Snoopy Amusement Park at Knott's Berry Farm in southern California. Based on these measurements and on reverberation calculations, spray-on sound-absorbing wall treatments were recommended. A sound system also was designed to provide the park with a background music system.

In addition to room acoustics modifications and sound-system design for the park, an acoustical analysis of the various enclosed theaters was undertaken. Recommendations for sound isolation and control of excessive ventilation system noise were coordinated with the project architect and mechanical engineer. The reverberation time was predicted for each theater, and various types of sound-absorbing material were recommended for installation on the wall and ceiling surfaces. The construction drawings were reviewed to identify locations where certain attractions, such as the roller coaster, might create excessive noise and vibration intrusion in the theaters. It was recommended that the roller coaster support structures be isolated from the theaters.

3 AQUARIUM

Monterey Bay Aquarium

Location Monterey, California •
Architect Esherick Homsey Dodge &
Davis, San Francisco, California

PRIOR to the design of this aquarium, the project developers traveled around the world visiting aquariums and noting the benefits and limitations of each. A number of the aquariums they visited had problems with excessive noise and vibration. Based on their experiences, the acoustical quality for this facility was made a high priority. The developers were concerned with noise control because aquariums are constructed mostly of concrete and glass. In this case, excessive reverberation was controlled by installing 76 mm (3 in.) thick faced fiberglass panels within the ceiling's concrete coffers.

Aside from the exhibit spaces, the aquarium also includes a multi-purpose auditorium and some divisible classroom space. The auditorium was shaped so as to optimize its room acoustics, and double-glazed laminated windows were specified to control traffic noise intrusion. For the classrooms, sound-rated operable partitions were specified, and all of the classroom's windows were fitted with laminated glass.

The mechanical noise emissions from the aquarium had to be limited to meet the city's noise ordinance requirements. Also, all of the vibration isolation and mechanical noise control devices specified for fans and pumps had to be specially treated so as to resist the corrosive salt air environment.

4 BOARDROOM

Airtouch

Location San Francisco, California • *Architect* Studios, San Francisco, California

THIS facility is located directly below the mechanical penthouse in a high-rise office building. Excessive structure-borne noise was transmitted into the area where the boardroom was to be located. Our analysis determined that the problem could be solved by adjusting and/or replacing the spring isolators for the building ventilation fans.

Based on industry standards, the maximum noise criteria for the boardroom was recommended to be NC 25. In order to achieve this noise level, it was necessary to specify the maximum airflow speeds throughout the ventilation system ductwork including minimum lengths of duct lining in both the supply and return ducts. Acoustical treatments helped create a quiet, effective meeting space. The rear projection screen was located behind acoustically treated doors, and sound-absorbing wall panels, acoustical ceiling tiles, and thick carpets were installed to minimize reverberation and echoes.

The rear projection equipment room design accommodates a computer graphics-capable video projector and an optical slide projector. The majority of the audio/visual equipment is stored in the projection room. Electrical conduit was installed between the boardroom and the rear projection room to connect source equipment, (VCRs, audio cassette players, and laser discs) to the video projector for display. This allows presenters to load media during presentations while remaining in the room. The room was provided with both a playback audio system and a separate speech-reinforcement system. Wall-mounted loudspeakers are located on each side of the projection screen for reproduction of video sound tracks. These loudspeakers were concealed behind removable fabric panels in the front projection wall to maintain the architectural appearance of the room. The speech-reinforcement system utilizes overhead ceiling-mounted loudspeakers. Tabletop microphones are used during audio conferencing.

5 BROADCAST

KQED

Location San Francisco, California •
Architect Gensler & Associates, San Francisco, California

THIS television and radio facility is located in a three-story former industrial building that is adjacent to a bus service and storage yard. The technical space includes 1580 m² (17,000 sq. ft.) for television, 280 m² (3,000 sq. ft.) for radio, and 85 m² (900 sq. ft.) for a boardroom. The initial phase of the project included conducting noise and vibration measurements. The building was exposed to noise generated by passing trains, trucks, and buses; the proposed studio space was also located above a parking garage.

To reduce the level of environmental noise intrusion, both vibration- and sound-isolating constructions were suggested. A floating floor was installed above the building's existing post-tensioned floor to isolate the studio from the parking garage. During the value engineering phase of the project, a 100 mm (4 in.) thick lift slab with a 50 mm (2 in.) air space was deemed the most financially practical means by which to achieve the necessary sound isolation. The walls in the studio were made highly absorptive using 150 mm (6 in.) thick faced glass fiber panels. The ceiling was treated with 50 mm (2 in.) thick faced glass fiber board. This treatment in the studio controls the reverberation time to less than 0.6 seconds.

Ventilation noise was controlled in the TV studios to NC 20 at full cooling load. The ventilation system for the two TV studios is supplied by a single packaged air handling unit mounted on the roof about 3 m (10 ft.) from the studios. Long, internally insulated ducts and acoustical expansion plenums are the primary noise reduction measures.

6 CLASSROOM

Dwinelle Distance Learning Center

Location University of California, Berkeley • *Architect* Savidge Warren & Fillinger, Berkeley, CA

THIS room was converted from a general assignment classroom to an interactive, instructional video classroom that serves approximately 30 students. It functions for both distance learning (the viewing of classes via video broadcast from another location) and video conferencing. During the design phase, various control room and seating layout configurations were evaluated so as to optimize the room's floor plan. The microphones, one for every two people, are mounted in the tabletop so as to provide adequate audio input. Loudspeakers are located adjacent to the video monitors for video presentation purposes; additional loudspeakers were installed in the ceiling for use with the speech reinforcement system.

It was required that the ventilation system be controlled to a maximum of NC 25. The existing windows were covered over with gypsum board to provide adequate sound isolation and for light control purposes.

Reverberation was controlled by carpeting the floor, installing an acoustical tile ceiling, and treating certain walls with 25 mm (1 in.) thick sound-absorbing panels.

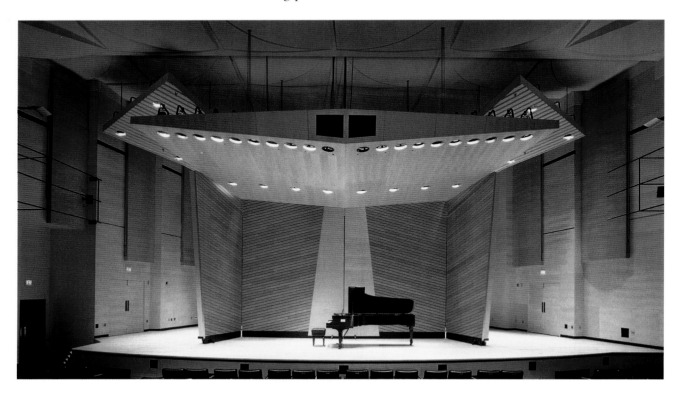

7 CONCERT HALL

Harris Concert Hall

Location Aspen, Colorado • *Architect* Harry Teague Associates, Aspen, Colorado • *Associate Acoustician* Elizabeth A. Cohen

BUILT by the Aspen Music Festival and School, the Joan and Irving Harris Concert Hall was designed to provide a performance and rehearsal space for the popular summer festival's students. The 500-seat, largely subterranean concert hall is used primarily for music performances, music recording, and film screenings. It has a reverse fan shape with a maximum ceiling height of 10 m (33 ft.) The canopy above the stage can be adjusted to accommodate the number of performers and the type of music that is being played. It is stepped and angled to help create a uniform early sound field. During music festivals, the first three seating rows can be removed to allow a pull-out stage to be extended, which accommodates over 150 musicians.

The suspended ceiling is 50 mm (2 in.) thick plaster to reflect low-frequency sound energy. In addition, all wall surfaces are angled and shaped to diffuse the sound throughout the audience area. The wall surfaces contain sliding 50 mm (2 in.) thick sound-absorbing panels that can be fully retracted into pockets. This allows for an adjustment of stage acoustics, particularly near the brass and percussion sections, which sometimes overpower quieter musical instruments. The overall reverberation time can be varied between 1.3 and 1.8 seconds when the Hall is empty. To avoid excessive noise, the mechanical equipment is isolated in an adjacent structure.

During the design phase, acoustical tests were conducted on a 6 to 305 mm ($\frac{1}{4}$ to 12 in.) scale model. In the physical model, certain wall and ceiling surfaces were faced with aluminum foil so that light reflections could be traced to the audience. This testing was conducted along with a computer ray tracing program. The objective of the model and computer studies was to achieve an adequate balance of side wall and ceiling reflections in the audience area.

Upon the hall's opening in August of 1993, its acoustics were almost unanimously extolled. Dubbed "The Carnegie of the Rockies" by the Denver Post, the hall was also called an acoustic "marvel" and a "gem."

8 CONFERENCE ROOM

BECAUSE this facility is adjacent to an alley, a furred sound-isolating wall was installed to isolate the studio from the alley noise. Double-stud sound-rated walls were also specified to provide appropriate sound isolation.

NC 20 was selected to be the maximum allowable background noise in the room. To control the noise produced by the ventilation system, internal duct lining and acoustical plenums were specified along with 950 mm (3 ft.) long silencers. The terminal air velocity was limited to a maximum of 1 m/s (200 fpm) for the supply ducts. The ceiling, which is made up of 38 mm (1½ in.) thick faced fiberglass tiles, has an NRC rating of 0.95. The wall panels are 38 mm (1½ in.) thick fabric wrapped over 38 mm (1½ in.) air space to help control excessive low-frequency sound buildup.

During the construction phase, site visits were conducted to verify the sealing of the electrical outlet boxes and the perimeters of sound rated walls.

Audio/Video Pacific Bell Corporate Television

Location San Francisco, California • *Architect* Garcia/Wagner & Associates, San Francisco, California

Cashman Complex

Location Las Vegas, Nevada • *Architect* Crosby Thornton Marshall, San Francisco, California • *Associate Architect* George Tate and Associates, Las Vegas, Nevada

THIS 9,290 m² (100,000 sq. ft.) facility is used for a variety of exhibits, including car and boat shows. The primary absorption in the space is provided by the sound-absorbing metal deck. The cost-effective deck provides both sound absorption and a roof structure for the facility.

Controlling mechanical equipment ventilation noise to a range of NC 40 to 45 was part of the program requirements. Because the facility was designed as an exhibit space, the results are less than ideal when it is used for other purposes, like large prayer meetings, for example. For exhibits, however, the room meets the client's needs.

Pleasant Hill City Hall

Location Pleasant Hill, California • *Architect of Record* Fisher/Friedman Architecture, San Francisco, California • *Design Architect* Charles W. Moore with Urban Innovative Group, Los Angeles, California

A TYPICAL meeting in this room includes five council members sitting at the dais, a city clerk, and an audience of between 6 and 200 people. The ceiling and wall shapes in this room help create optimum room acoustics in the audience area. The ventilation noise is controlled to NC 25 and various sound-absorbing wall treatments were installed to control reverberation and echoes.

To provide adequate sound reinforcement in the audience area, the city clerk lowers and raises the amplified sound as needed, with the use of electronic controls. To accommodate overflow crowds, a lobby is connected to the council chamber. Sound-rated walls reduce noise transfer from the lobby into the council chamber.

11 COURTROOM

Oakland Federal Building

Location Oakland, California • *Architect* Kaplan McLaughlin Diaz, San Francisco, California

THIS courtroom is part of a 92,940 m² (1,000,000 sq. ft.) federal office building complex. Federal courtrooms have specific acoustic criteria for background noise (NC 25 maximum), for reverberation time (0.8 seconds), and for partitions (STC 50 minimum).

Courtrooms always require a speech reinforcement system. In this courtroom, all the microphones are table-mounted. The judge's microphone has priority and mutes other courtroom microphones when it is in use. For privacy, the judge's main microphone and sidebar microphone can be muted. The audio system includes an infra-red listening system for those requiring hearing assistance. Loudspeakers are flush-mounted in the courtroom ceiling so as to be visually unobtrusive. The automatic microphone mixer system and a feedback suppressor minimize feedback. Individual microphones are automatically turned off when a person finishes speaking, thus minimizing background noise amplification. A remote control system provides overall volume control, along with control of the 8 individual microphones.

Skywalker Ranch Technical Building

Location Nicasio, California •
Architect Backen Arrigoni & Ross,
San Francisco, California

THE dubbing stage, along with the various recording, mixing, and editing rooms in this facility, was engineered to meet the owner's acoustic criteria for sound isolation, room acoustics, and background noise level. Floating floors, double-glazed windows, and masonry walls combined with furred drywall construction achieved the sound isolation requirements. The background noise level in the dubbing room was controlled to a maximum of NC 15 using in-duct silencers, plenums, oversized ventilation ducts, and a plaque air diffuser supply system. The reverberation time was controlled to 0.4 seconds. A portion of the Technical Building was constructed over a parking garage. Acoustical tests were conducted, and construction was designed so as to control the noise intrusion of car engines starting up.

It was desired that arches be part of the room's design. Cost studies conducted during the value engineering phase of the project dictated that the arches be constructed of glass fiber-reinforced gypsum rather than plaster. A 1-to-10 scale model was built as an aesthetic study model as well as an acoustical testing model. A 3 mm ($\frac{1}{8}$ in.) diameter microphone was used to receive the test signal in the model, and the sound reflection patterns in the model were displayed on an oscilloscope screen. The test indicated that the arches as designed would diffuse the sound, not create echoes. These test results were confirmed after the room was built.

During the construction phase, on-site field visits were conducted every two weeks to review the various sound-rated constructions and the installation of the ventilation system. Post-construction measurements of background noise were made in all noise-critical spaces to verify that the design criteria had been met.

The acoustical design of this building received an Honor Award from the American Consulting Engineers Council in 1988 in part because some of the recording spaces in the complex are among the quietest in the world.

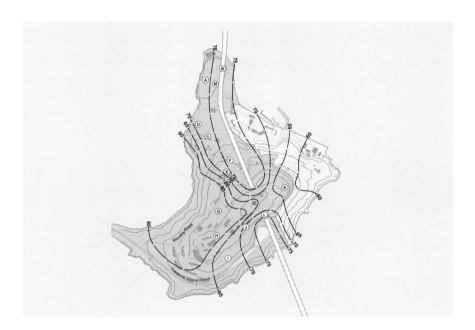

13 ENVIRONMENTAL
NOISE ANALYSIS

Treasure Island (Yerba Buena) Reuse Plan

Location San Francisco, California • *Architect* ROMA Design Group, San Francisco, California

TREASURE Island and Yerba Buena Island together make up a former San Francisco Bay Area military installation that was closed as part of the federal base closure program in 1995. The two islands, which are often referred to simply as Treasure Island, are joined by an isthmus, with Yerba Buena Island bordering the San Francisco Oakland Bay Bridge. The San Francisco Redevelopment Agency and the City and County Planning Department commissioned a study to evaluate alternative uses for the Yerba Buena Island and their environmental consequences. The environmental noise analysis is an essential part of the planning process because it helps determine the way in which Yerba Buena will be reused. Specifically, the report provides information about noise levels that will determine the most appropriate location for each proposed facility.

To quantify the traffic noise, measurements were conducted at strategic locations on the island. These locations included existing residential areas and potential future development sites. Based on the noise measurements, DNL contours were developed. The noisiest locations on the island are near the east- and west-side tunnel openings. The quietest areas are distant from the bridge and do not have line of sight to the traffic because of shielding from buildings or terrain.

The noise contours help determine the viability of proposed uses at specific locations. For example, a currently undeveloped site at the eastern tip of the island is being considered for residential use, a conference center, lodging, or a theme park. The bridge traffic generates a DNL of 70 to 76 dB in this area. For this noise exposure, new residential development should generally be discouraged according to the City's noise element. Traffic noise would have a significant effect on speech intelligibility in outdoor areas. Indoor noise can be reduced to acceptable levels by the building construction. Outdoor noise-reduction techniques, such as barriers, would be of limited use because traffic noise is generated by vehicles on the bridge, which is elevated 24 m (80 ft.) above the site.

14 HOSPITAL

Lucille Salter Packard Children's Hospital at Stanford

Location Palo Alto, California •
Architect Anshen + Allen Architects, San Francisco, California

ONE of the many acoustical design issues faced in this project was the control of emergency helicopter noise intrusion into patient rooms. On-site measurements determined that at a distance of 30 m (100 ft.), helicopters generated a maximum sound level of 100 dBA. The hospital administrator selected 55 dBA as the maximum allowable noise limit for patient room interiors, requiring a substantial reduction in noise levels. To achieve this reduction, windows with an STC rating of 45 were installed in those areas that were within 30 m (100 ft.) of the helicopter flight path. Within 60 m (200 ft.), windows rated at STC 39 were installed. The remaining rooms were furnished with windows rated at STC 33.

In addition to the helicopter noise control recommendations, acoustical analyses were conducted for other areas in the hospital. Dining rooms and banquet rooms were treated with sound-absorbing ceilings. Cleanable acoustic tile panels were specified for the hospital's treatment rooms, and sound-absorbing and waterproof panels were selected for the dishwashing room. In the on-call rooms, absorptive ceilings and walls were specified to minimize sound buildup. Also, sound-rated partitions were specified between patient rooms and between executive offices. Where sound-rated partitions intersect exterior window mullions, an aesthetically acceptable solution was developed to control sound transfer.

To meet Stanford University's exterior mechanical noise criteria of NC 45 (measured 3 m (10 ft.) from a building façade), acoustical louvers and other noise control means were implemented. Although the wall and door construction of the hospital's electrical rooms adequately controlled airborne noise, the concrete building structure transmitted electrical hum noise (generated by the 75 kVA transformers) to adjacent spaces. To control this vibration, rubber isolators were recommended.

15 HOTEL

Sutton Place

Location Chicago, Illinois • *Architect* MWM Architects, Oakland, California

BECAUSE each guest room in this urban luxury hotel was equipped with high-quality audio equipment, including televisions, CD players, and VCRs, the project developers expressed concern about noise transfer between guest rooms. In response, recommendations were made to limit the sound emissions level of the provided equipment. Single-stud insulated walls between guest rooms were also specified to control noise transfer.

During the construction phase, on-site acoustical measurements were made to help ensure that the party walls were sealed as required in the contract documents. A mock-up of a guest room fan coil unit was also tested. The three-speed fan generated sound levels between 32 and 46 dBA in the center of the guest room. This range, given that it varied according to the guest's adjustment of the fan's speeds, was judged to be acceptable. Traffic noise measurements were also conducted inside a guest room on the fifth floor. The measured noise intrusion varied between 35 and 41 dBA, which was deemed to be acceptable.

The City of Chicago has a property-line noise ordinance that limits the emissions of mechanical equipment. To meet these requirements, silencers for roof-mounted fans were recommended. In addition, it was advised that the plumbing pipes be isolated and that the vertical ventilation shaft be acoustically treated.

16 MULTI-FAMILY HOUSING

Portside

Location San Francisco, California • *Architect of Record* HKS, Inc., Houston, Texas • *Design Architect* Tower Architects, San Francisco, California

THIS condominium tower is located within 46 m (150 ft.) of the San Francisco-Oakland Bay Bridge. Because of the site's proximity to the Bridge, control of traffic noise intrusion was a central design concern. Various types of sound-rated glass were recommended. To assist the client's decision-making process, an audio/visual simulation was created to demonstrate what the rooms would sound like with the different types of window glass installed. Based on this simulation, the client chose an insulated, laminated window that reduced the interior noise level to below DNL 45 dB. Depending on the noise exposure the rooms received, the windows and exterior doors for this project were required to achieve between STC 31 and 41.

In addition to specifying windows, other acoustical design issues were addressed. The trash chute was treated with a damping compound and isolated from the building structure to prevent excessive noise transfer. A mock-up of a proposed garbage disposal was acoustically tested and found to be acceptable. Acoustical tests were conducted in the plumbing contractor's shop to quantify the noise produced by the tub/shower mixing valve. A mock-up of a ceramic tile installation on a floor was tested for impact noise isolation and found to meet project requirements.

Vibration isolation measures were also specified. It was advised, for example, that tenant storage lockers located above dwelling units be isolated. Pumps serving the facility's indoor swimming pool were also isolated since they were located over a dwelling unit. Garage door opening mechanisms were isolated from the structure using a detail similar to that shown in Chapter 18. It was also advised that exterior decks be isolated to prevent excessive footfall noise and vibration in the rooms below.

As part of quality control during the construction phase, a special acoustical inspection checklist was developed. The checklist, which was to be used by contractors as well as building inspectors, included plumbing noise control measures, caulking at sound-rated walls, and the insulation of sound-rated partitions.

17 SINGLE-FAMILY HOUSING

Private Residence

Location Northern California •
Architect Marcy Li Wong, Architect, Berkeley, California

IT was important to this homeowner that exterior noise intrusion be controlled to an acceptable level; in particular, this client required a quiet study room which would shield him from distracting noise, such as barking dogs, street traffic, and aircraft flyovers. The combined effect was measured, and sound-rated windows, entry doors, and skylights were selected in order to control the noise intrusion.

The client's former residence was tested and the client was interviewed in order to determine the acceptable noise levels for each part of the new house. Silencers and internal lining within the ventilation ducts were installed to control the ventilation noise to NC 15 in the bedrooms, study, and library. NC 20 was selected as the design criterion for the living room, while NC 25 was selected for the kitchen. The bedrooms, study, laundry room, and guest suite were fitted with solid core, acoustically gasketed doors. Quiet refrigerators, dishwashers, and garbage disposals were recommended for the kitchen. The clothes washer and dryer were isolated on a concrete inertia block to reduce vibration and noise propagation throughout the house. Also, the plumbing system and the roof drain pipes were vibration-isolated from the structure.

18 INDUSTRIAL

Oceanside Water Pollution Control Plant

Location San Francisco, California • *Architect* Simon Martin-Vegue Winkelstein Moris, San Francisco, California

SERVING the western half of San Francisco, this $220 million sewage treatment plant is a conceptually innovative, environmentally sensitive industrial facility. The plant is located on the coast adjacent to a residential community, the San Francisco Zoo, a recreation center, and other important public lands. Because of the project's size, the possibility of negative acoustical impact on the surrounding areas was a central design concern; this resulted in a design criteria that required that two-thirds of the building be installed underground.

The project's size and location presented some unique acoustical design concerns for the project designers. During the environmental analysis phase, for example, concern was expressed about vibration transfer from the construction/operations activities into the gorilla compound nearby. Ground vibration measurements were conducted to determine the response of the gorillas (as well as that of the zoo guests) to the expected level of vibration that the plant would produce. The anticipated vibration was hardly detectable and was judged not to be a problem.

During the environmental analysis stage, a stringent property line noise criterion of 50 dBA was established in order to minimize the acoustical impact of plant construction and operation. To achieve this level, the plant was designed so that noisier equipment was strategically located away from the zoo. To control noise emissions and yet allow access to the plant from the ground level, double roll-up doors were installed in a sound lock arrangement. Silencers and sound-absorbing treatments were used to control facility noise emissions from the plant's mechanical equipment. Ceiling areas in laboratories and offices were treated with sound-absorbing material. During the facility's construction, pile drivers were fitted with special mufflers and an earth berm was built to limit noise transfer from the construction site to the zoo and the recreation center.

19 LABORATORY

University of Washington Biomedical Sciences Research Building K Wing

Location Seattle, Washington • *Architect* MBT Architecture, San Francisco, California

THE University of Washington's Biomedical Sciences Research Building consists of individual laboratories, offices, and a teaching auditorium. Due to the university's laboratory ventilation requirements, interior sound control was a concern. The university had a fume hood noise criterion of 62 dBA measured at a distance of 305 mm (1 ft.). To meet this criteria, packless silencers were selected to control exhaust fan noise. Ceiling treatment was also required to prevent excessive reverberation. The facility's offices and auditorium were also considered during the design phase. Issues involved in their construction included: sound isolation, room acoustics, ventilation system, and noise control.

Researchers working in nearby laboratory wings were concerned that construction noise and vibration might affect their work. Measurements were made in the vicinity of the laboratory to establish an outdoor noise limit for mechanical equipment. Tests were conducted using heavy construction equipment to measure the vibration sensitivity of laboratory equipment in the adjacent labs. Based on these results, as well as on a survey of the researchers, recommendations were made to reduce the effects of internal and external vibration sources, including future construction activities. Specifically, optical microscopes, with up to 2,000 magnification or more, were installed on isolation tables.

To address noise intrusion into adjacent labs, the windows of the adjacent buildings were temporarily covered with plywood. An air space between the plywood and the window was created to achieve adequate sound isolation. A coordinated system of communication was established between the construction managers and the nearby scientists; this enabled certain researchers to conduct sensitive experiments lasting several hours while construction was inactive.

20 LIBRARY

Newport Beach
Public Library

Location Newport Beach, California
• *Architect* James Lawson Pirdy, AIA,
Inc., Newport Beach, California •
Design Architect Simon Martin-Vegue
Winkelstein Moris, San Francisco,
California

To help achieve sufficient quietness in this Newport Beach community facility, sound-rated wall constructions were installed around the air shafts, the elevator shaft, and the elevator equipment room. The doors to the elevator machine room and the mechanical rooms were specified to have an STC rating of 45. To minimize reverberation buildup and sound transfer to adjacent spaces, sound-absorbing materials were installed on the walls and ceiling in the mechanical rooms. In some areas, the specified acoustic wall panels doubled as tackable surfaces.

The acoustically rated structural metal deck shown above was specified for public areas as well as those parts of the library that did not have a suspended acoustic tile ceiling. The potential of aircraft noise intrusion was evaluated and deemed not to be a problem. Thus, the lightweight roof assembly was considered acceptable.

The maximum allowable background noise level was specified to be NC 30 in conference rooms and study rooms, NC 35 in reading rooms and private offices, and NC 40 in open plan office areas. For mechanical equipment noise control, silencers were recommended. Vibration isolation curbs were also recommended to isolate rooftop-mounted air-conditioning units.

THE original San Francisco Museum of Modern Art had difficulties with excessive reverberation, particularly in the large public lobby. This inspired the director of the new museum, completed in 1995, to seek acoustical counsel on this particular aspect of the design. The director was particularly interested in adequately controlling excessive reverberation in the atrium, which would periodically be used for parties and public gatherings. To achieve this control, portions of the main lobby's ceiling were treated with spaced-wood slats. Certain wall areas were made sound absorptive by installing fibrous material behind spaced-cement boards.

During the project's design phase, ventilation noise levels were specified for different rooms in the building. An acoustical analysis was conducted for all of the building's ventilation fans. The specifications required that the ventilation noise be controlled to the following maximums: auditorium, NC 20; boardroom and classrooms, NC 25; art galleries and exhibit areas, NC 30; and atrium, NC 35. Before the project manager accepted a substitute water-cooled chiller, measurements were conducted to verify the acceptability of the chiller-produced noise. Concern was expressed that high velocity airflow in the ventilation ductwork could create vibration in the art gallery ceiling, causing vibration in the ceiling-mounted light track and a vibrating light image on the walls. Experiments determined that the vibration would not generate excessive airborne noise nor cause movement of the lights.

Ten mm ($^3/_8$ in.) thick laminated glazing controlled airborne noise transfer from the rooftop-mounted mechanical equipment down through the skylights and into the atrium and top floor galleries. Acoustical louvers were specified for mechanical penthouse areas to control noise transfer to the out-of-doors. Also in the mechanical penthouse, a floating concrete slab was specified beneath the supply and return fans for both noise and vibration control.

Reverberation and echoes were controlled in the museum's multipurpose auditorium by making the sidewalls partially absorptive. This was achieved by rotating every third brick enough to expose the sound-absorbing material that was located behind the bricks. This technique was inspired by Bolt Beranek & Newman's acoustical design of the MIT chapel (designed by Eero Saarinen) in Cambridge, Massachusetts, which was completed in the mid-1950s.

The ceiling and walls of the loading dock, located directly outside of the auditorium, were treated with an acoustical plaster, and an STC 55 door was specified between the dock and the auditorium.

21 MUSEUM

San Francisco Museum of Modern Art

Location San Francisco, California • *Architect* Hellmuth Obata & Kassabaum, Inc., San Francisco, California • *Design Architect* Mario Botta, Lugano, Switzerland

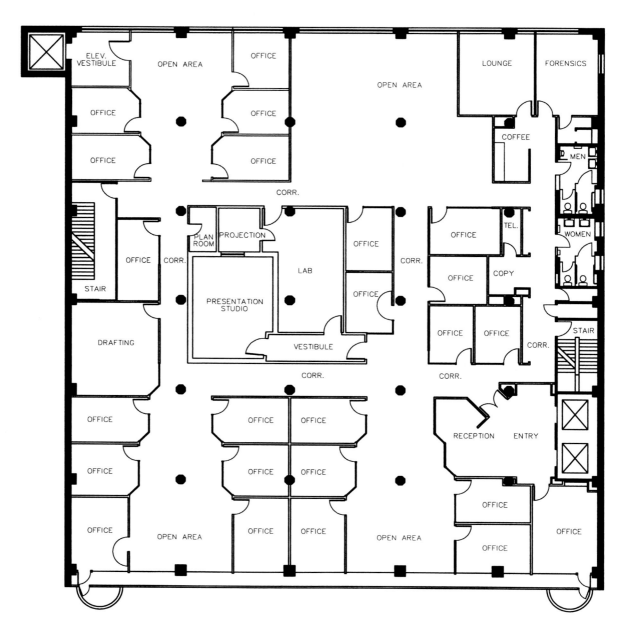

Charles M. Salter Associates, Inc. Office

Location San Francisco, Calif. •
Architect Aston Pereira & Associates,
San Francisco, California

OUR 930 m² (10,000 sq. ft.) office is located on the fifth floor of the historic Hallidie building in downtown San Francisco. The consultants are provided with private offices because we have found that enclosed offices encourage higher productivity. Our office does contain some open plan office areas, which are primarily used by secretaries, technicians, and marketing staff. Additional open space is used for staff meetings, client meetings, and periodic presentations to groups of students.

To control traffic-noise intrusion into the office areas, fixed glazing was installed 610 mm (2 ft.) from the building window/wall. Thus, single event traffic-noise intrusion is controlled to a maximum of 35 dBA.

In the private offices, gypsum board ceilings and single-stud insulated partitions are used for sound insulation between rooms. Acceptable room acoustics in these offices is achieved by the use of room furnishings such as books on bookshelves and fabric-wrapped acoustical panels. The ventilation noise in the offices is controlled to a maximum of NC 30.

In the open plan office area, a sound masking system was installed above the acoustic tile ceiling to provide a steady background noise level, thus minimizing distraction and annoyance caused by extraneous noise. The open plan office ventilation noise is controlled to a maximum of NC 35.

Charles M. Salter Associates, Inc. Office

Location San Francisco, California • *Architect* Backen Arrigoni & Ross, Inc. Architects, San Francisco, California

PRIOR to moving to our new offices in 1989, Charles Salter perceived the need for a room that would be used to demonstrate acoustical concepts, test audio equipment, and compare acoustical environments. This imagined room would provide clients and other interested people with the theoretical context in which to understand acoustical design issues relative to the array of potential design solutions. This space would be used to demonstrate acoustical situations to our clients so that they might better understand the issues that should be addressed in their own projects.

In June of 1991, Charles M. Salter Associates, Inc. completed this unique facility. The theater-like 5 m x 6 m (16 ft. x 20 ft.) room employs sophisticated multimedia technology, including computer animation. The surround sound system consists of 14 loudspeakers concealed behind the fabric-wrapped acoustical panels. The rear projection room contains a video projector and a slide projector along with control and amplification equipment including the multimedia computer.

The studio is acoustically isolated from the building with a floating concrete slab, 305 mm (12 in.) thick double stud walls and a spring isolated plywood and gypsum board ceiling. The ventilation noise in the room is controlled to a maximum of NC 5. Thousands of visitors have seen and heard presentations in our room.

Off Planet

(formerly Sega Music Group Studio) *Location* San Francisco, California • *Architect* Bottom/Duvivier, Redwood City, California

THIS intimate recording facility is composed of two control rooms, A and B, and a recording studio. Charles M. Salter Associates, Inc. provided acoustical consulting advice for the latter two room's acoustics. Control Room B also functions as a recording studio. Specific uses of the facility include sound effects design, tracking, mixing, and performance. The client wanted the facility's design to promote comfort, minimize fatigue, and encourage creativity for the artists who use the space. To achieve these qualities, the architect preserved an attractive high ceiling with a skylight to maintain a large open, uncluttered space. A resonant wood floor creates a wonderful acoustic environment for a grand piano or string quartet. Tilted acoustical panels suspended from the ceiling eliminate flutter echoes and create a desirable diffuse sound field. Room acoustics were made variable in the studio by providing quilted sound-absorbing blankets, which can be added or removed. The equipment is mounted on wheels that allow maximum flexibility in transforming the space for various uses.

In Control Room B, plaque air diffusers help control the supply air noise in the studio to less than NC 20. A double-glazed window isolates this space from the more reverberant and larger Studio A. Control Room B's design goal, to achieve a reverberation time of 0.4 seconds was also met. The reverberation was controlled by using a grid array of three materials on two adjacent walls, porous wood fiber board, fabric-wrapped glass fiber, and oriented strand board of compressed wood fiber. The remaining walls use these materials in angled wall panels. The client was particularly pleased with the design of Control Room B because the space was attractive and construction costs were kept within budget.

Mudd's

Location San Ramon, California • *Architect* Jacobson, Silverstein & Winslow Architects, Berkeley, California

OFTEN, restaurants contain hard surfaces, such as concrete floors, wood ceilings, and lots of windows. While these materials create an aesthetically appealing space, they can also cause undesirable side-effects, such as excessive reverberation. Built in 1980, Mudd's Restaurant presented similar acoustical problems. The floor is colored concrete, the walls are stucco, and the restaurant has many windows. To control reverberation, a spaced, kiln-dried, cedar slat ceiling with sound-absorbing material behind it was installed. In addition to effectively controlling reverberation and concealing necessary equipment (fire protection gear, HVAC, and sound system), the vaulted ceiling is aesthetically appealing.

A letter written by a San Francisco architect explains the way in which the ceiling has improved the quality of the Mudd's dining experience for him: "The interior of the dining room is truly delightful, intimate yet spacious. The acoustics are also excellent, which was important because I was with my 90-year-old mother who has a hearing aid that magnifies background noises, making conversation almost impossible in most restaurants."

Stanford Shopping Center

Location Palo Alto, California •
Architect Bull, Field, Volkmann,
Stockwell, San Francisco, California
• *Designer of sound system* KMK
Limited, White Plains, New York

THE Covered Mall No. 3 was designed as part of the expansion and remodeling work done at the Stanford Shopping Center in the late 1970s. Because this space was designed to be used occasionally for lectures, special events, and choral presentations, acoustical quality was a high priority. To provide this positive acoustical environment, the ceiling was made sound-absorbent by installing 190 mm (7½ in.) wide spaced wood slats with duct liner above. The openings into other circulation areas also absorb sound. As a result of the installed ceiling and the room's shape, the acoustics in the enclosed space are acceptably controlled. The room's good acoustics allow the speech reinforcement system, which is installed in the ceiling, to operate effectively.

Other acoustical design issues for this project included controlling rooftop mechanical equipment noise intrusion into retail spaces and limiting the distributed sound-system noise transfer into the neighboring residential areas.

Todd-AO

Location Los Angeles, California •
Architect Boto Architects, Inc.,
Venice, California

LIVE orchestra scoring for film and television is conducted in this room, which seats up to 150 musicians and has a reverberation time of 1.8 seconds. It is one of the largest scoring stages in the world, with a floor area of over 605 m² (6,500 sq. ft.) and a volume of approximately 5665 m³ (200,000 cu. ft.). The current room was created by remodeling a 50-year-old scoring stage that was too dry acoustically and had an objectionable bass resonance.

To blend the sound in the room, acoustical wedge diffusers were placed in the ceiling and fiberglass sound-absorbing panels are arranged on the walls. Some of the panels have an air space behind them to provide additional low-frequency sound absorption. The bass resonance problem was solved by applying additional mass to the existing poly-cylindrical shapes on the walls and ceiling.

28 SCREENING ROOM

Dolby

Location San Francisco, California • *Designer* Berry Reischmann, San Francisco, California

THIS room is used for film screening, presentation, audio recording, and training. New products for improving motion picture sound are demonstrated, evaluated, and developed here. It is located on the third floor of a building originally constructed in 1910. The size and shape of the room were optimized for motion picture presentations. The coffered ceiling creates a desirable aesthetic and helps to evenly diffuse sound throughout the room. To achieve Dolby's reverberation time criterion of approximately 0.3 seconds in the mid frequencies, about 70 percent of the wall and ceiling areas were made sound absorptive, using 25 mm (1 in.) thick sound-absorbing material over deep air spaces. The acoustical quality in the room can be varied using retracting sound-absorbing quilts in the side walls.

To develop structure-borne vibration control design standards for the 35 mm and 70 mm projectors in the projection room, vibration measurements were made on similar projectors at a nearby theater. The screening room projectors were mounted on a floating concrete slab, isolated from the surrounding floating floor.

Double stud walls, a sound-isolating gypsum board ceiling, and 75 mm (3 in.) thick acoustically gasketed doors control noise intrusion from the outside as well as from the theater to adjoining areas. Double glazing with a 200 mm (8 in.) air space was used to control projector room noise transfer into the screening room. The background noise in the room varies between NC 15 and NC 20 depending on the ventilation fan speed and thermal load. The office space below the screening room has an exposed ceiling for aesthetic reasons. The entire screening room is isolated from the office space by a concrete floating floor.

One Market Plaza

Location San Francisco, California • *Architect* Whisler-Patri, San Francisco, California • *Sculpture Design* Cesar Pelli

LOCATED in a high-rise building in downtown San Francisco, this sculptural 30 m (100 ft.) tower helps diffuse and absorb sound in the building's 28,320 m³ (over 1,000,000 cu. ft.) atrium. Given the atrium's size, reverberation could have presented numerous problems. The finished space has a reverberation time of 10 seconds. This fact called for a carefully designed, distributed sound reinforcement system which projects the sound signal to the audience, while not exciting the reverberant field of the room. The system was required to transmit both speech and music effectively. This system enabled the striking architectural form of the building to remain intact, while also increasing audibility for the atrium's visitors. The sound reinforcement system in the atrium is linked with a background music system in the adjacent food court area. Sound attenuators were specified to limit mechanical system noise in the atrium to a maximum of 45 dBA.

Fairfield Center for Creative Arts

Location Fairfield, California •
Architect ELS/Elbasani & Logan, San
Francisco, California

THIS 420-seat community theater accommodates both dramatic and musical productions. Multipurpose theaters like this typically present acoustical challenges to their designers. Such theaters should be able to effectively present a variety of programs, which necessitates versatility and often calls for an ability to adjust the room's acoustics.

The Fairfield stage was designed to be large enough for a 50-person orchestra, and the orchestra pit was sized to accommodate about 20 musicians. NC 20 was selected as the criteria for the theater and stage. A sliding glass window in the control room was provided so that the audio engineer responsible for mixing live shows could hear the sound being created in the theater by the sound reinforcement system. This window is sound-rated to control sound transfer from the control room into the theater when the window is closed.

The reverberation time in the theater was made variable by installing retractable velour draperies in an attempt to optimize the sound according to the program. These draperies, which were installed along the side walls, allow the reverberation time to vary by approximately 20 percent. When the theater is empty, for example, an average time of 1.2 to 1.4 seconds is achieved depending on the position of the drapes. Speech intelligibility measurements conducted at various seating areas determined that with the side wall curtains extended, the quality of speech intelligibility was measurably improved, as expected. When unamplified acoustic instruments are played, the drapery should be fully retracted so that the room can achieve its most reverberant condition. The stage house walls were treated with 25 mm (1 in.) thick faced sound-absorbing material to control excessive sound buildup.

In addition to the work done to improve the theater's multipurpose functions, roof drains over the stage and theater were isolated to control waterflow noise. Sound-rated doors and windows were also specified to control sound intrusion from the adjoining spaces, the lobby, and the busy street that borders the theatre.

Acoustical measurements were conducted to measure the level of noise intrusion due to aircraft flyovers from the nearby Travis Air Force Base. The theater's concrete block walls and insulated concrete roof, however, were sufficient to control the noise intrusion into the theater.

31 MOVIE THEATER

Stanford Theater

Location Palo Alto, California • *Architect* Hoover Associates, Palo Alto, California

THE design objective for this project was simply stated: to restore the 64-year-old, 1,200-seat Stanford Theatre to its original historic prominence as a movie house, and at the same time to bring the structure up to current seismic and safety standards, all without jeopardizing the building's antique authenticity.

Originally a vaudeville and silent movie theater, the Stanford was modified in the thirties when "talkies" were first developed. To make the room less reverberant at that time, the walls were pasted with wood fiber-board material, which was unattractive but nevertheless efficient. The organ was removed when the theater became a cinema, and a then-sophisticated sound system was installed.

In addition to the exacting renovation work that was done on the theater, the project managers, David and Pamela Packard of the Packard Foundation, wanted to improve the acoustical integrity of the space while restoring the theater to its original Greek-Assyrian splendor. The designers wanted the space to be acoustically suitable for both speech and organ music (an organ similar to the original was installed). Given the attention to detail that the renovation work required—uncovering old drawings of the original interior's design, working to match the exact color of the stage curtains, etc.—the acoustical objective called for coordination between all of the active disciplines. For example, a 25 mm (1 in.) thick semi-smooth acoustical plaster was selected because it approximated the acoustical quality of the wood fiberboard material that had been installed when the theater was converted to a cinema. In order to meet the Packard's aesthetic objective, the material was scored to make it like stone blocks.

The renovation also called for a public address system to facilitate presentations, talks, and panel discussions. The design of this system was particularly challenging because of the theater's natural reverberance, along with the requirement that the loudspeaker not be visibly intrusive. In addition, the original behind-the-screen cinema sound system was augmented by surround sound loudspeakers located on the side walls.

Mechanical ventilation equipment noise was controlled to a maximum of NC 30 in the theater's interior. Ceiling-mounted fan coil units were boxed in with gypsum board to limit casing-radiated noise. The roof-mounted compressor and condensing units were spring-isolated and mechanical noise transmission was controlled to meet the City of Palo Alto's property line ordinance.

32 WORSHIP SPACE

Moraga Valley Presbyterian Church

Location Moraga, California •
Architect Bull Stockwell Allen &
Ripley, San Francisco, California

THIS sanctuary is rhomboidal in shape, approximately 38 m (125 ft.) long, 32 m (105 ft.) wide, and 14 m (45 ft.) high at the mid-point. The 670 m² (7,200 sq. ft.) sloping main floor provides over 500 seats and additional choir seating for 64 people. The room's unusual shape, combined with the design objective that the space provide adequate acoustics for both speech intelligibility and organ music, made the acoustical design of this facility a distinct challenge.

CSA's recommended reverberation time for this space was 1.4 seconds. To achieve this level, a thin carpeting was installed in the aisles. Two layers of 16 mm (⅝ in.) thick gypsum board on the wall and ceiling surfaces provide a dense, reflective surface. Sound-absorbing wall treatment was applied on the room's rear wall surfaces to control echoes. The shapes of the ceiling and sidewalls provided desirable lateral early sound reflections to the pews for improved acoustics.

The ventilation noise was controlled to the maximum recommended level of NC 25. Noise intrusions due to exterior sources such as aircraft flyovers were controlled to a maximum of NC 30. The City had an exterior property line limit of 40 dBA between the hours of 10:00 PM and 7:00 AM, and 45 dBA between 7:00 AM and 10:00 PM. In order to achieve this level, it was recommended that most of the church ventilation equipment be located underground, and the equipment's outside air openings be acoustically treated.

The sound system provided speech reinforcement for clergy and music reinforcement, as well as playback and recording capability. The central cluster loudspeakers transmit the audio to the congregation at a dispersion angle of approximately 120 degrees. Reinforcement and playback were also provided to the choir facing the congregation.

After the sound system was adjusted and balanced, it was discovered that phase cancellation occurred in certain seating locations near the center aisle due to the symmetrical loudspeaker orientation. This problem was remedied by delaying the sound of one loudspeaker with respect to the other loudspeaker by 1.5 milliseconds.

18

Multi-Family Housing

Eric L. Broadhurst, P.E.

Governmental Regulations • Building Codes • City and County Requirements • HUD Requirements • Marketability • Privacy • Owner Expectations • Partition Design Issues • Room Layout • Plumbing Noise • Ventilation System Noise • Elevator Noise • Garage Door Opener Noise • Additional Noise Sources

As urban areas expand, so does the need for well-designed multi-family housing. Because people spend a significant amount of time at home, housing is an important area of acoustical design. The home should be a place of solace, a place to rest and relax.

In rural and suburban areas, people generally live in detached single-family homes. In urban areas, however, many people live in multi-unit buildings. The design of multi-family housing, which includes condominiums, townhouses, apartments, hotels, motels, and college dormitories, requires special attention because residences share "common" partitions.

This chapter explores issues that relate to multi-family housing, including governmental regulations, marketability, and design. Figure 18.1 illustrates a variety of noise sources that affect housing. Although this chapter focuses on multi-family housing, some of the design considerations are applicable to single-family residences, particularly for luxury homes, or homes that are located near highways or rail lines.

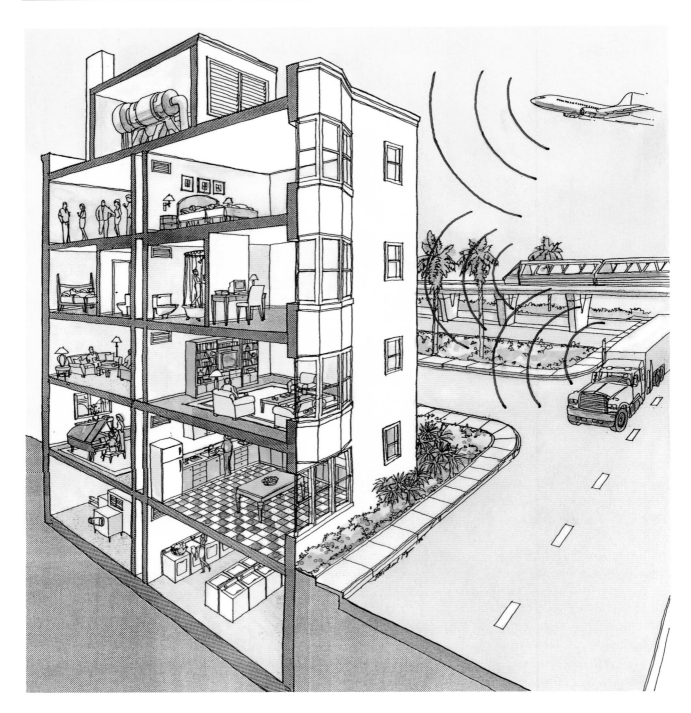

Figure 18.1 Various noise sources that affect housing.

Governmental Regulations

Governmental regulations that relate to multi-family housing are: (1) building codes, (2) municipal planning requirements, and (3) the United States Department of Housing and Urban Development (HUD) requirements.

Building Codes. Many building codes include acoustical requirements for multi-family housing. Tables 18.1 and 18.2 summarize California's minimum requirements for design and after construction. The floor/ceiling shown in Figure 18.2 is an example of a partition that would require an acoustical design that meets both STC and IIC standards.

Some states also have limits for interior noise levels due to environmental noise sources. In areas where the exterior DNL is 60 dB or more, California code requires that the exterior facade must provide sufficient sound isolation to reduce the interior DNL to 45 dB or less. In most cases, this can be accomplished solely by installing sound-rated windows. However, in areas where the DNL exceeds 75 dB, it may be necessary to upgrade the exterior walls and roof as well. Since windows need to be closed to meet these acoustical requirements, mechanical ventilation must be provided.

City and County Requirements. In addition to the indoor noise criteria in the building code, a municipal general plan often includes environmental noise limits. With housing, the goal is typically a DNL of 55 to 65 dB for outdoor use spaces.

	Minimum STC rating	Minimum IIC rating
Party/corridor wall	50	N/A
Floor/ceiling	50	50
Entry door	26	N/A

Table 18.1 Minimum California building code requirements during design phase.

	Minimum NIC rating	Minimum FIIC rating
Party/corridor wall	45	N/A
Floor/ceiling	45	45
Corridor wall w/ entry door	30	N/A

Table 18.2 Minimum California building code requirements after construction.

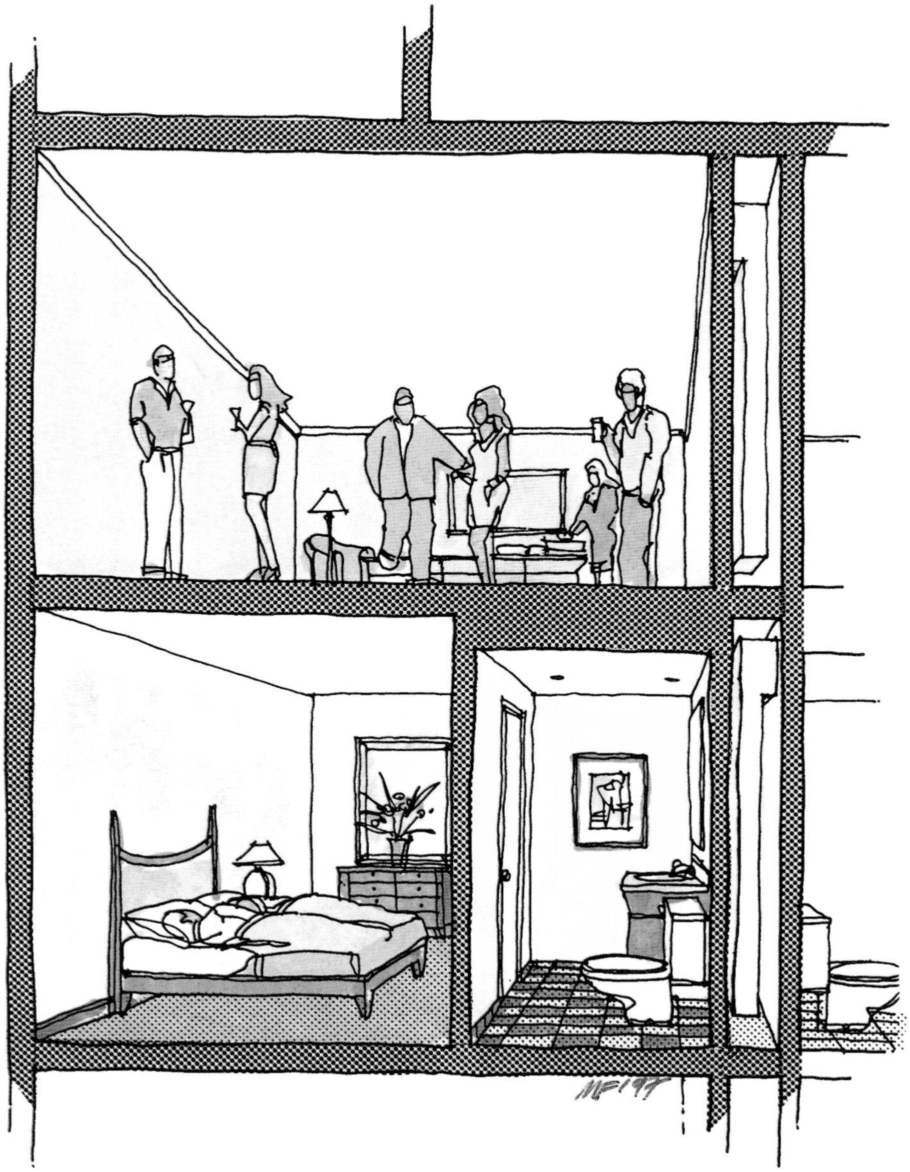

Figure 18.2 Floor/ceiling construction needed to meet building code standards.

	Minimum STC rating	Minimum IIC rating
Party/corridor wall	45*	N/A
Floor/ceiling	45	45
Entry door	33**	N/A

Table 18.3 Minimum noise isolation requirements for HUD project.

* Increase STC/IIC requirements by five points when living unit adjacent to uncarpeted corridor or high noise area (e.g., mechanical equipment room, elevator shaft, most commercial space)

** Where entry door is located opposite doors to other living units or service areas.

Additional limits are sometimes applied indoors since DNL may not be an accurate indicator of annoyance due to occasional loud events, such as train passbys or aircraft flyovers. Loud, short duration events can cause sleep disturbance and speech interference. Therefore, some communities include criteria for limiting indoor maximum noise levels where train, truck, and aircraft noise have been a problem. An example would be a maximum sound level of 50 dB in sleeping areas and 55 dB in other living spaces.

HUD Requirements. HUD has noise isolation requirements for certain projects that receive federal funds.[1] Table 18.3 summarizes these criteria. For environmental noise, HUD states that a DNL of 65 dB or less is acceptable, and indoor DNL should not exceed 45 dB.[2]

Marketability

While all projects need to meet the applicable building code and planning requirements imposed by government, there are also separate issues that relate to the property's marketability and the standard of care within the industry. The term standard of care refers to the legal standard that has been set for a specific professional service in a particular community at a given time. (This is discussed in Chapter 21.)

A quiet acoustical environment is likely to enhance the marketability of homes, but creating such a condition could result in higher construction costs. The acoustical engineer can play an important role in helping improve the marketability of a project in a cost-effective manner. The following sections discuss issues that affect marketability and standard of care.

Privacy. The ability to hear one's neighbor depends on both the partition noise reduction and the background noise level. If a room has a relatively high background noise level, it is difficult to hear your neighbor through a party wall. However, if a room is quiet, noise transmitted through the party wall is more audible. Rural areas are generally quieter than urban areas; therefore, for equivalent acoustical privacy, a partition in a rural area needs to provide more isolation than one in a city. Table 18.4 shows how the

Table 18.4 The relationship between STC of a partition and audibility of speech and music for various background noise levels.

20 dBA (very quiet) background noise	30 dBA (quiet) background noise	40 dBA (moderate) background noise	Speech & music audibility
STC 30	STC 20	—	Normal speech easily understood
STC 40	STC 30	STC 20	Loud speech easily understood
STC 50	STC 40	STC 30	Normal speech faintly heard
STC 60	STC 50	STC 40	Loud speech faintly heard but not understood. Normal speech inaudible.
STC 70	STC 60	STC 50	Loud speech inaudible. Amplified music audible.
STC 80	STC 70	STC 60	Amplified music at typical levels inaudible.

STC rating of a partition needs to be increased as the ambient noise level decreases, in order to maintain the same level of privacy between residences.

Owner Expectations. The sound isolation between dwelling units should be commensurate with the expectations of the residents. While an STC 50 party wall may be acceptable for an apartment or entry-level condominium, it might not be acceptable if a project is marketed as luxury housing. As the selling price of a project increases, so should the sound isolation in order to meet buyer expectations.

HUD has published an acoustical guide that discusses the various quality levels of construction.[3] This document suggests three levels of acoustical quality for multi-family housing corresponding to minimum standard, market-rate, and luxury. Our firm and others have adapted this information into a design approach that is summarized in Table 18.5.

These ratings are modified depending on the room adjacencies. For example, a partition between kitchens will typically not require STC and IIC ratings as high as a partition that separates a kitchen from a bedroom. In addition, there are technological limitations, particularly in wood frame construction, that may not allow a project to achieve the design values listed in Table 18.5.

In addition to increasing the sound isolation ratings as the quality of housing increases, interior noise criteria may also be made stricter. It is desirable to have an indoor noise level quieter than the code required level. Table 18.5 also includes typically recommended indoor noise goals for the various quality levels.

Even with careful design and construction of multi-family housing, partitions cannot block all noise; some noise transfer between units is inevitable. In order to avoid unreasonable expectations on the buyer's part, it is necessary to refrain from describing any dwelling unit as "sound-proof," "so quiet that you will never hear your neighbors," or "highest quality construction."

Partition Design Issues. In order for a partition to perform to its potential, care must be taken to avoid *sound leaks*. Often leaks occur via pipe penetrations, uncaulked partition perimeters, and electrical boxes. Sound leaks can be reduced by the following practices: (1) sealing of outlet boxes with pads (Figure 18.3); (2) caulking partition perimeters; (3) sealing penetrations through partitions (Figure 18.4); (4) gasketing entry doors (Figure 18.5).

If a partition is poorly caulked or a door is left ungasketed, the noise reduction could be 5 to 15 dB less than expected. This could cause a "luxury-grade" party wall (STC 60) to fall below minimum standards (STC 50).

For luxury multi-family housing projects, sound-rated interior partitions, in addition to common partitions, should be considered. For example, to protect bedrooms in a two-story townhouse, sound-rated walls and floor/ceilings are often appropriate.

Housing grade	Design STC and IIC rating	Design indoor DNL
Minimum standard	50	45 dB
Market-rate	55	40 dB
Luxury	60	35 dB

Table 18.5 Partition noise isolation and DNL criteria for various grades of housing.

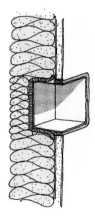

Figure 18.3 Sealed electrical outlet box.

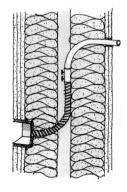

Figure 18.4 Sealed penetrations through sound-rated partition.

Figure 18.5 Gasketed entry door.

Figure 18.6 Desirable room adjacency between two dwelling units.

Room Layout. Situating noise-generating rooms of one dwelling unit next to noise-sensitive rooms of another unit should be avoided. For example, if a kitchen is located next to another unit's bedroom, the impact noise generated in the kitchen could be transmitted into the bedroom through the building structure, potentially causing sleep disturbance. If a kitchen or bathroom were instead placed adjacent to another kitchen, the occasional noise transfer would not likely be a problem. Similarly, problems can arise when a bathroom, kitchen, or living room with a hard-surfaced floor is placed over another unit's bedroom. The design solutions for these adverse juxtaposition can be difficult to implement and costly. Figure 18.6 shows room adjacencies that would create minimal conflict.

Plumbing Noise. While building codes in the U.S. do not address this issue, plumbing noise transfer generates a significant number of complaints in multi-family housing. This is due to the intermittent nature of the noise, its connotation, and its annoying quality.

It is difficult to estimate the level of noise that a plumbing system will generate due to the many variations in plumbing systems, including valve type, water pressure, pipe size, and piping layout. However, there are design practices and construction techniques that reduce the likelihood of excessive plumbing noise transfer.

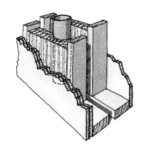

Figure 18.7 Plumbing piping separation.

Plumbing noise transmission can be both airborne and structure-borne. Airborne noise radiates from the piping, travels through the framing cavity, and passes through the gypsum board into a room. Structure-borne noise is conducted along the piping and transmits into building components such as framing and other piping prior to being radiated into a room.

To reduce airborne plumbing noise levels, the water and waste pipes should not touch the gypsum board (see Figure 18.7). When plastic waste/drain pipes are used, they should be wrapped with a putty-like material to suppress the radiated noise.

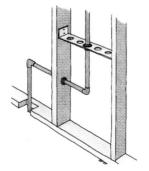

Figure 18.8 Isolated plumbing pipes.

To reduce structure-borne plumbing noise transfer, the piping can be isolated from the structure using resilient elastomeric/felt isolators (see Figure 18.8). While piping isolation is generally applied when the plumbing serving one unit passes through another unit, for luxury housing, it is often appropriate to isolate all plumbing piping.

Plumbing isolation during the construction phase is relatively inexpensive; conversely, retrofits made after the building is occupied are expensive. Due to people's sensitivity to plumbing noise and the relative ease and low cost of implementing isolation schemes, plumbing noise reduction should be considered for most multi-family housing projects and some single-family homes.

Ventilation System Noise. Housing often contains air-conditioning equipment, exhaust fans, and other noise- and vibration-generating mechanical equipment. The location of equipment should be addressed early during the design phase. Furnaces should be located away from noise-sensitive spaces such as bedrooms and living rooms, and building-wide mechanical equipment rooms should be located near as few residences as possible. The condition shown in Figure 18.9 should be avoided.

Once the mechanical equipment has been located, the noise transmitted to residences needs to be mitigated. Specific design issues include airborne noise, structure-borne noise, air turbulence noise, and vibration. To mitigate airborne noise, mechanical room partitions and doors should be designed so that they provide sufficient sound isolation. This entails determining the noise level generated by the mechanical equipment, comparing the level to the criteria for various types of rooms, and designing appropriate partitions. Table 18.6 lists some noise goals that are based on those published by ASHRAE.[4]

The most direct method to reduce duct-borne noise is to specify a quiet fan. Noise mitigation can also be provided within the ducting system by including long duct runs, internally lined ducts, and sound attenuating devices, such as plenums. (See Chapter 9 for more detailed information on mechanical noise mitigation.) Air registers should be selected at a sound rating that will achieve the design goal for each room and air volume dampers should not be located near the registers.

Where duct systems for two dwelling units are connected—as can occur in high-rise hotels—it is important to design the ducts so that sound does not transfer between rooms. Air registers should not be located near party walls, and ducts between units should be internally lined.

All mechanical equipment should be equipped with vibration isolation devices to control noise transfer through the structure. The isolation systems depend on the type of mechanical equipment, the noise sensitivity of the adjacent spaces, and the building structure.

Elevator Noise. Elevator noise transfer to dwelling units is largely structure-borne. Elevator shafts should be located adjacent to areas like storage rooms, public corridors, and mechanical rooms to reduce noise transfer into dwelling units. If an elevator shaft is adjacent to dwelling units, the shaft wall should be designed to provide adequate noise isolation. In addition, the power units in the elevator equipment room should be vibration-isolated.

Rooms	NC level
Private Residences	25–30
Hotel rooms	30–35
Corridors	35–40

Table 18.6 Typical noise criteria levels by room type.

Figure 18.9 Undesirable mechanical equipment location.

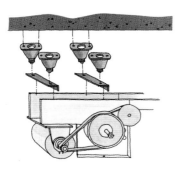

Figure 18.10 Isolated garage door opener.

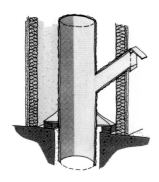

Figure 18.11 Isolated and treated trash chute.

Garage Door Opener Noise. For residents in a dwelling unit located above a garage, the structure-borne noise from a motorized garage door operator can be obnoxious. The noise is impulsive, intermittent, can occur at any hour, and is often initiated by one's neighbors. By selecting a "quiet" garage door opening system and installing the equipment on isolators, structure-borne noise can be significantly reduced (see Figure 18.10).

Additional Noise Sources. Noise from neighbors' appliances, cabinet door slams, and trash chutes often generate complaints. Appliances and cabinets should be located away from party walls. To minimize structure-borne noise of cabinet door slams, resilient materials at the closure contact points can be used. Noise from trash chutes can be mitigated by adding a damping material to the trash chute, installing isolators between the trash chute supports and the structure, and slowing the closing of the hopper door (see Figure 18.11).

Conclusion

Because people are sensitive to unwanted sounds in their home, proper acoustical design is critical to a successful residential project. This involves not only complying with governmental regulations, but also designing to meet the industry standard of care, which can include reducing noise intrusion into bedrooms, mitigating plumbing and ventilation system noise, and isolating elevators and garage door openers.

If appropriate acoustical treatments are specified prior to construction, they are relatively inexpensive to incorporate. Correcting acoustical deficiencies after a project is occupied can be difficult. The retrofit process is most often disruptive and expensive.

Notes

1 "HUD Minimum Property Standards: Multi-Family Housing," HUD, 1973, pp. 4–30.

2 "Environmental Criteria and Standards," HUD, 1984, p. 59.

3 "A Guide to Airborne, Impact, and Structure-borne Noise Control in Multi-Family Dwellings," HUD, 1963. This guide does not consist of regulations for HUD projects but rather is a collection of acoustical design ideas.

4 Heating, Ventilation, and Air-Conditioning Applications, American Society of Heating, Refrigeration, and Air-Conditioning Engineers Inc. (ASHRAE), 1996, Chapter 42.

19

Office Acoustics & Speech Privacy

Michael D. Toy, P.E.

Evaluation of Building Site • Assessment of Project-Generated Noise Sources • Ventilation Equipment Noise and Vibration • Other Noise Sources • Room Acoustics • Enclosed Offices • Open Plan Offices • Speech Privacy Prediction Method for Enclosed Offices • Sample Speech Privacy Calculation for Enclosed Offices • Speech Privacy Prediction Method for Open Plan Offices • Sample Speech Privacy Calculation for Open Plan Offices

AN acceptable acoustical environment is essential to productivity in most offices. The requirements for background noise and sound insulation vary according to the unique needs of each business. For example, low background noise levels are preferred in conference rooms to enhance speech intelligibility. In both open plan and enclosed offices, noise should be controlled to a level that reduces distractions. Counseling professionals require a high degree of speech privacy in order to reinforce their client's perceptions of confidentiality.

Part I of this chapter discusses the most important acoustical design issues for office buildings such as site selection, project-generated noise sources, and mechanical equipment noise control. Part II covers speech privacy in enclosed and open plan offices and Part III presents speech privacy calculation methods. With these calculations, the anticipated levels of satisfaction for various office design scenarios can be estimated.

I. General Acoustical Issues

DURING the development of an office building, the following acoustical issues should be considered: (1) the noise environment at the proposed building site; (2) potential project-generated noise impact on neighbors; (3) ventilation equipment noise and vibration; (4) miscellaneous noise issues associated with both base building and tenant improvements; and (5) design of appropriate room acoustics. These issues are discussed below.

Evaluation of Building Site

Acoustical analysis for an office development begins during the site selection process. Knowing the noise environment at a prospective location is essential for developers. Sometimes office buildings are located in noisy areas near major roadways or airports as a trade-off for either lower land costs or convenient accessibility.

Typically cities and counties have compatibility guidelines for commercial land uses. In California, for example, many of the community guidelines consider a site that has a DNL below 70 dB to be "normally acceptable." A DNL between 70 and 75 dB is considered "conditionally acceptable," meaning that construction should be preceded by an acoustical analysis that includes measures necessary to mitigate exterior noise.

The criteria for offices depend on the nature of the exterior noise source. If a project site is near a railroad or an airport, then the noise environment may fluctuate much more than if it were next to a highway. A reasonable interior criteria due to steady outdoor noise sources, such as freeway traffic, would range from an average sound level of 35 to 50 dBA. Alternatively, an interior maximum goal of 45 to 55 dBA could be used to address typical intermittent noise events such as train passbys and aircraft flyovers. The aforementioned average and maximum ranges depend on the anticipated office's usage. The lower ends of the ranges, average 35 dBA and maximum 45 dBA, should be considered for private offices and conference rooms. The higher ends of the ranges, average 50 dBA and maximum 55 dBA, should be acceptable for open plan offices.

An acoustical analysis for a proposed office building usually includes on-site measurements. From the measured data, the resultant indoor noise environment is calculated and the required sound-rated window/wall and roof assemblies are determined. Developers would want to avoid constructing a building using standard materials only to discover later that it cannot be leased without expensive retrofits. If a developer determines that the cost to meet prospective tenants' acoustical needs is too high, then an alternative site might be considered.

Assessment of Project-Generated Noise Sources

Many municipalities have noise ordinances that regulate the maximum allowable noise emitted by stationary sources at property lines. These regulations, or noise limits, vary according to the land use, the type of noise, the time of day, as well as the frequency and duration of the source.

Occasionally, ordinance requirements will be based on the existing ambient noise level at a site rather than on specific noise levels. In those cases, noise measurements are used to determine the ordinance limits. Some ordinances also establish maximum noise limits and allowable hours for construction activities.

Ventilation Equipment Noise and Vibration

The potential noise and vibration generated by a building's ventilation system should be analyzed during the design phase. Often, the ventilation system design is divided into two phases, the base building and tenant build-out. The base building includes the exterior building envelope and core areas such as elevators and toilet facilities. Tenant build-out areas include leased spaces. Once a tenant's requirements are known, the tenant part of the ventilation system is designed.

Base building ventilation system issues can include noise transmission to both neighbors and tenants. Vibration from rooftop equipment and from shafts are also concerns. The tenant improvement ventilation issues involve noise from air distribution systems serving the tenant spaces.

Other Noise Sources

Additional noise sources that could potentially impact offices include plumbing pipes and fixtures, electrical transformers, and elevators. The best way to deal with these noise sources is to locate them away from noise-sensitive areas. If this is not feasible, engineering solutions are available.

Room Acoustics

Public lobbies and common areas are often designed as a part of the base building phase. Adequate sound-absorptive treatment should be provided in these areas to control excessive reverberation. This is particularly important at reception areas in lobbies where people need to communicate.

II. Speech Privacy

DEGREES of speech privacy can be classified as *normal* or *confidential*. Normal speech privacy is defined as "the ability to comprehend an occasional word but never full sentences that are spoken in an adjacent room." Normal speech privacy is usually adequate for most offices. Confidential speech privacy means a neighbor is aware that a conversation is occurring in an adjoining room, but is not able to understand individual words. Confidential speech privacy is the recommended criterion for medical examination rooms, conference rooms, and executive offices. In both enclosed and open plan offices, the required level of speech privacy depends on both the sound isolation of the construction and the background noise levels. Figure 19.1 shows an enclosed office next to an open plan office.

Figure 19.1 (*left*) Enclosed office next to open plan office.

Figure 19.2 (*right*) Normal speech privacy provided by ceiling height partition and sound-rated ceiling.

Enclosed Offices

Frequently, enclosed offices are constructed with walls that extend only to the underside of a continuous suspended ceiling. Although this particular assembly is less expensive than extending the walls from slab-to-slab, speech privacy is compromised. To prevent the common ceiling plenum from being a "flanking path," an acoustical ceiling tile should be selected that has sound isolation rating at least 6 dB greater than the STC rating of the intervening wall. A rating scheme called CAC, *Ceiling Attenuation Class,* is used by acoustical tile manufacturers to rate the room-to-room sound transfer through ceilings based on ASTM E1264. Therefore, if an inter-office partition had an STC rating of 40, then the acoustical ceiling tile should be rated at a minimum of CAC 46 to minimize flanking, however, this may not always be feasible.

Normal speech privacy can typically be achieved with ceiling-height partitions if: (1) the intervening partition is internally insulated and sealed; (2) the background noise environment is at least 40 dBA; and (3) ceiling tiles have a minimum CAC rating of 35. The components that provide normal privacy conditions are illustrated in Figure 19.2.

Demountable wall assemblies are sometimes used to maintain flexibility in an office layout. Although these assemblies can achieve reasonable STC ratings in test laboratories, their field performance can be significantly lower than standard gypsum board partitions. A sound-masking system may be necessary to increase the background noise environment in order to achieve a "normal" speech privacy level. A sound-masking system electronically increases the background noise level, which in turn, improves speech privacy by making it more difficult to discern speech from adjacent areas. The sound-masking system was originally developed for use in open plan offices and is diagrammed in Figure 19.4.

A slab-to-slab full-height partition is typically required to achieve confidential speech privacy. When a partition extends from slab-to-slab, the noise transmitted through the ceiling does not affect the overall sound isolation.

Office designers need to be aware of miscellaneous flanking sound paths when utilizing sound-rated assemblies. Examples of flanking paths

 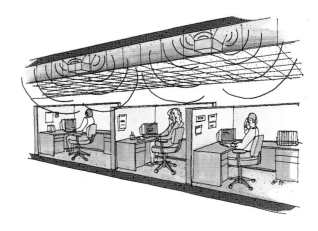

include doors, recessed electrical boxes, partition intersections at exterior window mullions, and continuous ceiling systems. Figure 19.3 indicates another path, through the ventilation system ductwork, called *crosstalk*.

Figure 19.3 (*left*) Example of *crosstalk* through flanking path.

Figure 19.4 (*right*) Open plan office with sound masking system.

Open Plan Offices

To achieve adequate speech privacy in open plan offices, it is often necessary to provide office screens (partial height partitions), sound-absorptive ceilings, sound-masking, sound-absorptive treatment on the walls, and adequate distance between work stations.

In open plan offices, the room's surfaces should be absorptive and include carpet, acoustical wall panels, and acoustical tile ceilings. The ceilings should have a Noise Reduction Coefficient (NRC) of at least 0.80. Absorptive treatments are especially beneficial on work station walls. The office screens should be both absorptive and sound-isolating, having minimum values of NRC 0.80 and STC 22, respectively. Figure 19.4 illustrates an open plan office.

The level generated by a sound-masking system is set so that it is not objectionable to occupants in the open plan area. The loudspeakers in the ceiling plenum are aimed up so their direct sound bounces off the structure above and is evenly distributed throughout the office as it passes through the suspended ceiling. Besides an electronic noise-generating system, adequate sound-masking can occasionally be provided by a ventilation system, vehicular traffic noise, and some types of office equipment. The advantage of an electronic sound-masking system is that the sound is steady and can be adjusted to achieve the optimum level and frequency spectrum. Ventilation and traffic noise are usually less reliable.

III. Speech Privacy Calculations

Speech Privacy Prediction Method for Enclosed Offices[1,3,4]

The calculation procedure shown as Figure 19.5 can be used to predict required STC rating for intervening partitions or the approximate level of speech privacy for enclosed offices. Factors required for calculation include:

Figure 19.5 Speech privacy analysis procedure for enclosed offices.

Note: For B and E a *furnished* room assumes an acoustical tile ceiling, carpeting, and typical office furnishings, while a *bare* room assumes hard-surfaced ceiling and floor with normal office furnishings. The room factors vary based on 10 log (floor area).

A.	Sound source level @ 1 m (3 ft.) from speaker	60 dBA	Conversational
		66 dBA	Raised voice
		72 dBA	Loud voice

B.	Effect of source room on sound level	Floor area of furnished room	Factor (dB)	Floor area of bare room	Factor (dB)
		5.6 m² (60 sq. ft.)	+9	5.6 m² (60 sq. ft.)	+15
		11.6 m² (125 sq. ft.)	+6	11.6 m² (125 sq. ft.)	+12
		23.2 m² (250 sq. ft.)	+3	23.2 m² (250 sq. ft.)	+9
		46.4 m² (500 sq. ft.)	0	46.4 m² (500 sq. ft.)	+6
		92.8 m² (1000 sq. ft.)	-3	92.8 m² (1000 sq. ft.)	+3

C.	Privacy criteria	20	Inaudible
		15	Confidential
		9	Normal

D. Partition STC

E.	Ratio of receiving room floor area to common partition area	for furnished room	Factor (dB)	for bare room	Factor (dB)
		1	0	1	-5
		1.5	+2	1.5	-3
	Floor area	2	+3	2	-2
	Partition area	3	+5	3	0
		4	+6	4	+1
		5	+7	5	+2
		6	+8	6	+3
		10	+10	10	+5

F. Background noise level in receiving room (in dBA)

(A) Speech effort: A conversational voice level is about 60 dBA at a distance of 1 m (3 ft.). When someone is using a speakerphone, the speech sound level might increase to 75 dBA. In an extreme case, such as a psychotherapist's office involved with primal scream therapy, the voice level could be as loud as 95 dBA.

(B) Source room effect: This is an adjustment for the amount of absorption in the source room based on the room size and furnishings; the larger the room is, the less the sound builds up.

(C) Privacy allowance: Select either normal or confidential speech privacy. These factors adjust the detectability of speech to the subjective response of listeners based on field tests as described in W. J. Cavanaugh et al.[1]

(D) Partition STC: The STC is the transmission loss of the partition separating two rooms.

(E) Noise reduction factor: This factor adjusts the sound isolation of the intervening construction to account for sound absorption in the receive room and the size of the intervening partition.

(F) Background noise: The ambient noise level in the receive room.

The computation procedure can be used to determine either the required STC of a partition between two rooms or the level of acceptability that would occur for a given partition STC and other relevant variables.

Required STC = A + B + C - E - F

Level of acceptability = A + B + C - D - E - F

Sample Speech Privacy Calculation for Enclosed Offices

To calculate the STC requirement for a wall separating executive offices, the factors and calculation procedures are listed below.

Source room (room where noise is generated) 3.66 x 4.57 m (12 x 15 ft.). Floor area is 16.7 m² (180 sq. ft.). Assume furnished room.

Receiving room (room where noise is received) 3.66 x 3.66 m (12 x 12 ft.). Floor area is 13.4 m² (144 sq. ft.). Assume furnished room.

Size of intervening wall 3.66 x 2.44 m (12 x 8 ft.). Wall area is 8.9 m² (96 sq. ft.)

Design for 65 dBA voice level and confidential speech privacy.

(A) Source sound level is estimated to be 65 dBA.

(B) Source room floor area is 16.7 m² (180 sq. ft.). This yields a factor of +4.

(C) Privacy allowance: 15 for confidential.

(E) The receive room effect can be calculated by dividing the common wall area 8.9 m² into the receive room floor area 13.4 m². The result is 1.5. The factor for a furnished room is +2 dBA.

(F) Background noise levels in enclosed offices typically vary from 25 to 45 dBA depending on the ventilation system and traffic noise intrusion through the exterior facade. A typical level is 33 dBA.

Therefore, the required STC rating is 65 + 4 + 15 - 2 - 33 = 49 to achieve a satisfactory speech privacy situation.

Alternatively, the level of acceptability can be calculated as follows: Level of acceptability = 65 + 4 + 15 - 49 - 2 - 33 = 0.

The anticipated human response to different speech privacy conditions is illustrated in Figure 19.6. In our example, the level of acceptability equals zero, thus leading to what is called *apparent satisfaction*. With apparent satisfaction, it is expected that approximately 9 out of 10 people will be satisfied.

If the speech privacy calculation yielded +3 because an STC 46 partition was used, then a condition called *mild dissatisfaction* would occur.

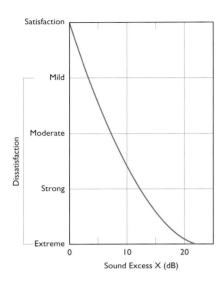

Figure 19.6 Level of acceptability chart.

With mild dissatisfaction, 8 out of 10 people would be satisfied. Sporadic complaints from the remaining 20 percent would still be expected to occur, although corrective action is rarely taken in these cases. These complaints would likely disappear with time as people became accustomed to the noise environment.

If the intervening wall assembly achieved an STC rating that is 7 points lower than STC 49, *moderate dissatisfaction* would occur. With moderate dissatisfaction, it is expected that 6 out of 10 people will be satisfied. Complaints registered by the 40% who were dissatisfied would most likely encourage some reparative effort in order to improve the level of speech privacy.

If the intervening wall assembly was 12 STC points lower than STC 49, only about 3 out of 10 people would be satisfied. Numerous, continual complaints would be expected and the dissatisfaction would likely be great enough to encourage significant improvement of the room's speech privacy.

If the construction were 20 points below the recommended rating, or STC 29, one would expect few, if any, people to be satisfied. The tenants under these circumstances would be expected to take drastic measures, such as to break the lease and find new offices if no remedy were provided.[1]

Speech Privacy Prediction Method for Open Plan Offices[1,2,3,4]

The open plan calculation procedure on Figure 19.7 assumes that the floor area of the office space is large compared to its height. Office space having enclosed areas (rooms with ceiling high walls) will have the acoustical characteristics of enclosed rooms and cannot be evaluated using this method.

Refer to Figure 19.7 while reviewing the calculation factors outlined below.

(A) Source Sound Level: This term describes the vocal levels that people typically use in an open plan office. When using this prediction method, it is necessary to select one of the four speech sound levels listed below.

Low: Telephone communication voice level typical in open plan offices that are treated with sound-absorbing materials.

Conversational: Typical voice level for people talking across a table.

Raised: People at a group meeting can be expected to talk using a raised voice level.

Loud: Workers using speakerphones may talk loudly to communicate.

(B) Privacy allowance: Select the level of desired privacy, either normal or confidential speech privacy.

(C) Distance from source to receiver: sound drops off at a rate of about 5 dB per doubling of distance in an open plan office with an acoustic tile ceiling and a carpeted floor.

(D) Barrier noise reduction: The width of the barrier should be at least twice its total height to reduce flanking around the sides.

(E) Room background noise level: This factor accounts for the masking sound available. Without sound-masking, the background noise can be as low as 30 dBA in a room, thus making it difficult to achieve adequate speech privacy.

(F) Directivity: The sound level of a voice is substantially louder in front of a person than behind. The data in F indicates that if a person is talking in a direction 180 degrees away from a receiver, the attenuation factor is 9 dBA. If a receiver is in front of a talker, the axis orientation is 0 degrees and there is no attenuation due to the directivity.

Figure 19.7 Speech privacy analysis procedure for open plan offices.

A.	Sound source level (@ 1 m (3 ft.) from speaker)	54 dBA	Low voice
		60 dBA	Conversational
		66 dBA	Raised voice
		72 dBA	Loud voice

| B. | Privacy criteria | 15 | Confidential |
| | | 9 | Normal |

C.	Distance from source to receiver (assume 5 dB per doubling of distance for rooms with sound absorbing ceiling and typical open plan office furnishings)	@ 1 m (3 ft.)	0 dB
		@ 2 m (6 ft.)	5 dB
		@ 3.7 m (12 ft.)	10 dB
		@ 7.4 m (24 ft.)	15 dB

D. Barrier noise reduction[5]

Distance between source and barrier (D_S)

Distance between receiver and barrier (D_r)	1 m (3 ft.)	2 m (6 ft.)	3.7 m (12 ft.)	1 m (3 ft.)	2 m (6 ft.)	3.7 m (12 ft.)
	Slight break in line-of-sight			.3 m (1 ft.) break in line-of-sight		
1 m (3 ft.)	5	5	5	8	7	7
2 m (6 ft.)	5	5	5	7	7	6
3.7 m (12 ft.)	5	5	5	7	6	6

Distance between receiver and barrier (D_r)	1 m (3 ft.)	2 m (6 ft.)	3.7 m (12 ft.)	1 m (3 ft.)	2 m (6 ft.)	3.7 m (12 ft.)
	.6 m (2 ft.) break in line-of-sight			1 m (3 ft.) break in line-of-sight		
1 m (3 ft.)	11	10	9	13	12	12
2 m (6. ft.)	10	9	8	12	11	10
3.7 m (12 ft.)	9	8	8	12	10	9

E. Background noise level in room (in dBA)

F. Speaker orientation relative to receiver

→ Diagram for use with D. Barrier noise reduction calculation

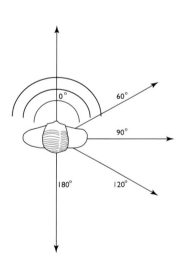

F. Directivity

Speaker axis relative to receiver	Attenuation (dBA)
0°	0
60°	3
120°	6
180°	9

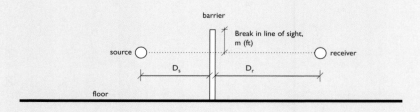

Sample Speech Privacy Calculation for Open Plan Offices

Assume that work stations at an advertising agency are arranged 3.7 m (12 ft.) apart. A sound-masking system is adjusted to provide 46 dBA of evenly distributed sound throughout the room. Each desk has a sound-absorbing tackable board that breaks the acoustical line of sight between work stations by 305 mm (1 ft.). The intervening barrier is 1 m (3 ft.) from the source and 3 m (9 ft.) from the receiver. The floor of the room is carpeted and there is a suspended acoustical tile ceiling.

(A) Source Sound Level. We expect a low speech level (54 dBA).

(B) Privacy allowance. We assume that normal privacy will be acceptable (9).

(C) Distance from source to receiver is 3.7 m (12 ft.) with an absorptive ceiling and floor. The estimated amount of attenuation is 10.

(D) Barrier provides 7 dBA of attenuation given a break in the line of sight of 0.3 m (1 ft.) and a spacing of 1 m (3 ft.) between source and barrier, and 3 m (9 ft.) between receiver and barrier.

(E) Room background noise level. The masking sound level is 46 dBA.

(F) Directivity. Assume that a talker is oriented toward a receiver like the desk arrangement in Figure 19.4.

Level of acceptability = A + B - C - D - E - F
= 54 + 9 - 10 - 7 - 46 - 0 = 0

The rating of 0 relates to the degree of satisfaction corresponding to Figure 19.6. We can expect apparent satisfaction from the scenario described above.

Conclusion

Various factors determine the acoustical qualities of an office—its location, construction, and layout are particularly defining factors. Speech privacy should be a central acoustical concern for office designers. Techniques exist to design offices so as to achieve adequate speech privacy.

Notes

1 W. J. Cavanaugh, W. R. Farrell, P. W. Hirtle, and B. G. Watters, Bolt Beranek & Newman, "Speech Privacy in the Buildings," *Journal of the Acoustical Society of America,* vol. 34 (1962) pp. 475–492.

2. Rein Pern, "Acoustical Variables in Open Planning," *Journal of the Acoustical Society of America*, vol. 49, no. 5 (Part 1), (May, 1971) pp. 1339–45.

3. Robert W. Young, "Revision of the Speech-Privacy Calculation," *Journal of the Acoustical Society of America*, vol. 38,. no. 4, (1965), pp. 524–530.

4. M. David Egan, *Architectural Acoustics* (McGraw-Hill Book Company, 1988).

5. Barrier calculations based on Z.E. Maekawa's method in "Noise Reduction by Screens," *Applied Acoustics*, 1, 1968, pp. 157–173.

20

Industrial Noise Control

John C. Freytag, P.E.

Industrial Noise Criteria • Predicting Noise Levels • Noise Mitigation

EFFECTIVE noise management should be a key element in the design and operation of all industrial facilities in order to help protect workers from noise-induced hearing damage. People in surrounding communities may also benefit from reduction of plant noise. Approaches to industrial noise control are evolving from quieter equipment, improved equipment enclosures, noise insulation materials, and active noise control. This chapter discusses criteria, prediction, and mitigation of industrial noise specifically for protection of workers. Minimizing noise effects in the community is addressed in Chapter 5.

Industrial Noise Criteria

Non-Regulatory. The non-regulatory goals of industrial noise mitigation are intended to promote adequate speech communication among workers and to reduce worker fatigue and discomfort. These goals are typically established by the plant operators and depend upon the type of work performed, company policy, and economic feasibility. Governmental regulations do not address speech communication or fatigue.

Regulatory. Hearing damage risk, on the other hand, is a health issue legally mandated in the United States by the Occupational Safety and Health Act of 1970 (OSHA). OSHA's noise exposure standards were formulated from extensive research into noise-induced hearing loss. The daily noise exposure limit is based on a model of a worker who experiences the same noise environment for twenty years and, as a consequence, suffers some permanent hearing loss.

The maximum daily noise exposure permitted under OSHA is shown in Table 20.1. The allowable duration T, for noise level L, is computed using Equation 20.1:

EQUATION 20.1

$$T = \frac{8}{2^{(L-90)/5}}$$

The allowable exposure (or daily dosage) is determined by a combination of noise level and duration throughout an eight-hour day. Eighty dB is defined as the *threshold level,* the minimum level used to calculate the daily dosage. From Table 20.1, it is obvious that the allowable exposure time is reduced by half for each 5 dB increase in noise level, or doubled for each 5 dB decrease in noise level. This relationship is called the *exchange rate.* The daily noise exposure is computed by summing the fractions of time spent at a given noise level relative to the time permitted at that same level as shown in Equation 20.2.

Duration per day (T)	A-weighted noise level (L)
8 hours	90 dB
6 hours	92 dB
4 hours	95 dB
3 hours	97 dB
2 hours	100 dB
1.5 hours	102 dB

Table 20.1 Maximum OSHA noise exposure.

EQUATION 20.2

$$D = \left[\frac{C_1}{T_1} + \frac{C_2}{T_2} + \frac{C_3}{T_3} + \cdots + \frac{C_n}{T_n}\right]$$

Where D = the total dose (not to exceed 1); C_i = the time spent at a measured sound level; T_i = the time permitted at that level, from Table 20.1.

For example, if a worker spends two hours in an area with a 90 dB noise level, the exposure at 90 dB constitutes $2/8$ of the maximum daily noise exposure, or 25 percent. If the worker moved to another location and spent three hours at 95 dB, the worker would have accumulated the maximum permitted noise dose as shown below:

EQUATION 20.3

$$D = \frac{2}{8} + \frac{3}{4} = 1$$

An alternative means of expressing the daily noise exposure is the eight-hour time-weighted average sound level (TWA). TWA is the equivalent eight-hour steady-noise level whose dosage is the same as one comprised of varying noise levels. The total noise dose can be converted to a TWA using the relation:

EQUATION 20.4

$$TWA = 16.61 \log(D) + 90$$

While the OSHA criteria apply to nearly all public and private industrial facilities throughout the United States, several other noise exposure criteria also exist. For example, the Department of Defense (DOD) and most international criteria have similar requirements except that they use a 3 dB increase for each doubling of duration in contrast to OSHA's 5 dB exchange rate. Likewise, the mining industry and NASA also have hearing damage risk criteria, which are the same as or even more restrictive than OSHA criteria.

Worker noise exposure in industrial facilities is routinely monitored by equipping the worker with a noise dosimeter that is worn throughout the day. The dosimeter comprises a calibrated microphone system and digital recording device that automatically computes the TWA.

A TWA of 85 dB is presently defined as OSHA's *Action Level* (AL). At this point, OSHA requires management to institute a hearing conservation program. Such a program involves employee training, audiometric testing, and availability of hearing protectors to affected workers. A TWA of 90 dB is OSHA's *Permissible Exposure Level* (PEL), at which point a citation is issued if employees are found without hearing protection. TWA values "significantly over the PEL" require engineering noise control. In these high-noise areas, the use of hearing protectors alone is not acceptable.

Predicting Noise Levels

Chapter 6 discusses some aspects of sound in rooms. One aspect that pertains to the prediction of noise levels is the acoustical properties within the room.

When moving away from a sound source, one eventually reaches a region where the sound level is influenced only by reflections from room surfaces. Here, the level remains constant regardless of distance from the source. This region is called the reverberant sound field or *far field*.

Predicting the noise level at some point in a room depends on three factors: (1) the sound power of the source (see sidebar); (2) the distance from the source; and (3) the acoustical properties of the room. Since reducing the amount of reflections in the reverberant sound field is an important means for engineering noise control, the next section will discuss some mathematical principles of room treatments.

Sound power cannot be measured directly; however, it can be calculated from several types of measurements under special acoustical conditions. One such condition is a reflection-free environment called a *free field*. For this condition, sound pressure levels are measured in a free field at various points over an imaginary surface around the source. The measured levels are averaged and the area of the imaginary surface is adjusted to 1 m² to yield the calculated sound power level. The sound power level of a device is a unique property not affected by room acoustics.

The total acoustical absorption in a room is:

EQUATION 20.5

$$A = S_1\alpha_1 + S_2\alpha_2 + S_3\alpha_3 + \cdots + S_n\alpha_n$$

Where A = the total number of metric sabins (sabins); $S_{1\ldots n}$ = the area of a building surface element in m^2 (sq. ft.); $\alpha_{1\ldots n}$ = the absorption coefficient of the building surface element.

For example, consider an untreated 6 m x 6 m (20 ft. x 20 ft.) room, 3 m (10 ft.) high, with an absorption coefficient (α) of 0.1 for all surfaces. The total surface area is 150 m^2 (1,600 sq. ft.). Therefore, A has a value of 15 (i.e., 0.1 x 150 m^2). If all surfaces were treated with a material having α = 0.8, A would be increased from 15 to 120 (i.e., 0.8 x 150 m^2).

In deciding whether to treat a room with sound-absorption, it is useful to determine if a particular worker location is in the reverberant field. Engineers can calculate this parameter, which has been named the *room radius*. Room radius is the distance from a source where the free field merges into the reverberant field. The room radius (r) can be approximated by using Equation 20.6.

EQUATION 20.6

$$r \cong \sqrt{\frac{A}{6\pi}}$$

Where: r is room radius, meters (ft.); A is number of metric sabins (sabins).

In the untreated room previously described, the room radius is 1 m (3 ft.). Thus, treating the room with sound absorption will reduce noise levels for a worker located more than 1 m (3 ft.) from the source.

As mentioned in Chapter 6, the reduction in noise level due to added absorption is:

EQUATION 20.7

$$LR = 10 \log\left[\frac{A_2}{A_1}\right]$$

Where: LR is the reduction in sound level (dB); A_1 is the number of sabins in the untreated room; and A_2 is the number of sabins after treating the room with sound absorption.

In our previous example, $A_1 = 15$ and $A_2 = 120$. Therefore, treating the room will reduce the reverberant noise level by 9 dB as shown below:

EQUATION 20.8

$$LR = 10 \log \left[\frac{120}{15} \right] = 9 \text{ dB}$$

The noise level close to a large sound source is fairly constant. As one moves away, the sound level begins to gradually diminish. Figure 20.1 shows how the sound level decreases as a function of the sound source dimensions and the distance from the source.[1] For instance, assume that a bank of cooling towers that is 3 m high and 9 m long has a sound level of 100 dB close in. As one moves about a meter away, the sound level will begin to decrease at the rate of 3 dB per doubling of distance. As one continues moving back to about 3 meters, the sound level will begin decreasing more rapidly—6 dB per doubling of distance. The latter is a characteristic of the far field.

Therefore, at a distance of 1 m, the sound level will be about 100 dB. At 2 m, the sound level decreases by 3 dB to 97 dB. As one moves further away, Figure 20.1 indicates that the sound level will be 95 dB at 3 m and 89 dB at 6 m.

Noise Mitigation

Two approaches are available for the mitigation of industrial noise: administrative controls and physical controls. Administrative controls limit the amount of time that workers are allowed in particular areas so their daily noise exposure limit (dosage) is not exceeded. Physical controls serve to reduce noise levels by addressing hearing protection for workers as well as modifying the source and path of noise.

Hearing protection devices such as ear plugs are an example of receiver controls. Although hearing protectors do reduce noise exposure, they are not always acceptable because (1) their performance depends on how well they fit the ears of individual workers; (2) their effectiveness varies with the spectra of the noise; and (3) workers may find them uncomfortable, thus cease wearing them. OSHA policy does not permit personal hearing protection in lieu of engineering noise controls except where the latter are deemed infeasible. The feasibility issue can be contentious; a costly equipment replacement may seem reasonable to an OSHA inspector, but can be perceived by management as an unacceptable expense.

Mitigation of the sound path generally employs equipment enclosures and/or barriers. Equipment enclosures are available from specialty equipment enclosure manufacturers. These enclosures are typically constructed of insulated dual metal panels. The interior panel surface is perforated to absorb sound and the exterior panel surface is solid. Enclosures typically provide an A-weighted noise reduction of 20 dB, depending on

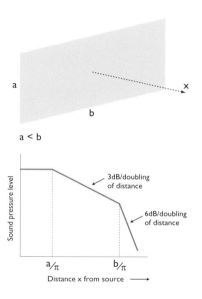

Figure 20.1

Notes

1. Rathe Theory described in Leo
L. Beranek, *Noise and Vibration
Control* (McGraw-Hill Book
Company, 1971), based on "Note on
two common problems of sound
propagation," *Journal of Sound &
Vibration*, vol. 10, 1969, pp. 472-477.

the spectrum of the noise source. Enclosures often require ventilation for cooling. Ventilation not only increases the complexity and cost of the enclosure, but can also decrease its sound attenuation.

Noise barriers can also be effective for factory floor operations. These barriers should be placed as close as possible to the noise source or to the sensitive receiver locations. Operations such as electronic assembly using pneumatic injection machines and conveyor-type operations are good candidates for acoustical barriers. Sound-isolating barriers should be faced with sound-absorptive material, whenever possible, to increase the noise reduction.

Source noise control consists of specifying quiet equipment. Usually this equipment is placed within a sound-isolating enclosure, fitted with vibration isolators, and treated with damping materials to minimize noise emissions.

Active Noise Control. Active noise has been used successfully to control ducted fan noise and is expected to gain wider use in the future (refer to Chapter 10). It is most effective on tonal noise. Low-frequency noise sources are good candidates for active noise control.

Conclusion

A variety of means are available to mitigate the sources of noise. Many of these may be implemented in the field, while others affect the particular design and operation of equipment, and therefore must be done by the manufacturer. Room acoustics treatment, barriers, and hearing protection for workers continue to be important issues for plant management.

21

Legal Issues

Jeffrey A. Leon, Esq. and
Charles M. Salter, P.E.

*Negligence • Breach of Contract/Warranty • Strict Liability • Nuisance and Inverse
Condemnation • Hearing Loss Claims • Case Studies*

WHEN a tree falls in the forest, does it make noise if there is no one present
to hear it fall? Today's world makes that long-debated question almost obsolete. On our increasingly crowded planet, it is difficult for many to imagine
that a place of true quiet exists. Highway and airport noises intrude into residential areas and the sounds of devices such as pagers, cell phones, and car
burglar alarms are pervasive. Proposed new highways and airports trigger
protracted battles over prospective noise. To escape this virtual cacophony,
society demands and expects greater protection from noise exposure.

The courts are increasingly being asked to resolve competing inter-
ests over sound. Part I of this chapter highlights the legal standards frequently considered when acoustical claims are made. A series of case studies in Part II illustrate how a variety of acoustical issues has been raised and
resolved through litigation.

I. Legal Standards for Evaluating Acoustical Claims

Acoustical issues arise in many different contexts and cast a wide net of
potential liability. As the case studies detail, a great deal of time and money
can be required to resolve claims against: employers, due to hearing loss
from noise exposure; landlords, as a result of noisy tenants or neighbors;
governmental agencies, due to disputes about land development and noise
from public transportation; contractors, architects, and engineers, for con-

struction and design flaws resulting in excessive noise for homeowners and other real property owners such as hotels, restaurants, hospitals, and schools; real property sellers and their real estate agents, resulting from alleged false promises concerning noise.

To provide insight into the complexity of such acoustical claims, the legal standards that are most commonly applied to such claims first should be considered.

Negligence

Some of the most frequently litigated acoustical claims are those filed against construction and design professionals who have worked on multi-family projects. The typical question asked when evaluating these claims is: Were the construction and/or design professionals negligent in the steps they took to reduce noise?

Negligence is determined by assessing whether a defendant has acted reasonably in light of the relevant community and professional standards or has breached the duty of care. As applied to construction and design defects, determining whether conduct has been reasonable often leads to an economic calculation. In assessing what standard of care should govern, courts will frequently compare the cost of the defect against the cost of the precautions that would have avoided the defect. The more that the cost of the defect exceeds the cost of avoidance, the more likely the construction or design professional will be found negligent.

Negligence is determined in part by consulting the appropriate building codes. For example, many local building codes specify the minimum sound reduction requirements for party walls in a multi-family project. If the project fails to satisfy the acoustic requirements of the local building code, then the developer and construction/design professionals may be found negligent per se, because they have violated the law in failing to meet the code acoustical requirements for the project. In this manner, the building code and the related concept of negligence establish the minimum—but not necessarily the maximum—legal standard for acoustical performance.

Even if the acoustical requirements of the applicable building code are satisfied, the developer as well as the construction and design professionals may still be held liable should the acoustic standard of care exceed the building code minimum. Relying solely on building code compliance may be a sufficient defense in jurisdictions where the courts look solely to minimum standards in determining reasonable acoustical performance. The modern trend, however, is to consider local practice as only one of many factors in determining the standard prevailing throughout the construction industry. Parties claiming negligence and demanding better acoustics may turn to a variety of sources for appropriate criteria, including other governmental standards, trade association guidelines, and expert witnesses. In addition, as the price of a housing unit increases from market rate to luxury, the reasonable expectations of the home buyer may buttress support for a higher than code-minimum standard for acoustical performance.

Before a more demanding acoustical standard can be adopted, however, due consideration in the negligence analysis must be given to such factors as technological limitations and cost. For example, no one can be expected to meet an acoustical standard that technology cannot achieve. Nor can anyone be expected to meet a higher acoustical standard if other design or construction specifications (e.g., necessitated by structural or fire-rating requirements) are likely to frustrate achievement of that standard. Furthermore, cost may limit what acoustical standards are reasonable. At a relatively high cost, a technologically sophisticated "quiet" room can be built to reduce noise significantly. Residential home builders, however, are unlikely to be able to afford such quiet; the sales price for such homes pushes most buyers out of the market. In contrast, perhaps a judge or jury would be more sympathetic to an argument that a very modest cost increase to provide for sound-rated windows would have significantly reduced noise intrusion into a residence from a nearby highway.

Thus, a variety of factors may contribute with different levels of significance to determinations of negligence. With societal demands for noise reduction on the rise, those determinations will likely become more hotly contested over time.

Breach Of Contract / Warranty

A contract or warranty may voluntarily set a higher acoustical standard than the standard that a negligence analysis would require. For example, builders and designers of buildings where multimedia events will be produced or where confidential work will occur may be required to satisfy more rigorous acoustical standards imposed by contract. Some companies in fact have developed their own acoustical standards, which may be incorporated into their construction contracts. Consequently, meeting the building code minimum or satisfying some other "reasonable" acoustical standard will not shield the developer or building professional from liability if the level of noise reduction promised in the contract or warranty is not delivered.

Frequently, and sometimes unintentionally, a developer of a residential project may also commit to a more demanding acoustical standard. Units may be offered for sale with marketing materials stating that the windows have been "acoustically rated" or the building has been "specially engineered" to control noise. More generally, but potentially as problematic for the builder defending a later acoustical claim, buyers may be advised that the units are of "highest quality construction" or are "luxurious." Such pronouncements may be viewed as a routine element of any sales program, but the builder must be keenly aware that these representations may readily support acoustical standards well above code minimums and may even support a fraud claim depending upon the specificity of the representation. Real estate agents and brokers also may incur legal exposure if they make undocumented and ultimately untrue assertions about the acoustical qualities of a unit to facilitate the sales process.

Although there is debate over whether a home purchaser "buys" the precise project plans and specifications, the potential also exists that those documents may be another basis for a warranty claim. If, for example, the plans call for 0.3 m (12 in.) thick sound-dampening insulation throughout the floor/ceiling construction of vertically adjoining units, but in fact the insulation is left out or 25 mm (1 in.) thick insulation is installed, then the home buyer may contend that the unit was not built to its warranted plans.

Strict Liability

Architects and engineers, as service providers, must be found negligent or in breach of a contract or warranty before they can be found liable. Most states apply the same rules to developers and contractors. Courts in those states reason that developers and contractors do not truly have sufficient control over what may be dozens of subcontractors and hundreds of products such that they should be held strictly liable for their buildings in the event of construction defects.

However, in a few states (including California), developers of mass-produced homes may be held strictly liable for construction defects even absent any negligence. As opposed to negligence, strict liability ignores whether an individual or business entity has exercised reasonable or even all possible care to avoid defective design or construction.

The doctrine of strict liability likely has contributed to the fact that 70 percent or more of all California condominium projects have had some sort of construction defects litigation. To show a "defect," as opposed to negligence, a party would likely point to non-compliance with building codes or industry standards, or a deviation from the project plans or specifications. In general, it is easier to prove that a party was strictly liable for a defect as opposed to having been negligent because even a party's extensive precautions to avoid a defect will not insulate the party from being found strictly liable if a defect exists.

Nuisance and Inverse Condemnation

The other major area of litigation in which acoustical issues arise is when a governmental agency or private industry engages in activities that generate an allegedly unreasonable level of noise. In these situations, property owners can attempt to stop the alleged nuisance and/or recover damages to mitigate the nuisance's effect. When the government blocks a private landowner from engaging in activities due to the potential for excessive noise or engaging in activities that generate an unreasonable level of noise, the private landowner may seek compensation for the alleged condemnation of the property.

A "nuisance" has been very flexibly viewed as any sort of injury to health, offense to the senses, or other obstruction to the free use of property. The touchstone for nuisance claims is whether the activity in dispute is a reasonable use of the property.

As applied to claims for acoustical nuisance, the first question is whether there is an unreasonable level of noise. The next question is what costs would be incurred to reduce the noise. For example, is the noise inside a home from San Francisco cable cars any higher than the levels of noise experienced in other houses on the same street? And even if cable car noise as a single event were significantly louder than any other traffic noise, what steps could be taken to mitigate that noise? No court is likely to shut down cable car operation if, for example, laminated window glass could significantly reduce the cable car noise within the home. Similarly, the external cost of noise on many residents near an airport that seeks to bring in bigger and noisier planes can be internalized through a nuisance action. The airport and the airlines would be forced to weigh whether it would be less costly to pay the hundreds or thousands of residents to upgrade their window glass to reduce interior sound or to require the airlines to buy sound-dampening equipment for their airplane engines. As described in Chapter 5, the FAA has chosen to mandate quieter planes and limit local planning options for operational control of airports.

A case of inverse condemnation involves a variant of the legal approach taken in nuisance cases. The standard applied is that the landowner is entitled to "just compensation" for the condemned land. In the case of inverse condemnation (in contrast to eminent domain in which the government agrees that it is taking the property), an important issue is whether the government has truly subjected the property to an encroachment or restricted its use in a manner which impairs the property value sufficiently to be a "taking" of the property. For example, will a proposed highway and its resulting additional noise make any difference to the value of a property that is already in the flight path of a major airport? Does a government agency's refusal to allow construction of a housing development due to noise concerns deprive the property of any reasonable economic use warranting just compensation to the landowner? Or are limitations on development simply reasonable use restrictions on the property which land-owners must bear?

"No Fault" Compensation for Job-Related Hearing Loss

Federal and state laws provide comprehensive systems of medical and other remedial treatment to reduce the effects of job-related injuries including hearing loss. These workers' compensation and other administrative systems generally trade off prompt compensation, regardless of the employer's fault, against smaller financial awards than would be recovered if the worker could sue the employer and show the employer's fault for the injury. Typically, workers under these no-fault compensation systems are not prevented from suing non-employers for their contribution to the workers' injuries (e.g., suits by shipyard and construction workers against asbestos manufacturers). No different than any other injury, a worker claiming job-related hearing loss can apply for temporary and permanent disability. Temporary disability provides wage replacement during the injured worker's healing period. Permanent disability compensates for the remaining handicap or impairment of function and facilitates the injured worker's return to the labor market.

Hearing loss claims may be somewhat more complicated than many other industrial injuries, however, because they often arise from the cumulative effects of noise exposure over a considerable time and not usually from a discrete event. As a result, questions may be raised concerning other causes for the hearing loss ranging from the natural effects of age to noise exposure either unrelated to work or related to work for another employer. Acoustical and medical experts therefore may be required to assess such claims for hearing loss.

II. Case Studies

THE following section presents 18 legal case studies wherein employees of Charles M. Salter Associates, Inc. have served as expert witnesses. These cases were selected to illustrate the range of legal issues that involve acoustics and vibration.

Case Study 1 State of Nevada v. Property Owner.

1. State of Nevada v. Property Owner

As part of Nevada's freeway expansion program in Reno, a portion of a trailer park was taken by the State via eminent domain. Following this, the trailer park owner filed a lawsuit against the State of Nevada, claiming that future freeway noise would substantially reduce the property value of the remaining park property. During the trial, the plaintiff's acoustical expert played a tape recording of generic freeway noise for the jury to illustrate how annoying the future noise could be.

On behalf of the State of Nevada's Highway Legal Department, Charles M. Salter Associates, Inc. measured the existing ambient noise in the trailer park and predicted the future environmental noise that the freeway expansion would create. We concluded that the overall noise level at the park would not change.

The trailer park was located next to the Reno Municipal Airport. National Guard Phantom Jets taking off generated over 100 dB at the trailer park. The created sound level was so high that we had to bring in a consultant's personal loudspeakers from San Francisco to accurately reproduce the sound in the courtroom—we could not locate rental loudspeakers in Reno that were powerful enough.

To convey our acoustical findings to the jury, we presented an audio/visual demonstration. One of our audio presentations was of freeway noise recorded in Milpitas, California, adjacent to a freeway which had similar traffic volumes and characteristics to that anticipated for the Reno freeway. After deliberation, the jury ruled in favor of the State's position.

Several years after the conclusion of the case, the attorney for the trailer park owner met with Charles Salter. The attorney confessed that when he heard the tape recording of the freeway noise in Milpitas (where the sounds of birds chirping and children playing were evident) during our audio/visual simulation, he knew that he had lost the case.

Case Study 2 Homeowner v. Developer.

2. Homeowner v. Developer

A woman purchased a townhouse from a developer who assured her that the home was built to meet the highest quality construction standards. After moving in, the homeowner was disturbed to find that her neighbor's noise transferred easily into her home. The developer ignored the woman's complaints for months. Eventually, she filed suit against the developer, claiming that her home did not meet a reasonable standard of care. After reviewing the design, Charles M. Salter Associates, Inc. concluded that the party wall design was inadequate and was known to be acoustically defective. As designed, the partition was rated at STC 47, three points below the minimum code requirement. The developer's earlier statement that the design met the "highest quality standards" could not be supported. It also was determined during the investigation that the wall as constructed did not meet fire and structural code requirements. Eventually, the developer agreed to rescind the transaction and to pay for the plaintiff's legal and consulting costs.

Case Study 3 Homeowner v. City of San Francisco.

3. Homeowner v. City of San Francisco

In the late 1980s, a couple purchased a mansion next to the San Francisco cable car line on Hyde Street. Several years after this purchase, they filed suit against the City of San Francisco over the cable car noise. They claimed nuisance, inverse condemnation, and a violation of the City's Noise Ordinance.

The City argued that the San Francisco Noise Ordinance did not contemplate controlling transportation noise sources such as the cable car system. The purpose of the ordinance was to control noise sources such as mechanical equipment or amplified music.

Charles M. Salter Associates, Inc. (CMSA) was hired by the City of San Francisco to conduct exterior and interior noise measurements at the plaintiff's home. Our measurements documented both the average noise level as well as the single event noise generated by the cable car line and street traffic. The exterior measurements indicated that the traffic on the streets surrounding the home would be considered "loud," over 70 CNEL.

The issue of whether the noise constituted a nuisance went to trial before a jury. In order to educate the jury about the case, CMSA prepared a calibrated acoustical simulation of cable car noise within a bedroom in the home. Our simulation included the following elements: (1) Noise in the bedroom as is (L_{eq} 49 dB); (2) Noise in the bedroom assuming the windows were gasketed normally (L_{eq} 44 dB); (3) Noise in the bedroom assuming the window glass was upgraded to 10 mm ($\frac{3}{8}$ in.) laminated glass (L_{eq} 35 dB); (4) Noise in the bedroom with double laminated glass windows (L_{eq} 28 dB).

The plaintiff's solution for the alleged noise problem would have cost the City of San Francisco hundreds of thousands of dollars and caused an end to the cable car system. CMSA testified that installing sound-rated windows throughout the home would cost approximately $15,000. The jury ruled in favor of the City, concluding that this cost should be borne by the homeowners, who knowingly purchased a home on a noisy street. The California Court of Appeals upheld the judgment, and the California Supreme Court declined to review the case.

Case Study 4 Homeowner v. Condominium Association Board of Directors.

4. Homeowner v. Condominium Association Board of Directors

A homeowner in a high-rise condominium requested permission from the Board of Directors of the Condominium Association to install a hardwood floor throughout most of his unit. The owner of the unit located directly below expressed his concern to the Board about noise transfer and suggested that the proposed floor design be reviewed by an acoustical engineer. The Board dismissed the downstairs neighbor's concern and approved the installation of the hardwood floor.

Following the floor's installation, the downstairs neighbor filed suit against the Board of Directors for not meeting their fiduciary responsibilities. Charles M. Salter Associates, Inc. was hired by the attorney of the downstairs neighbor to quantify the resulting noise. We found that the wood floor was glued down onto the concrete slab without any consideration for impact isolation. Consequently, the hardwood floor achieved impact isolation ratings in various rooms of IIC 40 to 44, below code minimum standards.

In our professional opinion, the noise created in the downstairs apartment was unacceptable to people of "normal sensibilities." The defense's expert witness measured virtually identical results when he conducted his tests at another time when we were not present. After a three-week trial, the jury ruled in favor of the plaintiff and rendered a $162,000 judgment. The jury required that two-thirds of the judgment be paid by the Condominium Association Board of Directors and that one-third be paid by the upstairs neighbors.

Case Study 5 Home Buyers v. Real Estate Broker and Home Seller.

5. Home Buyers v. Real Estate Broker and Home Seller

A couple purchased a home on a busy street in Northern California. After moving in, they became upset about the level of traffic noise intrusion in their new home. The homeowners did not hire a contractor, architect, or acoustical engineer to help them determine how to reduce the noise in their bedroom located near the street. The homeowners instead attempted to reduce the noise by applying fiberboard over the bedroom windows. This repair proved to be ineffectual. After spending a few more nights in the noisy bedroom, the couple decided that they wanted to rescind their purchase. They sued the real estate broker and the previous owner of the house when their request for recession was rejected.

One of the plaintiff's claims was that the broker and the previous owner should have disclosed that the home was on a busy street. The couple also claimed that the broker purposefully took them to the house between noon and 2:00 PM, when the traffic noise was at its lowest. Charles M. Salter Associates, Inc. (CMSA) conducted three days of continuous acoustical measurements at the home. Our measurements indicated that the hourly noise levels were about the same from 6:00 AM to 10:00 PM every day. The hourly L_{eq} during the day and evening was 60 ± 2 dB. The sound level impinging on the home was DNL 65 dB; thus the noise exposure at the home met the city standards for single-family homes. During the court proceedings, a CMSA consultant explained to the judge how easy it would be to reduce the noise intrusion into the bedrooms by upgrading the bedroom windows. The judge rendered a defense verdict.

Case Study 6 Homeowners' Association v. Developer.

6. Homeowners' Association v. Developer

A fifty-four-unit wood frame residential project was constructed atop a bluff overlooking the Pacific Ocean near San Diego. To allow for good ocean views, the living rooms were oriented to the west and extended deep into the building. The arrangement of the living room required wood joists spanning 7 m (23 ft.). The length of the span, in combination with the need for thin floors/ceilings, dictated the choice of doubled 38 mm x 286 mm (2 x 12) wood joists to support the living room floor. This unusual floor design satisfied the structural requirements of the building code.

After construction was completed, the homeowners complained about "feelable" floor vibration generated by people walking within the same unit and also by their neighbors walking in adjacent units. Along with being "feelable," these disturbances caused china dishes to rattle in their cabinets. Guests who experienced the vibration in these units would ask their hosts if an earthquake had just taken place.

A litigation ensued and the floors were found to have been properly constructed; hence, the focus of the investigation became the structural engineer's design of the floor itself. The floor vibration situation was investigated by several acoustical engineers. Field measurements indicated that the floor system had a resonance around 10 Hz—a typical value for light frame construction. Since American building codes do not address floor vibration, the measured floor stiffness was compared to the requirements in other countries. We concluded that such a floor system would be acceptable according to most design guidelines and building codes in Canada and Europe.

The plaintiff's expert proposed a repair which added steel angles to the joists in order to increase the bending stiffness. The added weight of the steel angles would probably have reduced the resonant frequency of the floor system, negating the improvement in "feelability." Despite the fact that the proposed repair did not appear to solve the problem, the defendant's insurance company decided to pay $600,000 to settle the case. This amount represented the bulk of the monetary settlement.

Case Study 7 Homeowner v.
Municipality.

7. Homeowner v. Municipality

A couple bought a house next to a town square and sued the municipality over the noise generated by a pump that operated the square's lawn
sprinkler system. They claimed that the noise made sleep impossible, caused
the house to vibrate, and induced stress that resulted in diminished sexual
appetite. Charles M. Salter Associates, Inc.'s vibration measurements indicated that there was no detectable vibration. At 3:00 AM with the sliding
glass door wide-open in the master bedroom, the loudness of various
acoustic events were as follows: Crickets = 13 dBA; Pump = 21 dBA;
Breathing = 25 dBA; Birds = 26 dBA; Refrigerator = 32 dBA; Stomach
gurgle of plaintiff 4.6 m (15 ft.) from the microphone = 35 dBA; Dog bark
= 42 dBA; Rain = 45 dBA; Car passby = 45 dBA; Jet flyover = 48 dBA.

With the sliding glass door open 460 mm (18 in.), the noise of the
pump was inaudible. Eventually the case was settled out of court.

Case Study 8 Seaman v. Shipping
Company.

8. Seaman v. Shipping Company

Workers exhibiting hearing loss sometimes sue their employer, claiming that their deafness was caused by the noise they encountered at work. This legal case involved hearing loss claims of a retired merchant seamen. Similar cases involve the hearing loss claims of retired railroad worker. These types of cases are complicated because of the diverse variables that must be considered before a verdict is rendered. These variables include: (1) *The age of the worker*, because people lose their hearing as they age, determining the percentage of hearing loss that was caused by aging as opposed to noise exposure is difficult; (2) *Noise-induced hearing loss* due to activities that are not directly job-related: workers often lose hearing due to flying aircraft, operating chain saws, firing guns, listening to loud music, and engaging in other similar activities; (3) *The economic loss due to demonstrated hearing loss;* For example, people can lose 5 dB of hearing acuity at 4,000 Hz and still have no difficulty in carrying on a conversation.

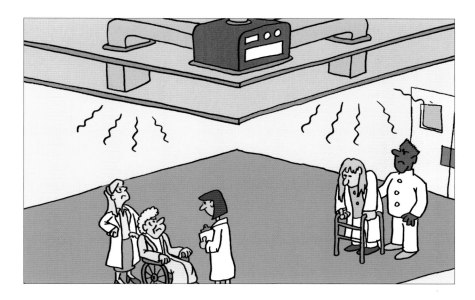

Case Study 9 Medical Center v. Architect and Engineer.

9. Medical Center v. Architect and Engineer

After a three-story medical office building was constructed and before it was occupied, the facility's owners complained that the ventilation system generated excessive noise and vibration. The hospital hired an acoustical engineer to develop solutions. The hired consultant speculated as to the potential causes of the problems, but never developed practical solutions. Already frustrated that their newly constructed building was not usable, the hospital became more frustrated at the inactivity and lack of helpfulness demonstrated by the architect and the mechanical engineer. The hospital filed suit against both, demanding money not only for the cost of "fixing" the problem, but also for the loss of income during that period of time when the building could not be occupied due to the excessive noise and vibration.

Hired by the architect's defense, Charles M. Salter Associates, Inc. conducted a design review, which confirmed that the architectural design met the standard of care. It appeared that the noise and vibration problems were caused by inadequate implementation of the noise and vibration control requirements, particularly by the incorrect installation of the specified vibration isolation devices. This case was eventually settled out of court.

Case Study 10 Lodge Owner v. Lodge Builder et al.

10. Lodge Owner v. Lodge Builder et al.

A lodge was built in Northern California in a very quiet location. The owner's objective was to create a five-star facility—a relaxing getaway.

Following construction, the acoustics were deemed unacceptable by the lodge owner. The owner complained of plumbing noise as well as sound transfer through the floor/ceilings and party walls. Both the design and the construction of the lodge were questioned. It was discovered that the architects specified party walls and floor/ceiling constructions that met minimum code standards, which are well below those that are expected at a luxury hotel. In addition, the architects neglected to detail the party walls and floor/ceiling constructions so that code minimum standards could be achieved. They also completely neglected plumbing noise control in their design. Thus, even if our client (the builder) had built the constructions perfectly, a serious plumbing noise problem would still have existed.

This case exemplifies the situation that occurs in many projects with inadequate construction budgets. A client's understandable desire to save money can lead to inadequate constructions that require significantly more money to reconstruct or repair. This lawsuit involved the owner, general contractor, architect, subcontractor, manufacturers, and material suppliers. Eventually the case was settled through arbitration.

11. Property Owners v. Municipality

With the help of an experienced city planner, this city developed an environmental assessment rating system that evaluated the feasibility of construction for each piece of land. Several years before the city planner was hired, Charles M. Salter Associates, Inc. (CMSA) created noise contours for the city, which were later used by the planner in the rating system. By assigning values to a variety of factors: noise exposure, degree of land slope, drainage, fire safety and others, the system averaged these numbers for each piece of property to determine the allowable extent of development. With a freeway running through the town, a substantial portion of its land was exposed to high traffic noise levels. Under the new rating system, many pieces of land were pronounced unsuitable for construction. Landowners in this community filed suit against the city because they felt that the system was unfair and overly restrictive.

Hired by the city's attorneys, CMSA defended the appropriateness of the noise contours. The landowners argued that the contours were too general in nature. They claimed that the contours did not account for the attenuation due to intervening terrain. The owners argued that the contours therefore represented an excessively conservative estimate of the noise level on each property, which should be adjusted on a property-to-property basis as a result of on-site measurements.

In an attempt to settle the lawsuits, a series of public hearings was held between the city officials, the landowners' attorneys, and the general public. At these meetings, CMSA explained the overall planning purposes of the city's noise contours and answered other questions that were posed by various attendees. Eventually, the city and the land owners agreed to a modification in the planning procedures and the cases were settled out of court.

Case Study 12 Store Owners v. Landlord.

12. Store Owners v. Landlord

A retailer located on the first floor of a wood-framed building was exposed to high levels of impact noise created by people walking above in an architect's office. The building's floor/ceiling construction consisted of only 28 mm (1⅛ in.) plywood on wood joists with no ceiling. Every footstep was audible in the quiet store below. When a chair was dragged across the floor above, it sounded as if it were occurring in the store. The retail establishment, believing that the landlord should solve the acoustical problem so that the noise intrusion would not be disruptive to their business, complained to the landlord. The landlord ignored the complaints of the store owners and they filed suit against him. We were hired by the store owners to measure the noise levels in the store, which we found to be disturbing and intrusive. This case was heard before a judge and eventually a settlement was reached.

Case Study 13 Real Estate Developer v. Architect and Contractor.

13. Real Estate Developer v. Architect and Contractor

An office building was designed for a site that bordered a freeway. After the building's construction, the owner sued the architect and the contractor, claiming that revenue was lost due to a delay in the construction because an adequate sound-rated window facade had not been specified in a timely fashion.

During the design phase, the architect solicited the advice of an acoustical consultant with respect to the appropriate STC ratings for the office building windows. Without conducting acoustical measurements or visiting the site, the acoustical consultant recommended STC 30. As a result, STC 30 rated windows were specified.

At the beginning of construction, the owner expressed concern about sound control and hired the architect's acoustical consultant to discuss the noise issue with him. The acoustical consultant, again without conducting measurements, suggested that STC 40 windows would be appropriate for the project. Because of the confusion that ensued during construction, substantial delays occurred.

We were hired by the architect's defense council to conduct acoustical measurements in the building, which had been fitted with the specified STC 30 glazing. We found that the noise level within the building was L_{eq} 40. At an arbitration, Charles M. Salter Associates, Inc. testified that this environment was acceptable for the tenants in the building. Eventually this case was settled.

Case Study 14 Mall Owner v. Discotheque.

14. Mall Owner v. Discotheque

A bar playing loud disco music was located in a shopping mall. The owner of the mall planned to build a luxury hotel directly adjacent to the bar. The sound levels that were typically generated within the bar were 110 dBA. The mall owner sued the bar owner for the amount of money the mall owner's acoustical consultant estimated it would take to upgrade the wall and window constructions to block the music noise emissions from the bar into the proposed hotel. Charles M. Salter Associates, Inc. was hired by the bar owner to conduct an investigation of the sound level and to develop cost- effective means by which to adequately control the music noise.

Our analysis concluded that the exterior walls and windows of the hotel would need to be sound rated anyway because of the single event noise intrusion from cars and trucks in the parking lot. This level of sound isolation would adequately control the music noise intrusion into the hotel. Eventually, this lawsuit was settled out of court.

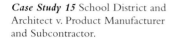

Case Study 15 School District and Architect v. Product Manufacturer and Subcontractor.

15. School District and Architect v. Product Manufacturer and Subcontractor

A school district retained an architect to design and specify an operable partition that would divide a multipurpose room into several sections. The architect specified an STC rating for the partition but did not provide the necessary noise control construction surrounding the partition. This error made it impossible to control sound transfer between the divided class- rooms reliably.

The general contractor selected and installed an operable partition that did not meet the architect's minimum STC rating for the product. Following this installation, the school district complained about sound trans- fer between classrooms to the architect. The architect blamed the subcon- tractor who selected and installed the partition. Charles M. Salter Associates, Inc. was hired by the partition manufacturer to evaluate the situation. Our analysis indicated that even if the specified partition had been installed as directed, the acoustical problem that the school district complained about would still have occurred. Eventually this case was settled out of court.

16. Restaurant Owner v. Architect

A restaurant was designed to have a sound-absorbing, aesthetically accept-able ceiling. During the construction phase, the contractor offered to sub-stitute another material that he claimed was acoustically equivalent to the specified product. Because of the time constraints for the project and because the substituted material was aesthetically acceptable, the architect approved the substitution without requesting authentication of its acousti-cal value.

The substituted material turned out to be sound reflecting, rather than sound absorbing, due to the manufacturer's fraudulent representation of the acoustical qualities of the product and the subcontractor's incorrect installation of the product. In addition, live music was played in the bar adjacent to the restaurant, which added to the restaurant's ambient noise level. The combination of the sound-reflecting ceiling and the live music, created an unacceptable acoustical environment in the restaurant. Restaurant patrons told the restaurant owner that they would not come back because they could not carry on a conversation during their meal. The restaurant owner sued the architect for over $300,000 due to loss of business because of excessive noise. After four years of legal maneuvering, the case was settled out of court.

Case Study 17 Relative of Deceased
v. Landlord.

17. Relative of Deceased v. Landlord

In 1986, a fire in East Oakland burned an apartment building to the ground
and took the life of one woman. Later, fire experts determined that the fire
originated in the apartment of the deceased, who was smoking in bed. The
daughter of the deceased filed suit against the landlord, alleging his partial
responsibility because he failed to have a smoke detector and alarm system
installed in the apartment as required by the City Code. During the trial,
the landlord's fire expert alleged that such alarms are not always heard. The
presence of a fire alarm, he argued, would likely not have allowed the plaintiff's mother, who was incapacitated and incapable of leaving the building
by herself most of the time, to be rescued by other tenants.

Hired by the plaintiff, Charles M. Salter Associates, Inc. conducted
acoustical tests to determine if a code-approved fire alarm would have been
heard, had it been in the apartment. Fortunately, a virtually identical apartment building existed nearby the burned site. Because this building had the
same background noise environment, we were able to conduct accurate
tests to demonstrate the audibility of a smoke detector throughout the
building. A smoke detector was placed in a prominent location of an apartment physically identical to that in which the fire originated. The alarm
was activated and acoustical measurements were made throughout the
building. This demonstration made it clear that the alarm could have been
heard throughout the entire building, alerting other building tenants who
might then have been able to rescue the fire victim. The convincing results
of this study were presented during the trial. Still, given the fact that (1)
the victim started the fire by smoking in bed, and (2) the daughter had not
been in contact with her mother for several years, the jury ruled in favor
of the landlord, despite the established audibility of the required fire alarm.

Case Study 18 Association v. Developer.

18. Association v. Developer

A real estate developer purchased a multi-family building site near a municipal airport. The city planning department required that prescriptive constructions be used for exterior walls, windows, and roof/ceilings if a building site was exposed to noise levels that are DNL 65 dB and above. The city-approved designs consisted of double stud walls and double windows. The noise level at the building site was well above DNL 65 dB. Regardless, the developer, acoustical engineer, architect, and contractor did not follow the city's prescriptive constructions. The construction as designed and built was below the requirements of the city.

The sound level measured in the homes due to single-event aircraft noise was clearly unacceptable. It was obvious to everyone that the completed construction was inadequate. The homeowners association filed suit against the project's acoustical engineer and the architect. The defense hired Charles M. Salter Associates, Inc. in attempts to lessen the settlement levied against them. Eventually, the case was settled out of court and the homeowners received enough money to reconstruct their buildings to control aircraft noise intrusion adequately.

Glossary of Acoustical Terminology

Absorption Coefficient: The fraction of the incident sound power that is absorbed by a material on a scale from 0 to 1. (See *Sound Absorption, Sabin.*)

Accelerometer: A transducer for measuring acceleration levels associated with low-frequency vibration.

Acoustic, Acoustical: Qualifying adjectives that describe something containing, producing, related to, or associated with sound. Acoustic usually pertains to hearing, while Acoustical pertains to the science of acoustics.

Acoustics: (1) The science of sound, including its production, transmission, and reception. (2) The physical qualities that determine how a room sounds.

Ambient Noise: The background noise, including sounds from many sources near and far, associated with a given environment.

Anechoic Chamber: A laboratory test room where all surfaces are covered by sound absorbing material in an attempt to eliminate sound reflections. (See *Dead Room.*)

Articulation Index: A calculated coefficient used for rating the intelligibility of speech.

Aspect Ratio: The ratio of dimensions usually normalized to the smaller dimension.

Average Sound Pressure Level (L_{eq}): The average sound pressure level occurring in a specified period (e.g., 1 hour).

A-weighting: A standard frequency weighting that de-emphasizes low-frequency sound similar to average human hearing response and approximates loudness and annoyance of noise. A-weighted sound levels are frequently reported as dBA.

Bandwidth: A specific range of frequencies.

Bel: 10 dB, named after Alexander Graham Bell.

Binaural: Auditory reception associated with two ears.

Breakout Noise: Noise associated with fan or airflow noise that radiates through the walls of a duct into the surrounding area.

Broad-band Noise: Noise comprised of a wide frequency range and not characterized by any tonal component.

Ceiling Attenuation Class (CAC): A laboratory rating system that results in a single value to rate the sound transmission properties of ceiling systems (usually between offices where the wall does not extend above the ceiling). Specified in ASTM E1264.

Codec: An electronic device that compresses and decompresses digital signals. Codecs usually perform A-to-D (analog-to-digital) and D-to-A conversion.

Coefficient: A unitless quantity generally scaling from 0 to 1.0.

Coloration: Colloquial term for timbre (quality of tone).

Community Noise Equivalent Level (CNEL): A descriptor for the 24-hour average A-weighted noise level that accounts for a human's increased sensitivity to noise during the evening and nighttime hours. Noise levels between the hours of 7 PM to 10 PM are penalized 5 dB; noise levels between the hours of 10 PM to 7 AM are penalized 10 dB. (See also *Day/Night Average Sound Level.*)

Compressor: The part of an audio system that is used to compress the intensity range of signals at the transmitting or recording end of a circuit.

CRT (three-gun) Video Projector: A projector that uses three colored cathode-ray tube guns (red, green, and blue), that when properly combined form a full-color image.

Damping: The dissipation of sound or vibration. Damping reduces the velocity of vibrating molecules in an elastic medium at resonant frequencies, resulting in the more rapid dissipation of energy.

Day/Night Average Sound Level (DNL): A descriptor established by the U.S. Environmental Protection Agency to describe the 24-hour average level A-weighted noise with a ten decibel penalty applied to noise that occurs between the hours of 10 PM and 7 AM to account for a human's increased sensitivity to noise during sleeping hours. (See *Community Noise Equivalent Level.*)

dBA: See *A-weighting.*

Dead Room: A room that is characterized by a large amount of sound-absorbing material and with little reverberation.

Decay: The decrease of sound or vibration with time. Decays are most readily observable subsequent to an impulse. (See *Damping, Reverberation Time.*)

Decibel (dB): The measurement unit used in acoustics for expressing the logarithmic ratio of two sound pressures or powers. Typically used to describe the magnitude of a sound with respect to a reference level equal to the threshold of human hearing.

Diffraction: The process that bends a wave due to an obstacle or aperture.

Diffuse Sound Field: A sound field where the intensity is the same in all directions. Also used in room acoustics to describe the property of being "immersed" in a uniform reverberant sound field.

Dipole Surround Speaker: A loudspeaker system in a single enclosure that radiates from both the front and back. Often used in a home theater to provide a diffuse sound field, using only two surround loudspeakers.

Direct Sound: A wave radiated directly from a sound source that reaches a receiving point without reflections.

Distance Learning: A specific video conference application in which an instructor is in video conference with students at a distant location.

Driver: A device that supplies energy to another system or circuit (as a loudspeaker).

Echo: A distinctly audible sound reflection, usually considered undesirable.

Electroacoustics: The conversion of acoustic energy into electrical energy, or vice versa.

Equal Loudness Contours (Fletcher-Munson Curves): A set of frequency curves that indicates the loudness levels of pure tones heard monaurally.

Equalizer: A device that adjusts the frequency response characteristics of an electronic signal.

Far Field: The portion of the acoustic radiation field of a noise source where sound pressure level decreases at a rate of 6 dB per doubling of distance.

Feedback: A part of the output of a system (as an electronic circuit) that is returned to the input.

Filter: An electronic device for separating specific frequency components of a signal.

Flanking Transmission: The transmission of airborne sound from a source room to an adjacent receiver room by a path other than the common partition.

Flutter: Distortion that occurs in sound reproduction as a result of undesirable speed variations during the recording, duplicating, or reproducing process.

Flutter Echo: Echoes occurring repetitively between a pair of surfaces.

Fourier Analysis: Frequency and phase analysis of a signal based on the addition of sine waves.

Free Field: The sound field or portion of a sound field unaffected by reverberation or reflected sound.

Frequency: A descriptor for a periodic phenomenon. The frequency is equal to the number of times that the pressure wave repeats in a specified period of time. In the case of sound and vibration, frequency is measured in units of Hertz (Hz), which correspond to one cycle per second (cps).

Fresnel Lens: A lens often used on the projector side surface of a rear projection screen. The lens functions as a condenser by refracting all projected light perpendicular to the screen surface, which improves image brightness. (Also used in lighthouse lenses.)

Fundamental Frequency: The "first" harmonic or resonance within a sound spectrum often having the greatest amplitude.

Gain: The ratio of increase in output compared to the input in an amplifier, frequently measured in units of decibels referenced to a voltage.

Haas Effect: See *Precedence Effect*.

Harmonic: A single component of a sound's spectrum that has a frequency that is an integer multiple of the fundamental frequency.

Hertz (Hz): A unit of measure for describing the frequency of a wave phenomenon, such as sound.

Impact Insulation Class (IIC): A single number rating derived from one-third octave band values of impact noise levels measured through a floor/ceiling system using a standard tapping machine. The IIC describes the impact noise insulating properties of a floor/ceiling assembly and approximates footfalls on hard-surfaced floors. ASTM E989 and E1007 govern the measurement and calculation of IIC.

Impulsive Noise: Usually a brief noise characterized by an instantaneous sound pressure that significantly exceeds the ambient noise level.

Inertia Base: A concrete slab which serves as a base for mechanical equipment such as fans and pumps. The base is supported on vibration isolators to reduce the transmission of vibration to the building structure.

Insertion Loss: The difference in sound levels before and after a noise barrier or sound attenuator is constructed or installed.

Inverse Square Law: The sound attenuation due to distance from a source based on the reciprocal of the square of the distance. For a point source in the far field, this is 6 dB per doubling of distance.

ISDN (Integrated Services Digital Network): A digital means for moving large amounts of information over phone lines, often used in conjunction with video conferencing equipment.

Kilo-: Prefix representing 10^3 or thousand.

Lambert's Law: The angle of a reflected sound "ray" from a surface is equal to its angle of incidence.

LCD Panels: A liquid crystal display (LCD) panel used with a conventional light-based overhead projector to display computer images on a screen. The semi-transparent LCD panel is connected to the computer video output and sits on the illuminated overhead projector.

LCD Video Projector: A projector that uses a liquid crystal display (LCD) panel and single high-intensity light to project a colored image.

Lenticular Surface: A lens often used on the audience-side surface of a rear projection screen. The lens refracts projected light to minimize the projection of light to the ceiling and floor while providing a wide viewing angle, which improves image brightness.

Light Valve Projector: A projector that uses an image light amplifier consisting of a solid film liquid crystal display and high intensity light to project a colored image.

Limiter: An electronic circuit used to prevent the amplitude of an audio signal from exceeding a specified level while preserving the shape of the wave at amplitudes less than the specified level.

Loudness: The magnitude of the physiological sensation produced by sound, which varies directly with sound intensity and is also dependent on both frequency and time.

Masking: The process by which sensitivity to a sound is decreased by the presence of another (masking) sound. Masking noise can be used to reduce the intelligibility or distraction of an intruding sound, such as speech.

Micro-: Prefix representing 10^{-6} or one-millionth.

Mode: Any of various stationary vibration patterns of which an oscillatory system is capable; for example, a standing wave in a room or a fundamental on a guitar string.

Monaural Sound: A single channel of sound reproduced through one or more loudspeakers.

Near Field: The acoustic field that is near a sound source, where the sound level drops off at less than 6 dB per doubling of distance.

Newton: Unit of force that will impart acceleration of one meter per second squared to a kilogram mass.

Noise: (1) Any undesired sound. (2) Any statistically random wave.

Noise Criteria (NC) Curves: A set of spectral curves used to obtain a single number rating describing the "noisiness" of environments for a variety of uses. NC is typically used to rate the relative loudness of ventilation systems.

Noise Isolation Class (NIC): A single-number rating representing the sound isolation between two enclosed spaces separated by a partition. The sound paths include the partition and flanking paths. (See *Sound Transmission Class*.)

Noise Reduction (NR): The difference (reduction) in sound pressure level of sound transmitted through a building partition, usually measured in octave or one-third octave frequency bands.

Noise Reduction Coefficient (NRC): A single-number rating of the sound-absorption of a material equal to the arithmetic mean of the sound-absorption coefficients in the 250, 500, 1,000, and 2,000 Hz octave frequency bands rounded to the nearest multiple of 0.05.

Octave: A separation in frequency constituting doubling or halving of frequency.

Octave Band: A frequency band with the highest frequency being twice the lowest frequency defining the frequency range.

Pascal (Pa): Unit of pressure equal to one newton per square meter. Twenty micro pascals is the reference pressure used in determining sound pressure level.

Phase: A time-point reference in a periodic wave, usually measured in degrees (angular), with 360 degrees representing one complete cycle. Phase is used to measure the amount a wave leads or lags another.

Phon: A unit of loudness of a sound that relates sound pressure at different frequencies to the loudness of a reference sound pressure level of a 1,000 Hz tone.

Phoneme: A member of a set of the smallest units of speech that serve to distinguish one utterance from another in a language or dialect.

Pink Noise: Noise whose spectrum is filtered so that there is equal power in each octave band. For example, the sound power in the 63 Hz octave band (45 to 89 Hz) is equal to the sound power in the 1,000 Hz octave band (707 to 1,414 Hz).

Pitch: The perceived frequency of a sound determined primarily by the fundamental frequency of the sound and also upon its harmonics and their relative sound pressure levels (e.g., spectrum).

Plasma Display: A video display in which the image is created by the glow of electrically excited neon and xenon gases arranged in a flat panel matrix.

Preamplifier: An amplifier designed to amplify weak low level signals from a device (as a microphone) before the signals are fed to additional amplifier circuits.

Precedence Effect: The ability of the human auditory system to suppress perception of echoes, primarily up to 40 msec after a direct sound.

Psychoacoustics: The scientific study of human auditory perception.

Quadraphonic Sound: A four-channel system for reproducing sound.

Rarefaction: The instantaneous, local reduction in density of a gas resulting from passage of a sound wave. For example, this occurs in the environment when sound travels through an atmosphere which changes temperature with elevation above the ground.

Reflection: The phenomenon by which a sound wave is re-radiated ("bounced") from a surface.

Reflectogram: A graph showing the measured or computed amplitude and time of arrival of acoustical reflections.

Refraction: The phenomenon by which the direction of a sound wave is changed due to the spatial variation in the speed of sound. For example, this occurs when sound travels through an atmosphere in which temperature changes with elevation above the ground.

Resonance: (1) The large amplitude vibration in a mechanical system caused by a relatively small stimulus having the same period as the natural vibration period of the system. (2) An increase in the sound pressure of one or more harmonics of sound.

Reverberation Time: The time (in seconds) required for the sound pressure level to decrease 60 dB in a room after a noise source is abruptly stopped. Reverberation time relates to a room's volume and sound absorption.

Reverberance: The subjective impression produced by reflected sound.

Reverberation: Collection of time-delayed sounds following a direct sound that result from reflections indoors.

Reverberant Field: The sound field in a room that is dominated by reverberation. Usually the sound level no longer is reduced by moving further from the sound source.

Room Constant (Absorption): The total acoustic absorption provided by objects and surfaces, including air, in a specific room. For a given frequency, the single number value equal to the product of the average absorption coefficient of the room and the total internal area of the room.

Sabin: A unit of acoustic absorption equal to the absorption provided by a unit area of a perfect absorber.

Seismic Snubber: Device used in conjunction with vibration isolation to stabilize system during earthquakes.

Shock: An impulsive force capable of exciting mechanical resonances; for example, a blast produced by explosives.

Sine Wave: A wave whose amplitudes vary sinusoidally as a function of time. The simplest possible wave without harmonics (pure tone).

Sone: A subjective unit of loudness for an average listener equal to a 1,000 Hz tone that is 40 dB above the listener's threshold of hearing. The scale is subjectively linear where a doubling of the sone value represents a doubling of loudness.

Sound: (1) An oscillation in pressure, resulting from molecular motion, in a viscous or elastic medium such as air, water, wood, steel, etc. (2) Sound is an auditory sensation evoked by air molecules vibrating in a frequency range between 20–20 kHz.

Sound Absorption: The property of absorbing sound energy possessed by objects and surfaces, including air. (See *Sabin, Absorption Coefficient.*)

Sound Exposure Level (SEL): A descriptor used for expressing the noise dosage of a transient event. The SEL is expressed as a sound level containing an equivalent amount of sound energy lasting 1 second.

Sound Insulation: (1) The capacity of a structure to prevent sound from being transmitted from one space to another. (2) Insulation used in a wall, floor, or ceiling cavity to add damping and decrease transmitted sound. (See *Sound Transmission Loss.*)

Sound Intensity (I): The rate of flow of energy per unit area in a specified direction.

Sound Isolation: Sound attenuation through a material or assembly. (See *Noise Reduction.*)

Sound Masking System: An electronic system that generates noise to help render speech less intelligible.

Sound Power (W): The acoustical energy of a sound source independent of distance.

Sound Power Level (PWL): Sound power radiated by a sound source reported as a level in decibels referenced to 1 pico-watt (without regard for distance from the source).

Sound Pressure: The time varying pressure exerted in an elastic medium such as air, generally in the 20–20,000 Hz frequency range.

Sound Pressure Level (SPL): Sound pressure measured as a level in decibels referenced to 20 micropascals generally at a specific location or distance from a sound source.

Sound Transmission Class (STC): A single-number rating derived from laboratory measurement of sound transmission loss. STC is calculated in accordance with ASTM E413, "Classification for Rating Sound Insulation." The STC describes the sound-insulating properties in the 100–4 kHz frequency range, primarily for assessing speech transmission through a structure, such as a partition. (See *Noise Isolation Class.*)

Sound Transmission Loss (TL): A laboratory measure of sound insulation indicative of the sound-intensity flow transmitted through a partition without regard to the partition size, usually measured in one-third octave bands.

Spectrum: The relative magnitude of different frequency components that comprise a single wave.

Speech Interference Level (SIL): A single-number rating used to evaluate interference based on the background noise level and voice level. The SIL is the arithmetic mean of the noise levels at 500, 1,000, and 2,000 Hz.

Speech Privacy Potential: A single-number rating used to evaluate the degree of speech privacy between two adjacent work stations.

Speech Reinforcement System: An electronically amplified audio system designed to reproduce speech at a sufficient level for intelligibility to overcome distance and other acoustic limitations.

Structure-borne Noise: Noise propagated through a structure and re-radiated as airborne noise.

Subsonic or infrasonic: Pertaining to signals involving frequencies below the range of human hearing (20 Hz).

Timbre: Perceived auditory sensation of "tone color" mainly dependent on sound spectrum.

Tone: A pure tone is a singular frequency produced by a sine wave.

Transducer: A device that is actuated by energy from one system and supplies energy, usually in another form, to a second system (as a microphone or loudspeaker).

Tripole Surround Speaker: A loudspeaker system in a single enclosure that radiates from the front, back, and side. Often used in home theaters to produce a diffuse sound field using only two surround loudspeakers.

Tungsten-Halogen Lamps: An incandescent lamp that utilizes a tungsten filament and a halogen gas to produce a high-intensity light with high-color temperature. Often used with LCD and other video projectors.

Tweeter: A high-frequency transducer component of a loudspeaker system often used in conjunction with a woofer (low-frequency transducer).

Ultrasonic: Pertaining to signals involving frequencies above the range of human hearing (20 kHz).

Variable Air Volume Boxes (VAV): A unit that regulates the flow of air with a motor-operated damper in a ventilation system.

Vertical Keystone Correction: Correction for the undesirable effect caused when a projector is not perpendicular with the center of a screen. An uncorrected "keystone" image looks narrower at the top or bottom depending on the location of the projector relative to the screen.

Vibration: A periodic motion of molecules in an elastic medium with respect to equilibrium. Vibration by mechanical equipment can be a factor in structure-borne noise radiation.

Vibration Isolation: The methods used to reduce vibration in a structure caused by vibrating equipment, including the use of springs and elastomeric materials.

Video Conference Room: A room designed for simultaneous audio and video communication between two groups at different locations.

Wavelength: In a periodic wave, the distance between two points having the same phase in consecutive cycles along a line in the direction of propagation (e.g., the distance between two adjacent peaks of a sine wave).

Weighting: A prescribed frequency response commonly provided by an electronic filter in a sound level meter.

White Noise: Noise having a frequency spectrum with equal power at each frequency. For example, the sound power between 100 and 200 Hz is equal to the sound power between 1,100 and 1,200 Hz.

Suggested Reading

Michael Barron, *Auditorium Acoustics and Architectural Design* (London: E & FN Spon an Imprint of Chapman & Hall, 1993).

Durand R. Begault, *3-D Sound for Virtual Reality and Multimedia* (Boston: AP Professional. 1994).

Leo L. Beranek. *Concert and Opera Halls, How They Sound.* Woodbury, NY: Acoustical Society of America through the American Institute of Physics, 1996.

Noise and Vibration Control Engineering Principles and Applications. Leo L. Beranek and Istvan L. Ver, editors. (John Wiley & Sons, Inc., 1992).

David A. Bies and Colin H. Hansen. *Engineering Noise Control Theory and Practice* (Academic Division of Unwin Hyman, Ltd., 1988).

Lothar Cremer and Helmut A. Muller. Translated by Theodore J. Schultz. *Principles and Applications of Room Acoustics,* vols. 1 and 2. (Applied Science Publishers, 1978).

David M. Egan, *Architectural Acoustics, Third Edition.* (McGraw-Hill Book Company, 1988).

Cyril M. Harris, Ph.D., Editor-in-Chief. *Handbook of Acoustical Measurements and Noise Control* (McGraw-Hill, Inc., 1991).

Tomlinson Holman, *Sound for Film and Television.* (Focal Press, 1997).

John W. Kopec, *The Sabines at Riverbank.* (N.Y. Acoustical Society of America, 1997).

Latest applicable issue of *ASHRAE Guide and Data Book*, Sound and Vibration Control Chapter.

M.E. Schaffer, "A Practical Guide to Noise and Vibration Control for HVAC Systems" (ASHRAE, 1991).

List of Contributors in Photograph

(1) Ross Jerozal; (2) Tim Der; (3) Julie Malork;
(4) Harold Goldberg; (5) Karen Decker; (6) Al Rosen;
(7) Charles Salter; (8) Jason Duty; (9) Cristina Miyar;
(10) Michael Flynn; (11) Brenda Yee; (12) Eric
Broadhurst; (13) Kathie Leavy; (14) Ken Graven;
(15) Jeff Leon; (16) Phil Sanders; (17) Michael Toy;
(18) Maya Van Putten; (19) Durand Begault; (20) Tom
Schindler; (21) David Schwind; (22) Jack Freytag;
(23) Tony Nash; (24) Eva Duesler; (25) Claudia
Kraehe. (Not pictured: Rachel Dangermond, Dylan
Jhirad, Henry Rollmann.)

Begault, Durand R., Ph.D.
Affiliated Consultant

Ph.D., Computer Audio, University of California,
    San Diego, 1987
M.F.A., Electronic Music and Recording Media
    Mills College, Oakland, 1981
B.A., Music, University of California, Santa Cruz,
    1979

Visiting researcher at the Spatial Auditory Display
    Laboratory (Flight Management and Human
    Factors Division) at NASA Ames
    Research Center, since 1988.
    Lecturer positions at San Diego State
    University (1987–8), the University of
    California, San Diego (1986-8) and San
    Francisco State University (1996–7), and the
    University of California, Santa Cruz Science
    and Technology Extension (1996–present)

3-D Sound for Virtual Reality and Multimedia
    (Academic Press Professional)
The Sonic CD-ROM for Desktop Audio Production
    (Academic Press Professional)
Two U.S. patents related to spatial audio

Acoustical Society of America
Audio Engineering Society
Human Factors and Ergonomics Society
Institute of Noise Control Engineering

Broadhurst, Eric L., P.E.
Principal Consultant
Acoustical Consultant since 1989

B.S. Mechanical Engineering, University of California,
    Berkeley, 1989

Registered Mechanical Engineer, California, 1993
Institute for Noise Control Engineering, Board
    Certified, 1997

Corbett, Thomas J.
Senior Consultant
Audio/Visual Consultant since 1982
Communications Technology Specialist (CTS), 1991

M.A. Speech, Theater, University of Wyoming,
    Laramie, 1967
B.S., Theater and Psychology, St. Cloud University,
    1965

20 Years as College/University Instructor
ICIA Institute for Professional Development

Society of Motion Picture and Television Engineers
United States Institute for Theater Technology

Dangermond, Rachel E.
Writer/Editor
Writing by Design since 1990

B.A., English Literature and History, University of New
    Orleans, 1989
M.A. coursework, English Literature, Mills College,
    Oakland, 1991–92

Fiction Editor, *Ellipsis Literary Journal*, 1987–88

"FM Talk Radio," *City Business*, New Orleans, 1996
Norco Museum Interpretive Plan, Shell Oil Company,
    Louisiana, 1996
California Lottery Scripts, Spanish & English,
    1990–1995
New Orleans Air, sitcom pilot for *WINGS*, 1989
"Drama in the Hills," *Louisiana Life*, May 1988

Decker, Karen E., P.E.
Senior Consultant
Acoustical Consultant since 1990

B.A., Engineering, Pennsylvania State University,
    1990

Registered Mechanical Engineer, California 1994
    Construction Specifications Institute

Der, Timothy M.
Principal Consultant
Acoustical Consultant since 1987

B.S. Mechanical Engineering, University of California,
    Berkeley, 1987

Duty, Jason R.
Consultant
Acoustical Consultant since 1996

B.E., Electrical Engineering, Thayer School of
    Engineering, Dartmouth College, 1996
B.A., Engineering and Music, Certificate in Envir-
    onmental Studies, Dartmouth College, 1995

Freytag, John C., P.E.
Director
Acoustical Consultant since 1972

M.S., Engineering, Stanford University , 1976
B.S.M.E., Arizona State University, 1972
Graduate Business Studies, Golden Gate University,
    1980

Registered Mechanical Engineer, California, 1981
Institute of Noise Control Engineering, Board
    Certified, 1989
Acoustical Society of America
Audio Engineering Society, Forensic Audio Working
    Group

Transportation Research Board, Aviation Noise
    Committee
American Institute of Physics
American Institute of Aeronautics and Astronautics
American Institute of Architects, Associate Member

Goldberg, Harold S., P.E.
Principal Consultant
Acoustical Consultant since 1985

B.S., Mechanical Engineering, University of California,
    Berkeley, 1985

Registered Mechanical Engineer, California 1991
Institute for Noise Control Engineering, Board
    Certified, 1997

Graven, Kenneth W., P.E.
Principal Consultant
Acoustical and Audio/Visual Consultant since 1989

B.S., Electronic Engineering Technology, Music Minor,
    California Polytechnic State University, San
    Luis Obispo, 1988

Registered Electrical Engineer, California, 1996
    Acoustical Society of America
Audio Engineering Society

Jerozal, Ross A.
Consultant
Audio Video Consultant since 1996

B.S. Electrical Engineering, California Polytechnic
    State University, San Luis Obispo, 1996
Audio Engineering Society

Jhirad, Dylan R.
Graphics
Graphic Designer at EHDD since 1995

B.A. Comparative Literature, University of California,
    Berkeley, 1990

Leon, Jeffrey A., Esq.
President, The Law Offices of Leon & Leon
Practicing Law since 1983

B.A., Economics, Phi Beta Kappa & Highest Honors,
    University of Virginia, 1978
J.D., Order of the Coif., University of Virginia Law
    School, 1981

Directed a major First Amendment battle on behalf of
    an award-winning journalist and The San
    Francisco Chronicle against the Federal Bureau
    of Prisons, appeared as a guest on Ted
    Koppel's Nightline, Attorney of the Year by the
    Bay Area Society for Professional Journalists.

District of Columbia Bar, 1981
State Bar of California, 1982

Malork, Julie A.
Consultant
Acoustical Consultant since 1996

M.Arch., University of California, Berkeley, 1995
B.A., Architecture and Art, Yale University, 1989

Organization of Women Architects
American Institute of Architects, Associate Member

Miyar, Cristina L.
Consultant
Acoustical Consultant since 1995

M.Arch., University of California, Berkeley, 1995
B.A., Music, Concentration in experimental acoustics,
    Wesleyan University, 1990

American Institute of Architects, Associate Member

Murray, Rachel V., P.E.
Principal Consultant
Acoustical Consultant since 1983
M.S. Mechanical Engineering, Pennsylvania State
    University, 1984
B.S. Mechanical Engineering, Pennsylvania State
    University, 1983

R.V. Murray & S.I. Roth, "A Computerized Fan Noise
    Prediction Method," Noise-Con, 1985
Registered Mechanical Engineer, California, 1990

Nash, Anthony P., P.E.
Vice-President
Acoustical Consultant since 1972

B.S., Electrical Engineering, Oregon State University,
    1966

Registered Electrical Engineer, California, 1977
Institute of Noise Control Engineering, Board
    Certified, 1978
Acoustical Society of America
ASTM, Acoustics Standards Committee
Structural Engineers Association of Northern
    California

Rollmann, Henry D.
Graphics
Practicing architecture at EHDD since 1996

B.A., Architecture, University of California, Berkeley,
    1993

Rosen, Alan T.
Principal Consultant
Acoustical Consultant since 1984

B.S., Electrical Engineering, University of Illinois at
    Urbana-Champaign, 1984

Association of Environmental Professionals (Chapter
    President 1990)
American Planning Association
Acoustical Society of America

Salter, Charles M., P.E.
President
Acoustical Consultant since 1968

M.B.A., Finance, Boston College, 1972
B.S. Art and Design, Architecture, City Planning
    Minor, MIT, 1969
B.S.C.E., Structural Engineering, Economics Minor,
    Tufts University, 1965
Adjunct Professor, University of California, Berkeley,
    1994-present
Lecturer in Acoustics, University of California,
    Berkeley, 1973-1994

Registered Mechanical Engineer, California and
    Nevada, 1974
Institute of Noise Control Engineering, Board
    Certified, 1975
Western Construction Consultants Association
    (President, 1995-97)
Acoustical Society of America
American Institute of Architects, Associate Member
Association of Environmental Professionals

Sanders, Philip N.
Principal Consultant
Acoustical Consultant since 1989

M.A., Linguistics, Stanford University, 1989
B.A., Linguistics, Stanford University, 1985

Society for Marketing Professional Services (SMPS)
Acoustical Society of America
American Institute of Architects, Associate Member

Schindler, Thomas A., P.E.
Principal Consultant
Acoustical Consultant since 1984

B.S., Ocean Engineering, Florida Atlantic University,
    1982

Registered Electrical Engineer, California, 1996
Audio Engineering Society
Society of Motion Picture and Television Engineers

Schwind, David R., FAES
Senior Vice-President
Acoustical Consultant since 1975

B.S.E., Interdisciplinary Engineering, Purdue
    University, 1974

Patent on device to disperse sound from a high
    frequency loudspeaker.

Institute of Noise Control Engineers
Acoustical Society of America
Audio Engineering Society, Fellow
The League of Historic American Theaters
Society of Motion Picture and Television Engineers
United States Institute for Theater Technology
American Society of Mechanical Engineers

Skye, Robert B.
Principal Consultant
Acoustical and Audio/Visual Consultant since 1976

The National Academy of Recording Arts and
    Sciences
Gold and Platinum Records for engineering the
    recording of "AC/DC Live," and a Grammy
    Award for engineering the live recording of
    "Tribute to Miles" with Herbie Hancock, Wayne
    Shorter, Tony Williams, Ron Carter, and
    Wallace Roney

Audio Engineering Society
Musician's Local #292

Toy, Michael D., P.E.
Senior Consultant
Acoustical Consultant since 1987

B.S. Mechanical Engineering, University of California,
    Berkeley, 1986
Registered Mechanical Engineer, California, 1991

Yee, Brenda R.
Manager, Graphics Department
CAD drafter since 1993

Certificate, Interior Design, University of California,
    Berkeley, Extension, 1998

B.A., History, English Minor, University of California,
    Davis, 1993

Acknowledgments

Charles M. Salter, P.E.

In a sense, the preparation of this book began two decades ago when I first started my own practice. I was fortunate to have waited till now to give it solid form, because the book has benefited from the advice and experience of many of my associates, employees, and colleagues.

Any work of scholarship is a cooperative venture and mine is no exception. I wish to thank all the people whose assistance and cooperation helped bring this book to completion. It is a pleasure to thank the authors of each chapter, along with the people who edited, wrote, and provided support in all aspects of the book: The historical data for Chapter 1 was compiled by Brenda Yee, Nikko Wenner, David Schwind, Durand Begault, and Cynthia Flewelling. In addition, John Kopec, of Riverbank Labs, offered valuable historical suggestions and supplied appropriate references for this chapter. Chapter 5 benefited from the input of Jack Freytag, Harold Goldberg, and Lynn Alexander, an environmental planning consultant. Jim Abbott helped with the math in Chapter 6. Jason Duty assisted in the development of the Appendices in Chapter 6. Robert Alvarado helped develop the information in Chapter 7. Steve Curry, S.E., and Hratch H. Kouyoumdjian, S.E. both commented on Chapter 8. Rachel Murray, Kathie Leavy, and Tim Der contributed to Chapter 9. David Malman of Architectural Lighting wrote a section on lighting for Chapter 12. The following individuals and organizations assisted in collecting relative cost information for Chapter 16: Scott Lewis, Oppenheim Lewis, Inc.; Mark Pearcy & Jamie Rusin, ELS/Elbasani & Logan Architects; Mike Rubio,

Norman Wright Equipment Company; Joe Mirando, Fry Metals, Inc.; Steve Saarman, Saarman Construction, Ltd.; Joe and Rick Wester, Wester Acoustics; Joan Blackmon, The Finish Line; Howard Podolsky, Pyrok, Inc.; and Al Spaulding, Eckels Spaulding Company. There are many others who have provided comparative cost information, both during the course of our day-to-day consulting activities, and in research geared specifically toward writing this chapter, we apologize for not mentioning your names specifically, but thank you for your contributions. Steven Winkle, FAIA, contributed building code research for Chapter 18. Cristina Miyar helped develop the speech privacy information in Chapter 19. My father, Leonard Salter, Esq., and son, Joshua Salter Berezin, Esq., both reviewed Chapter 21. In addition, Cristina Miyar, Julie Malork, and Claudia Kraehe worked on many chapters, while Jason Duty provided technical support throughout. Kim Emanuel assisted with the equations and fractions throughout the book. David Schwind and Michael Toy compiled the glossary. I am particularly indebted to Eva Duesler, for being the gatekeeper of the information as it was processed into text.

A special thanks to Jon Tomlinson, who edited the first draft, and to Maya Van Putten for her coordination and editing during the development of the manuscript. The editor of the book was Rachel Dangermond of Writing by Design.

The design of the book is credited to Dylan Jhirad and Henry Rollmann of Pick Design, who, with their graphic assistants Viola Rouhani and Mary Hayano, designed the book and cover, and created the tables, charts, and computer illustrations. Pick Design would like to thank Esherick Homsey Dodge & Davis for their patience.

Many thanks to Michael Flynn, of Michael Flynn Illustrations, who drew most of the illustrations. To Raul Viceral, whose cartoon illustrations added a nice touch to the collection. And to Peter Hasselman, for his beautiful illustration of the Cathedral San Marco.

An enormous thank you to my wife, Trudy, and to my family, colleagues, and clients, who encouraged this endeavor and supported the various stages of its development. Please see our website at **www.cmsalter.com** for errata. Comments about this book should be directed to our email address **acoustics@cmsalter.com**.